Ergebnisse der Mathematik und ihrer Grenzgebiete 90

A Series of Modern Surveys in Mathematics

R.E. Edwards G.I. Gaudry

Littlewood-Paley and Multiplier Theory

Springer-Verlag
Berlin Heidelberg New York 1977

R. E. Edwards
Institute of Advanced Studies, Australian National University, Canberra

G. I. Gaudry
Flinders University, Bedford Park, South Australia

AMS Subject Classifications (1970):
Primary 42–02, 42A18, 42A36, 42A40, 43–02, 43A15, 43A22, 43A70, 60G45
Secondary 42A44, 42A56, 42A68, 46E30, 60B15

ISBN 3–540–07726–X Springer-Verlag Berlin Heidelberg New York
ISBN 0–387–07726–X Springer-Verlag New York Heidelberg Berlin

Library of Congress Cataloging in Publication Data. Edwards, Robert E. Littlewood-Paley and multiplier theory. (Ergebnisse der Mathematik und ihrer Grenzgebiete; 90). Bibliography: p. Includes indexes. 1. Fourier analysis. 2. Multipliers (Mathematical analysis). I. Gaudry, G.I., 1941-joint author. II. Title. III. Series. QA403.5.E38. 515′.2433. 76–12349

Printed in Germany
Typesetting: in Japan. Printing: fotokop wilhelm weihert KG, Darmstadt. Binding: May & Co, Darmstadt.

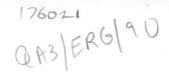

Preface

This book is intended to be a detailed and carefully written account of various versions of the Littlewood-Paley theorem and of some of its applications, together with indications of its general significance in Fourier multiplier theory. We have striven to make the presentation self-contained and unified, and adapted primarily for use by graduate students and established mathematicians who wish to begin studies in these areas: it is certainly not intended for experts in the subject.

It has been our experience, and the experience of many of our students and colleagues, that this is an area poorly served by existing books. Their accounts of the subject tend to be either ill-suited to the needs of a beginner, or fragmentary, or, in one or two instances, obscure. We hope that our book will go some way towards filling this gap in the literature.

Our presentation of the Littlewood-Paley theorem proceeds along two main lines, the first relating to singular integrals on locally compact groups, and the second to martingales. Both classical and modern versions of the theorem are dealt with, appropriate to the classical groups $\mathbb{R}^n$, $\mathbb{Z}^n$, $\mathbb{T}^n$ and to certain classes of disconnected groups. It is for the disconnected groups of Chapters 4 and 5 that we give two separate accounts of the Littlewood-Paley theorem: the first Fourier analytic, and the second probabilistic.

Some central results about multipliers of $\mathfrak{F}L^p(1 < p < \infty)$ are established, either collaterally with the Littlewood-Paley theorem, or as deductions from that theorem; for instance the famous theorems of M. Riesz, Marcinkiewicz, and Stečkin. In proving these concrete results, we have had also to develop or use certain portions of the *general* theory of Fourier multipliers. We think that the mix thus produced is a healthy one and that our book can therefore serve as a balanced introduction to the study of Fourier multipliers of L^p on LCA groups.

The applications, in the last chapter, to lacunary sets and Fourier multiplier theory, are meant to illustrate the importance of the Littlewood-Paley theorem as a tool in harmonic analysis. This is an idea which has been exploited with considerable success in recent years.

In addition to the general developments and applications just mentioned, our book contains a few results which, as far as we know, are new.

There are places where some readers may accuse us of pedantry. The fact is that we have merely tried to provide some details—possibly routine for the expert but troublesome for some others—which are almost always brushed aside with something close to contempt. The stock instance is the distinction between functions and the corresponding function-classes modulo negligible functions. Very often the vagueness and the cure are apparent even to a beginner. This is not always the case, however, and in such instances we have tried to replace the familiar hand-waving by something a little more convincing.

We have, in the main text, deliberately ignored historical and bibliographical matters. This is because we wished to pursue the mathematics, without undue distraction, to the goals we had set ourselves. Since, however, some bibliographical indications of the original sources of the main theorems are desirable, we have added a few comments of the kind in the Historical Notes at the end of the book. While we hope these Notes will be useful to some of our readers, we want to make it plain that they should be regarded as no more than a rough and incomplete guide to the literature.

We are indebted to many friends for encouragement and assistance in this enterprise.

Through his collaboration with the second of us, Alessandro Figà-Talamanca contributed indirectly in many ways to the present book. Even though it would be impossible to specify precise instances where his outlook, enthusiasm and ideas have made themselves felt in our writing, we (especially G^2) are well aware of his influence and are pleased to acknowledge it.

Edwin Hewitt has served on numerous occasions as a source of encouragement. It is due in no small measure to his good influence on us that our early plans for the book have now come to fruition.

We appreciate the exceedingly generous assistance of Jeff Sanders with the proof reading. Warm thanks are also due to Jan May for her expert typing of an early draft of our work, and to Cheryl Vertigan for producing the beautifully typed final version and for help with marking it up for the printers.

Contents

Prologue

Among the best known and most useful theorems in harmonic analysis is the Plancherel-Riesz-Fischer theorem. Stated for the circle group $\mathbb{T}$, this reads as follows.

(a) *The series*

$$\sum_{n\in\mathbb{Z}} c_n e^{inx} \tag{1}$$

is the Fourier series of a function in $L^2 = L^2(\mathbb{T})$ *if and only if*

$$\sum_{n\in\mathbb{Z}} |c_n|^2 < \infty; \tag{2}$$

in this case (1) *is the Fourier series of a function f in* L^2 *for which*

$$\|f\|_2^2 = \sum_{n\in\mathbb{Z}} |c_n|^2;$$

moreover, the series (1) *converges unconditionally in* L^2 *to f.*

Notice that the condition (2) is equivalent to the condition

$$\left\| \left(\sum_{n\in\mathbb{Z}} |c_n e^{int}|^2 \right)^{1/2} \right\|_2 < \infty$$

and (thanks to the orthogonality relations) to the boundedness with respect to N of the L^2 norms of the partial sums

$$\sum_{|n|\leq N} c_n e^{inx}.$$

Expressed rather loosely, the theorem affirms that we can build up an L^2 function from the series (1), provided only that we control (by keeping them bounded) how the sums

$$\sum_{|n|\leq N} |c_n|^2 = \left\| \left(\sum_{|n|\leq N} |c_n e^{int}|^2 \right)^{1/2} \right\|_2^2.$$

grow.

It is a standard fact that due to the orthogonality relations and the completeness of L^2, (a) is equivalent to the Parseval formula:

(b) *If $f \in L^2$, then*

$$\|f\|_2 = \left(\sum_{n \in \mathbb{Z}} |\hat{f}(n)|^2 \right)^{1/2},$$

$\hat{f}(n)$ denoting as usual the n-th Fourier coefficient of f.

It is natural to seek analogues of (a) and (b) which will be applicable with L^p in place of L^2 throughout. It turns out that fairly close analogues are available when $p \in (1, \infty)$.

An analogue of (b) is expressed by the Littlewood-Paley theorem, namely:

(b′) *To each p in $(1, \infty)$ corresponds a pair (A_p, B_p) of positive numbers such that*

$$A_p \|f\|_p \leqslant \left\| \left(\sum_{j \in \mathbb{Z}} |S_j f|^2 \right)^{1/2} \right\|_p \leqslant B_p \|f\|_p$$

for every f in L^p, where $S_j f$ is the j-th dyadic partial sum of the Fourier series of f, defined by the formulas

$$S_j f(x) = \begin{cases} \sum\limits_{2^{j-1} \leqslant |n| < 2^j} \hat{f}(n) e^{inx} & \text{if } j > 0 \\ \hat{f}(0) & \text{if } j = 0 \\ \sum\limits_{-2^{|j|} < n \leqslant -2^{|j|-1}} \hat{f}(n) e^{inx} & \text{if } j < 0. \end{cases} \tag{3}$$

It will be seen in 1.2.6(iv) that (b′) implies, among other things, the following partial analogue of (a):

(a′) *Suppose $p \in (1, \infty)$. The series* (1) *is the Fourier series of a function in L^p if and only if*

$$\left\| \left(\sum_{j \in \mathbb{Z}} |s_j|^2 \right)^{1/2} \right\|_p < \infty,$$

where s_j denotes the trigonometric polynomial obtained from formula (3) *by replacing $\hat{f}(n)$ by c_n throughout; in this case, the series* (1) *is the Fourier series of a function f in L^p satisfying the inequality*

$$A_p \|f\|_p \leqslant \left\| \left(\sum_{j \in \mathbb{Z}} |s_j|^2 \right)^{1/2} \right\|_p \leqslant B_p \|f\|_p;$$

moreover, the series $\sum_{j \in \mathbb{Z}} s_j$ converges unconditionally in L^p to f.

We can loosely paraphrase the Littlewood-Paley theorem (b′) by saying that the L^p norm of a function f can be computed, up to equivalence, by breaking the Fourier series of f into its dyadic partial sums, putting these together in an ℓ^2 fashion, and calculating the L^p norm of the resulting function. Similarly, statement (a′) tells us that we can manufacture an L^p Fourier series (1) by controlling the growth of its dyadic partial sums. Although the computations involved are more complex than the corresponding ones in (a) and (b), this is not too much of a drawback. See Chapter 9 for a number of important applications of these ideas.

The preceding remarks may help the reader to "place" the Littlewood-Paley theorem. It is without doubt one of the fundamental results in L^p harmonic analysis. In spite of this, it is difficult to find an exposition of the theorem and its proof which is well designed as an introduction, i.e., which assembles carefully the essential ideas and ingredients and which seeks to exhibit how these come together to make the theorem "work". This is our explanation of and apology for the existence of this monograph.

We present proofs of several versions of the Littlewood-Paley theorem for the groups $\mathbb{T}$, $\mathbb{Z}$ and $\mathbb{R}$ and their finite products, using so-called Hadamard decompositions of their duals, and for certain totally disconnected groups. At the same time we establish the equivalence of these Littlewood-Paley theorems with companion theorems about multipliers, the original versions of which are due to Marcinkiewicz, and prove the classical multiplier theorems of M. Riesz and Stečkin, so that the monograph may be regarded as dealing with some specialised aspects of multiplier theory.

Chapter 5 deals with a martingale version of the Littlewood-Paley theorem. It permits us to give an alternative and more general account of the Littlewood-Paley theorems for disconnected groups treated by other methods in Chapter 4. The basic necessities concerning conditional expectations and martingales are incorporated in Chapter 5 itself.

It is necessary to add, however, that by no means all aspects of Littlewood-Paley theory or of multiplier theory (even for the groups mentioned above) are dealt with; see 1.1.5 below. Besides this, we have not attempted to cover the recent work on Littlewood-Paley theory for certain noncommutative groups; see [7], [24] and [39].

With the notable exception of Chapter 5, already mentioned, the proofs of the Littlewood-Paley and multiplier theorems are based on the material in Chapter 2 concerning singular integrals, which is a known development of ideas due to Calderón and Zygmund.

Throughout the book, we have endeavoured to present the necessary ingredients of our account in a systematic fashion, and to keep matters reasonably self-contained; in particular, a number of auxiliary topics are treated briefly in the appendices. There is nothing very original in what we present since the proofs, or the ideas behind them, are well known to all "workers in the field". But we hope that the presentation is more accessible than existing ones and reasonably unified as far as it goes.

Chapter 1. Introduction

1.1. Littlewood-Paley Theory for $\mathbb{T}$

1.1.1. Our aim is to discuss certain portions of so-called Littlewood-Paley (= LP) theory purely as a part of abstract harmonic analysis. To give a preliminary idea of what this involves we will in this section describe the appropriate central results for the case in which the underlying group is the circle group $\mathbb{T}$: this is the original setting for the theory.

For this purpose we identify the character group of $\mathbb{T}$ with the additive group $\mathbb{Z}$ of integers, the integer n being identified with the character

$$\chi_n : e^{it} \to e^{int}$$

of $\mathbb{T}$ which it generates.

The so-called dyadic intervals in $\mathbb{Z}$ play a fundamental role: these are the intervals Δ_j ($j \in \mathbb{Z}$) defined as follows:

$$\Delta_j = \begin{cases} \{2^{j-1}, 2^{j-1}+1, \ldots, 2^j - 1\} & \text{if } j > 0, \\ \{0\} & \text{if } j = 0, \\ -\Delta_{|j|} & \text{if } j < 0. \end{cases}$$

If f is any integrable function on $\mathbb{T}$ and $\hat{f}$ its Fourier transform, we write

$$S_j f = \sum_{n \in \Delta_j} \hat{f}(n) \chi_n,$$

a partial sum of the Fourier series of f.

The first basic theorem is as follows.

1.1.2. The Littlewood-Paley theorem for $\mathbb{T}$. *To each p in $(1, \infty)$ correspond positive numbers A_p and B_p such that*

$$A_p \|f\|_p \leqslant \left\| \left(\sum_{j \in \mathbb{Z}} |S_j f|^2 \right)^{1/2} \right\|_p \leqslant B_p \|f\|_p$$

for (say) all trigonometric polynomials f on $\mathbb{T}$.

(It is easy to see that the same inequality can be derived for more general functions f on $\mathbb{T}$, but we do not wish to dwell on this point at the moment.)

1.1.3. The second basic theorem refers to the concept of *multiplier*.

Let ϕ be a bounded function on $\mathbb{Z}$ and p be in the range $[1, \infty]$. We say that ϕ is an *L^p multiplier* if and only if, for every f in L^p, the series

$$\sum_{n \in \mathbb{Z}} \phi(n)\hat{f}(n)\chi_n$$

is the Fourier series of some function g in L^p. In that case there is a number M such that $\|g\|_p \leqslant M\|f\|_p$, and the smallest such number M is denoted by $\|\phi\|_{p,p}$.

This can be expressed in another way. Introduce the operator T_ϕ defined initially as follows

$$(T_\phi f)^\wedge = \phi\hat{f}$$

for every trigonometric polynomial f on $\mathbb{T}$. Then ϕ is an L^p multiplier if and only if there is a number B such that

$$\|T_\phi f\|_p \leqslant B\|f\|_p$$

for every trigonometric polynomial f; and the smallest such B is $\|\phi\|_{p,p}$. (Here again the inequality will continue to hold for more general functions f.)

The second basic theorem is as follows.

1.1.4. The weak Marcinkiewicz multiplier theorem for $\mathbb{T}$. *To every p in* $(1, \infty)$ *corresponds a number K_p such that every bounded function ϕ on $\mathbb{Z}$ with a finite support and constant on every Δ_j is an L^p multiplier and*

$$\|\phi\|_{p,p} \leqslant K_p\|\phi\|_\infty.$$

This is a weak version of a multiplier theorem due to Marcinkiewicz; the strong version will be discussed in Chapter 8.

1.1.5. The Littlewood-Paley theorem for $\mathbb{T}$ for functions of power series type is included in Theorem (2.1) (cf. formulas (2.4) and (2.7)) of [40]. We emphasise the fact that the proof of the result given by Zygmund is quite different from the one presented here.

In the case of $\mathbb{T}$ and $\mathbb{R}$ and finite powers of these groups, there is another aspect of LP theory which is concerned with certain so-called "g-functions." These are quadratic functionals involving derivatives of the Poisson integral of f, arising through connections with analytic and harmonic functions. As such, they are somewhat more remotely connected with pure harmonic analysis and we shall leave them aside. For details see [38], Chapter IV and [39].

Also missing is all discussion of those aspects of LP theory for $\mathbb{T}$ relating to matters of pointwise convergence a.e. of subsequences of partial sums of the Fourier series of functions in L^p. Such matters are discussed in [40], Chapter XV. (Problems of convergence pointwise a.e. seem at present to be of relatively little significance in abstract harmonic analysis; quite possibly this is an aspect, the study of which lies in the future.)

1.1.6. In this exposition we shall be concerned with the Littlewood-Paley and Marcinkiewicz theorems themselves and with analogues of them for cases where the underlying group $\mathbb{T}$ is replaced by other groups, such as finite products of $\mathbb{T}$, $\mathbb{R}$ and $\mathbb{Z}$ and certain disconnected groups. The classical results for $\mathbb{T}$, $\mathbb{R}$ and $\mathbb{Z}$ will be found in Chapters 6, 7 and 8. Rather surprisingly, perhaps, the case of certain disconnected groups, discussed in Chapter 4, is in some respects technically a good deal simpler than the cases of $\mathbb{T}$, $\mathbb{R}$, and $\mathbb{Z}$ and leads to some new results even for the group $\mathbb{T}$. A martingale version of the Littlewood-Paley theorem is presented in Chapter 5. This approach leads to alternative proofs for, and more general versions of, the theorems in Chapter 4.

In Section 1.2 we shall formulate analogues of 1.1.2 and 1.1.4 for general families (Δ_j) and discuss an equivalence between them for the case of so-called decompositions.

1.2. The LP and WM Properties

1.2.1. Throughout this section G denotes a Hausdorff LCA group and X its character group. The Haar measures on G and X are assumed to be adjusted so that the Fourier transformation $f \to \hat{f}$ is an isometry of $L^2 = L^2(G)$ onto $L^2(X)$. (We take for granted the basic properties of the Fourier transformation as defined on L^1 and L^2. The reader familiar with the first two chapters of [35] will have an adequate background in abstract harmonic analysis for everything that we do.)

The following notation will be employed whenever G is a Hausdorff LCA group: if g is an extended-real-valued or complex-valued measurable function on G, we write

$$\|g\|_p = \left(\int |g|^p \, dx\right)^{1/p} \quad \text{if} \quad p \in [1, \infty),$$

$$\|g\|_\infty = \text{loc ess sup } |g|.$$

Thus $\|g\|$ may be ∞; and $g \in \mathscr{L}^p(G)$ if and only if g is real- or complex-valued and measurable and $\|g\|_p < \infty$. If h is an equivalence class (modulo negligible functions if $p \neq \infty$, and modulo locally negligible functions if $p = \infty$) of measurable functions, $\|h\|_p$ will denote the common value of $\|g\|_p$ for every function g belonging to the class h. Thus $h \in L^p$ if and only if g is measurable and $\|g\|_p < \infty$ for some one (and hence every) function g belonging to the class h.

1.2.2. Multipliers. If $\phi \in \mathscr{L}^\infty(X)$, we define T_ϕ to be the continuous linear mapping of L^2 into itself for which

$$(T_\phi f)^\wedge = \phi \hat{f}$$

for every f in L^2 (the equality is, of course, one between equivalence classes of functions). Compare this with 1.1.3: if G is noncompact, we cannot use trigonometric polynomials at this point, as we did in 1.1.3.

Note that T_ϕ depends only on the class of ϕ modulo locally negligible functions.

Given p in $[1, \infty]$ and ϕ in $\mathscr{L}^\infty(X)$, ϕ is said to be an L^p *multiplier* if and only if there is a number B such that

$$\|T_\phi f\|_p \leqslant B\|f\|_p$$

for every f in $L^2 \cap L^p(G)$. For this it is enough (see Appendix A.1, the notations of which are being used here) that the same inequality should hold for f in $L^1 \cap L^\infty$. The smallest admissible B is then denoted by $\|T_\phi\|_{p,p}$, $\|\phi\|_{M_p(X)}$ or $\|\phi\|_{p,p}$ and termed the (p-) *multiplier norm of* ϕ. The operator T_ϕ is called a *multiplier operator*.

When G is compact, ϕ is a multiplier of L^p in the above sense if and only if for every f in $L^p(G)$ the series

$$\sum_{\chi \in X} \phi(\chi)\hat{f}(\chi)\chi$$

is the Fourier series of some g in L^p; cf. 1.1.3 again.

We denote the set of L^p multipliers by $M_p(X)$, or sometimes more briefly by M_p when X is understood. Occasionally we say that an element Φ of $L^\infty(X)$ belongs to $M_p(X)$, meaning thereby that some one (and hence every) function belonging to the class Φ is a member of $M_p(X)$.

It is immediate from the Plancherel theory that $M_2(X) = \mathscr{L}^\infty(X)$ and that $\|\phi\|_{2,2} = \|\phi\|_\infty =$ loc ess sup $|\phi|$ for every ϕ in $\mathscr{L}^\infty(X)$.

We now record three fundamental properties of multipliers which will be used frequently in the sequel.

(i) Use of the Parseval formula, together with Hölder's inequality and its converse, shows that $M_p(X) = M_{p'}(X)$ with equality of the corresponding (semi-) norms (here, as always, p' denotes the exponent conjugate to p: $1/p + 1/p' = 1$).

(ii) The Riesz convexity theorem shows in particular that

$$\|\phi\|_\infty \leqslant \|\phi\|_{p',p'}^{1/2}\|\phi\|_{p,p}^{1/2} = \|\phi\|_{p,p}.$$

It can also be shown that $\phi \in M_1(X) = M_\infty(X)$ if and only if ϕ is equal l.a.e. to a Fourier-Stieltjes transform $\hat{\mu}$, where μ is a bounded measure on G, in which case $\|\phi\|_{1,1} = \|\phi\|_{\infty,\infty} = \|\mu\|$. [This is essentially a reformulation of [25], Theorem 0.1.1 or [3], Corollary 2.6.2.]

Since $\mu * f \in L^p(G)$ and $\|\mu * f\|_p \leqslant \|\mu\|\,\|f\|_p$ whenever μ is a bounded measure on G, $f \in L^p(G)$, and $1 \leqslant p \leqslant \infty$, it follows that $\hat{\mu} \in M_p(X)$ for all such μ and p; furthermore $\|\hat{\mu}\|_{p,p} \leqslant \|\mu\|$.

(iii) Suppose that $\phi \in \mathscr{L}^\infty(X)$ and $1 \leqslant p \leqslant \infty$. Then $\phi \in M_p(X)$ if and only if there is a constant C such that

$$\left|\int_X \phi(\chi)\hat{f}(\chi)\hat{g}(\chi)\, d\chi\right| \leqslant C\|f\|_p\|g\|_{p'}$$

for all integrable functions f and g with compactly supported Fourier transforms.

For more details concerning these and other aspects of multipliers, see [9], Chapter 16 and [25].

If Δ is a measurable subset of X and ξ_Δ denotes its characteristic function relative to X, we shall usually write S_Δ in place of T_{ξ_Δ}; S_Δ is a Fourier partial sum or partial integral operator.

1.2.3. Decompositions of X. In what follows, J denotes a countable (i.e., finite or countably infinite) index set and $(\Delta_j)_{j\in J}$ a family of measurable subsets of X.

Such a family is termed a *decomposition of* X if and only if

(i) the Δ_j are pairwise disjoint;

(ii) $X\backslash(\bigcup_{j\in J}\Delta_j)$ is locally negligible.

The LP and WM properties are to be formulated for families (Δ_j) of measurable sets. When it comes to proving theorems about the LP and WM properties, we shall frequently assume that the family (Δ_j) is actually a *decomposition*, even though the stronger assumption is not, in some instances, strictly necessary. On the contrary, the R property, to be introduced in 1.2.12, is formulated for and mainly used in cases where the sets of the family (Δ_j) may overlap.

Whatever the family (Δ_j) we shall for simplicity write S_j in place of S_{Δ_j}.

We consider two statements, each of which may be regarded as expressing a property of the family (Δ_j).

1.2.4. The LP **(Littlewood-Paley) property.** *To every* p *in* $(1, \infty)$ *corresponds a positive number* B_p *such that*

$$\left\|\left(\sum_{j\in J}|S_jf|^2\right)^{1/2}\right\|_p \leqslant B_p\|f\|_p \tag{1}$$

for every f *in* $L^2 \cap L^p(G)$.

The term on the left of (1) is defined to mean $\|g\|_p$ where $g = (\sum_{j\in J}|g_j|^2)^{1/2}$ and g_j is a representative function chosen from the class S_jf of L^2; the result does not depend on the choice of the g_j.

As will appear in 1.2.6(ii), to assert the LP property of a *decomposition* (Δ_j) amounts to saying that a direct analogue of the Littlewood-Paley theorem 1.1.2 is valid, the Δ_j taking the place of the dyadic intervals.

1.2.5. The WM **property: the weak Marcinkiewicz multiplier property.** *To every* p *in* $(1, \infty)$ *corresponds a positive number* K_p *such that every* ϕ *in* $\mathscr{L}^\infty(X)$ *which is constant on each* Δ_j *and zero off finitely many of them is an* L^p *multiplier* (*i.e., belongs to* $M_p(X)$) *and*

$$\|\phi\|_{p,p} \leqslant K_p\|\phi\|_\infty. \tag{2}$$

Obviously, to affirm the WM property amounts to saying that the direct analogue of the weak Marcinkiewicz multiplier theorem 1.1.4 is valid, the Δ_j again taking the place of the dyadic intervals. Stronger versions of this theorem will be discussed later.

The main purpose of this monograph is to exhibit non-trivial examples of groups and decompositions having the LP and WM properties. In the rest of this chapter, we add diverse comments on the LP and WM properties and show them to be equivalent.

1.2.6. Remarks concerning the LP property.

(i) Suppose that the LP property obtains for a family (Δ_j). It follows at once from (1) that, for each index j,

$$\|S_j f\|_p \leqslant B_p \|f\|_p$$

for f in $L^2 \cap L^p(G)$. This implies that ξ_{Δ_j} is an L^p multiplier, and that S_j can be extended uniquely from $L^2 \cap L^p$ into a continuous endomorphism of L^p. We continue to use the symbol S_j to denote this extension, and note that then the last inequality continues to hold for every f in L^p. If $1 < p \leqslant 2$, or if G is compact and $1 < p < \infty$, the relation $(S_j f)^\wedge = \xi_\Delta \hat{f}$ continues to hold for every f in L^p.

We claim that furthermore the inequality (1) *continues to hold for every f in L^p.*

Proof. Take any f in L^p and a sequence (f_n) of elements of $L^2 \cap L^p$ converging in L^p to f. If F is any finite subset of J we have by hypothesis that

$$\left\| \left(\sum_{j \in F} |S_j f_n|^2 \right)^{1/2} \right\|_p \leqslant B_p \|f_n\|_p$$

for every n. We can extract a subsequence $(S_j f_{n_i})_{i=1}^\infty$ such that

$$\lim_i S_j f_{n_i} = S_j f$$

holds pointwise a.e. for every j in F, and then Fatou's lemma applied to the last inequality yields the estimate

$$\left\| \left(\sum_{j \in F} |S_j f|^2 \right)^{1/2} \right\|_p \leqslant \liminf_i B_p \|f_{n_i}\|_p = B_p \|f\|_p.$$

Finally, if we let F expand to J, the monotone convergence theorem leads to (1). □

(ii) On comparing 1.1.2 with 1.2.4, the reader will note that the left-hand inequality is missing from the latter. The reason for this is that, as we shall now show, *the* LP *property for a decomposition (Δ_j) already implies the existence of a positive number A_p for each p in* $(1, \infty)$ *such that*

$$A_p \|f\|_p \leqslant \left\| \left(\sum_{j \in J} |S_j f|^2 \right)^{1/2} \right\|_p \tag{3}$$

for all f in $L^2 \cap L^p(G)$. Thus the LP *property implies the Littlewood-Paley theorem for the decomposition (Δ_j).*

Proof. The hypotheses (i) and (ii) in 1.2.3 combine with the Parseval formula to show that the series

$$\sum_{j \in J} S_j f \tag{4}$$

is unconditionally convergent in L^2 to f for every f in L^2. The Parseval formula shows also that

$$\int S_j f . \overline{S_r g} \, dx = 0 \tag{5}$$

if $j \neq r$ and $f, g \in L^2$. Take any increasing sequence (F_n) of finite subsets of J whose union is J. Applying the Cauchy-Schwarz inequality, Hölder's inequality, and (1) we then obtain (in view of the convergence of (4) and of the corresponding series with g in place of f):

$$\begin{aligned}
\left|\int f\bar{g}\, dx\right| &= \lim_n \left|\int \Big(\sum_{j \in F_n} S_j f\Big)\Big(\sum_{r \in F_n} \overline{S_r g}\Big) dx\right| \\
&= \lim_n \left|\int \Big(\sum_{j \in F_n} S_j f . \overline{S_j g}\Big) dx\right| \\
&\leqslant \liminf_n \int \Big(\sum_{j \in F_n} |S_j f|^2\Big)^{1/2} \Big(\sum_{j \in F_n} |S_j g|^2\Big)^{1/2} dx \\
&\leqslant \liminf_n \left\|\Big(\sum_{j \in F_n} |S_j f|^2\Big)^{1/2}\right\|_p \left\|\Big(\sum_{j \in F_n} |S_j g|^2\Big)^{1/2}\right\|_{p'} \\
&\leqslant \left\|\Big(\sum_{j \in J} |S_j f|^2\Big)^{1/2}\right\|_p B_{p'} \|g\|_{p'},
\end{aligned}$$

provided $f \in L^2 \cap L^p$ and $g \in L^2 \cap L^{p'}$. The converse of Hölder's inequality now gives the inequality

$$\|f\|_p \leqslant B_{p'} \left\|\Big(\sum_{j \in J} |S_j f|^2\Big)^{1/2}\right\|_p,$$

whence (3) with $A_p = B_{p'}^{-1}$. □

(iii) *It is in fact the case that, if the* LP *property holds for a decomposition* (Δ_j), *then* (3) *is true for every* f *in* $L^p(G)$.

Proof. If the LP property holds then, by (i), the operators S_j are well-defined on all of L^p, and the right side of (3) is finite when $f \in L^p$. Now suppose we take a sequence (f_n) of elements of $L^2 \cap L^p$ which converges to f. If we show that

$$\|(\sum |S_j f_n|^2)^{1/2}\|_p \to \|(\sum |S_j f|^2)^{1/2}\|_p,$$

the proof will be complete. Using twice the fact that

$$|\|u\| - \|v\|| \leqslant \|u - v\|,$$

it follows that

$$\begin{aligned} &\left|\|(\sum |S_j f_n|^2)^{1/2}\|_p - \|(\sum |S_j f|^2)^{1/2}\|_p\right| \\ &\leqslant \left\||(\sum |S_j f_n|^2)^{1/2} - (\sum |S_j f|^2)^{1/2}|\right\|_p \\ &\leqslant \left\|(\sum |S_j(f - f_n)|^2)^{1/2}\right\|_p \\ &\leqslant B_p\|f - f_n\|_p, \end{aligned}$$

which converges to zero. □

(iv) **The Plancherel-Riesz-Fischer theorem for L^p.** *Let G be compact and metrisable. Suppose that (Δ_j) is a decomposition of X possessing the* LP *property and such that every Δ_j is finite. Let c be a complex-valued function on X and define the trigonometric polynomials*

$$s_j = \sum_{\chi \in \Delta_j} c(\chi)\chi.$$

Suppose that $1 < p < \infty$. Then $c = \hat{f}$ for some f in $L^p(G)$ if and only if

(a) $$\left\|\left(\sum_{j \in J} |s_j|^2\right)^{1/2}\right\|_p < \infty,$$

in which case

(b) $$A_p\|f\|_p \leqslant \left\|\left(\sum_{j \in J} |s_j|^2\right)^{1/2}\right\|_p \leqslant B_p\|f\|_p.$$

Proof. Suppose first that $c = \hat{f}$ for some f in L^p. By 1.2.6(i), $s_j = S_j f$ for every j in J and the left hand side of (a) is majorised by $B_p\|f\|_p$, which is finite. Hence (a) holds. Moreover, (b) follows on combining 1.2.6(i) and 1.2.6(iii).

Conversely, suppose that (a) holds. Let (F_r) be any increasing sequence of finite subsets of J with union equal to J, and put $\Phi_r = \bigcup_{j \in F_r} \Delta_j$, so that the Φ_r form an increasing sequence of finite subsets of X with union equal to X. Write also

$$f_r = \sum_{j \in F_r} s_j,$$

which is a trigonometric polynomial. Since the Δ_j are pairwise disjoint, it is evident that $S_j f_r = s_j$ or 0 according as $j \in F_r$ or $j \in J \backslash F_r$. Hence 1.2.6(ii) entails that

$$A_p\|f_r\|_p \leqslant \left\|\left(\sum_{j \in F_r} |s_j|^2\right)^{1/2}\right\|_p,$$

which by (a) is bounded with respect to r. As a result, the sequence (f_r) has a weak limiting point f in L^p. Then $\hat{f}$ is a limiting point of $(\hat{f}_r)$ for the topology of pointwise convergence on X. Since (Δ_j) is a decomposition of X, $\hat{f}_r = \xi_{\Phi_r} c$, which converges pointwise on X to c. It therefore follows that $\hat{f} = c$, as we had to show. □

Remarks. Suppose (a) of 1.2.6(iv) holds. It will be shown in 1.2.9 below that then the series $\sum_{j\in J} s_j$, obtained from the series $\sum_{\chi\in X} c(\chi)\chi$ by grouping terms according to the decomposition (Δ_j), converges unconditionally in L^p to f.

On the other hand, if G is an infinite compact Abelian group, $p \in [1, \infty]$, and p is different from 2, it is false to assert that the Fourier series $\sum_{\chi\in X} \hat{f}(\chi)\chi$ is unconditionally convergent in L^p for every f in L^p. Indeed, in the contrary case, the numerical series $\sum_{\chi\in X} \hat{f}(\chi)\hat{g}(\chi)$ would be unconditionally convergent, and hence absolutely convergent, whenever $f \in L^p$ and $g \in L^{p'}$. From this and [10], Theorem 7.7.9 and Corollary 6.2.3, it would follow that there is a constant B such that if ω is any function from X into $\{-1, 1\}$, then

$$|\sum \omega(\chi)\hat{f}(\chi)\hat{g}(\chi)| \leqslant \sum |\hat{f}(\chi)\hat{g}(\chi)| \leqslant B\|f\|_p\|g\|_{p'}.$$

Since $\|f\|_p \leqslant \|f\|_\infty$ and $\|g\|_{p'} \leqslant \|g\|_\infty$ (G is compact), it would follow that both series $\sum_{\chi\in X} \omega(\chi)\hat{f}(\chi)\chi$ and $\sum_{\chi\in X} \omega(\chi)\hat{g}(\chi)\chi$ are at least Fourier-Stieltjes series whenever $f \in L^p$ and $g \in L^{p'}$. Hence ([11], Theorem (1.1)) we should have both that $\hat{f} \in \ell^2(X)$ whenever $f \in L^p$ and $\hat{g} \in \ell^2(X)$ whenever $g \in L^{p'}$, and hence both that $L^p \subseteq L^2$ and $L^{p'} \subseteq L^2$. Since G is infinite, the first conclusion is false if $1 \leqslant p < 2$, and the second is false if $2 < p \leqslant \infty$.

It is possible to formulate a Plancherel-Riesz-Fischer theorem for L^p in case G is noncompact, but the details are inevitably more complicated. In particular, if $p > 2$, a distributional-style Fourier transformation is involved.

1.2.7. Remarks concerning the WM **property.**

(i) *Suppose that the* WM *property holds for a decomposition* (Δ_j). *If* $\phi \in \mathscr{L}^\infty(X)$ *and is constant on every* Δ_j, *then* $\phi \in M_p(X)$ *and*

$$\|\phi\|_{p,p} \leqslant K_p\|\phi\|_\infty.$$

Proof. Let ϕ be as described in the hypothesis. Take an increasing sequence (F_n) of finite subsets of J having union equal to J and define ϕ_n to agree with ϕ on $\bigcup_{j\in F_n} \Delta_j$ and to be zero elsewhere on X. Since the WM property holds,

$$\begin{aligned}\|T_{\phi_n} f\|_p &\leqslant K_p\|\phi_n\|_\infty\|f\|_p \\ &\leqslant K_p\|\phi\|_\infty\|f\|_p\end{aligned}$$

for every f in $L^2 \cap L^p$ and every n. On the other hand, the Parseval formula shows that $\lim T_{\phi_n} f = T_\phi f$ in L^2, and so a subsequence $(T_{\phi_{n_i}} f)_{i=1}^\infty$ converges a.e. to $T_\phi f$. The last inequality then combines with Fatou's lemma to show that

$$\|T_\phi f\|_p \leqslant K_p\|\phi\|_\infty\|f\|_p$$

for every f in $L^2 \cap L^p$, and the result follows. □

(ii) *Conversely, suppose that every* ϕ *in* $\mathscr{L}^\infty(X)$ *which is constant on every* Δ_j *belongs to* $M_p(X)$ (*no a priori bound on* $\|\phi\|_{p,p}$ *being assumed*). *Then the* WM *property holds.*

Proof. Since alteration of ϕ on a locally negligible set does not affect T_ϕ, every ϕ in $\mathscr{L}^\infty(X)$ which is equal l.a.e. on every Δ_j to a constant belongs to $M_p(X)$. The set M of such ϕ is a closed subspace of $\mathscr{L}^\infty(X)$. By hypothesis, then, N: $\phi \to \|\phi\|_{p,p}$ is a seminorm on M. Since

$$\begin{aligned} N(\phi) &= \sup\{\|T_\phi f\|_p : f \in L^2 \cap L^p, \|f\|_p \leqslant 1\} \\ &= \sup\left\{\left|\int T_\phi f . \bar{g}\, dx\right| : f \in L^2 \cap L^p, g \in L^2 \cap L^{p'}, \|f\|_p \leqslant 1, \|g\|_{p'} \leqslant 1\right\} \\ &= \sup\left\{\left|\int \phi \hat{f}\hat{\bar{g}} d\chi\right| : f \in L^2 \cap L_p, g \in L^2 \cap L^{p'}, \|f\|_p \leqslant 1, \|g\|_{p'} \leqslant 1\right\}, \end{aligned}$$

and since the mapping

$$\phi \to \int \phi \hat{f}\hat{\bar{g}} d\chi$$

is plainly continuous on M for each fixed f in $L^2 \cap L^p$, and g in $L^2 \cap L^{p'}$, it follows that N is lower semicontinuous on M. Therefore ([10], 7.1.2) N is continuous on M, i.e. there exists a number K_p such that

$$\|\phi\|_{p,p} \leqslant K_p \|\phi\|_\infty$$

for all ϕ in M. This plainly implies that the WM property holds. □

Note. Hypothesis (i) of 1.2.3 is not used in the proofs above.

1.2.8. The equivalence of the LP and WM properties for decompositions.

(i) *The* WM *property implies the* LP *property*. Let $j \to j^*$ be an injection of J into the set of nonnegative integers, and let $r_0, r_1, r_2, \ldots$ denote the Rademacher functions on [0, 1]. Assume the WM property obtains. Since the Δ_j are pairwise disjoint, we have for any finite subset F of J that

$$\left\|\sum_{j\in F} r_{j^*}(t)\xi_{\Delta_j}\right\|_{p,p} \leqslant K_p$$

for every t in [0, 1]. This signifies that

$$\left\|\sum_{j\in F} r_{j^*}(t) S_j f\right\|_p \leqslant K_p \|f\|_p$$

for every t in [0, 1] and every f in $L^2 \cap L^p$, i.e., that

$$\int_G \left|\sum_{j\in F} r_{j^*}(t) S_j f(x)\right|^p dx \leqslant K_p^p \|f\|_p^p \tag{6}$$

for every t in [0, 1] and every f in $L^2 \cap L^p$. Integrate (6) with respect to t over [0, 1] and use the Fubini-Tonelli theorem to write the result in the form

$$\int_G \left\{ \int_0^1 \left| \sum_{j \in F} r_{j*}(t) S_j f(x) \right|^p dt \right\} dx \leqslant K_p^p \|f\|_p^p.$$

Since the set of Rademacher functions forms a set of type $(2, p)$ ([9], 14.2.1), there is a number $a_p > 0$ such that the inner integral in the last inequality is not less than

$$a_p^p \left(\sum_{j \in F} |S_j f(x)|^2 \right)^{p/2},$$

so that we are led to the estimate

$$\left\| \left(\sum_{j \in F} |S_j f|^2 \right)^{1/2} \right\|_p^p \leqslant a_p^{-p} K_p^p \|f\|_p^p.$$

If we now let F expand to J, the monotone convergence theorem leads to (1) with $B_p = a_p^{-1} K_p$. □

(ii) *The* LP *property implies the* WM *property*. Assume that the LP property holds and begin by observing that, as a consequence of Parseval's formula,

$$\int S_j f . \bar{g}\, dx = \int S_j f . \overline{S_j g}\, dx \tag{7}$$

for each j in J and all f and g in $L^2(G)$.

Let ϕ in $\mathscr{L}^\infty(X)$ be as stipulated in the statement of the WM property. Redefine ϕ to be zero on Δ_j whenever Δ_j is locally negligible; this leaves T_ϕ unaltered. Suppose that then ϕ takes the constant value c_j on Δ_j, where $c_j = 0$ for each j in $J \backslash F$ and F is a finite subset of J. Then $|c_j| \leqslant \|\phi\|_\infty$ whenever $j \in J$ and, thanks to 1.2.3(i),

$$T_\phi = \sum_{j \in F} c_j S_j.$$

So, using (7) and the Cauchy-Schwarz inequality, we obtain the estimate

$$\begin{aligned} \left| \int T_\phi f . \bar{g}\, dx \right| &= \left| \sum_{j \in F} c_j \int S_j f . \bar{g}\, dx \right| \\ &= \left| \sum_{j \in F} c_j \int S_j f . \overline{S_j g}\, dx \right| \\ &\leqslant \|\phi\|_\infty . \sum_{j \in F} \int |S_j f| . |S_j g|\, dx \\ &\leqslant \|\phi\|_\infty . \int \left(\sum_{j \in F} |S_j f|^2 \right)^{1/2} \left(\sum_{j \in F} |S_j g|^2 \right)^{1/2} dx \end{aligned}$$

for f in $L^2 \cap L^p$ and g in $L^2 \cap L^{p'}$. If we now change g into $\bar{g}$ and apply Hölder's inequality and (1) to the integral on the right, it appears that

$$\left|\int T_\phi f.g\,dx\right| \leqslant \|\phi\|_\infty B_p\|f\|_p B_{p'}\|g\|_{p'}$$

whenever $f \in L^2 \cap L^p$ and $g \in L^2 \cap L^{p'}$. The converse of Hölder's inequality applied to the integral on the left now shows that

$$\|T_\phi f\|_p \leqslant \|\phi\|_\infty B_p B_{p'}\|f\|_p \tag{8}$$

for all f in $L^2 \cap L^p$, which shows that the WM property holds. □

1.2.9. The LP property and unconditional convergence. *Consider the following three statements:*

(i) *the family* (Δ_j) *possesses the* LP *property;*

(ii) *for every j in J and every p in $(1, \infty)$, S_j is extendible into a continuous endomorphism of L^p and the series $\sum_{j\in J} S_j f$ is unconditionally weakly convergent in L^p for every f in L^p;*

(iii) *for every j in J and every p in $(1, \infty)$, S_j is extendible into a continuous endomorphism of L^p and the series $\sum_{j\in J}\gamma(j)S_j f$ is unconditionally convergent in L^p whenever $f \in L^p$ and $\gamma \in \ell^\infty(J)$.*

We claim that these statements are equivalent.

Proof. It is enough to show that (iii) implies (ii), that (i) implies (iii), and that (ii) implies (i).

(a) (iii) *implies* (ii). This is trivial.

(b) (i) *implies* (iii). Assume (i). It has been seen in 1.2.6(i) that, if $j \in J$ and $p \in (1, \infty)$, S_j is extendible into a continuous endomorphism of L^p and (1) continues to hold for all f in L^p. Therefore, if $f \in L^p$ and $\varepsilon > 0$, it follows from the dominated convergence theorem that there is a finite subset $F_0 \subseteq J$ for which

$$\left\|\left(\sum_{J\setminus F_0} |S_j f|^2\right)^{1/2}\right\|_p < \varepsilon. \tag{9}$$

Now if $\gamma \in \ell^\infty(J)$, and F_1 and F_2 are finite subsets of J such that $F_2 \supseteq F_1 \supseteq F_0$, it follows from Höldeı's inequality that

$$\begin{aligned}
\left\|\sum_{F_2\setminus F_1}\gamma(j)S_j f\right\|_p &= \sup_{\|g\|_{p'}\leqslant 1}\left|\int \sum_{F_2\setminus F_1}\gamma(j)S_j f.\bar{g}\,dx\right| \\
&= \sup_{\|g\|_{p'}\leqslant 1}\left|\int \sum_{F_2\setminus F_1}\gamma(j)S_j f.\overline{S_j g}\,dx\right| \\
&\leqslant \sup_{\|g\|_{p'}\leqslant 1}\int \sum_{F_2\setminus F_1}|\gamma(j)S_j f.\overline{S_j g}|\,dx \\
&\leqslant \|\gamma\|_\infty \sup_{\|g\|_{p'}\leqslant 1}\int\left(\sum_{F_2\setminus F_1}|S_j f|^2\right)^{1/2}\left(\sum_{F_2\setminus F_1}|S_j g|^2\right)^{1/2} dx
\end{aligned}$$

$$\leqslant \|\gamma\|_\infty \sup_{\|g\|_{p'}\leqslant 1} \left\| \left(\sum_{F_2 \setminus F_1} |S_j f|^2 \right)^{1/2} \right\|_p \left\| \left(\sum_J |S_j g|^2 \right)^{1/2} \right\|_{p'}$$

$$\leqslant \|\gamma\|_\infty \, \varepsilon \sup_{\|g\|_{p'}\leqslant 1} B_{p'} \|g\|_{p'} = \|\gamma\|_\infty B_{p'} \varepsilon,$$

by (9), and (1) applied to the pair (g, p'). Hence the series $\sum \gamma(j) S_j f$ converges unconditionally in L^p.

(c) (ii) *implies* (i). If (ii) holds, the numerical series

$$\sum_{j \in J} \int S_j f . g \, dx$$

is unconditionally convergent for every f in L^p, and g in $L^{p'}$. Hence this series is absolutely convergent if $f \in L^p$ and $g \in L^{p'}$. Put

$$N(f, g) = \sum_{j \in J} \left| \int S_j f . \bar{g} \, dx \right|.$$

For fixed f in L^p, the mapping $g \to N(f, g)$ is a seminorm on $L^{p'}$ which is evidently lower semicontinuous (as the supremum of $\sum_{j \in F} \ldots$, F finite). Hence ([10], 7.1.2) it is continuous on $L^{p'}$. Thus

$$N'(f) = \sup \{N(f, g) : \|g\|_{p'} \leqslant 1\} < \infty$$

is a seminorm on L^p. It is lower semicontinuous, being the supremum of all sums

$$\sum_{j \in F} \left| \int S_j f . \bar{g} \, dx \right|,$$

F ranging over all finite subsets of J and g varying over the set of functions in $L^{p'}$ such that $\|g\|_{p'} \leqslant 1$. Hence N' is continuous on L^p. This signifies that there is a number B'_p such that

$$\sum_{j \in J} \left| \int S_j f . g \, dx \right| \leqslant B'_p \|f\|_p \|g\|_{p'}$$

for all f in L^p and all g in $L^{p'}$. This, together with the converse of Hölder's inequality, shows that a fortiori

$$\left\| \sum_{j \in F} \gamma(j) S_j f \right\|_p \leqslant B'_p \|f\|_p \tag{10}$$

for every f in L^p, γ in $\ell^\infty(J)$ such that $\|\gamma\|_\infty \leqslant 1$ and any finite subset F of J. By using the properties of the Rademacher functions as in the proof of 1.2.8(i), we deduce from (10) that

$$\left\| \left(\sum_{j \in F} |S_j f|^2 \right)^{1/2} \right\|_p \leqslant B_p \|f\|_p,$$

and so, by the monotone convergence theorem, the inequality

$$\left\| \left(\sum_{j \in J} |S_j f|^2 \right)^{1/2} \right\|_p \leqslant B_p \|f\|_p,$$

for all f in L^p. This proves that (i) holds. □

Note. If the hypothesis (i) is replaced by the inequality (1) in 1.2.4 for both the single fixed index p in $(1, \infty)$ and its conjugate p', the proofs can proceed. Thus 1.2.9 can be formulated for a single fixed, but arbitrary, p rather than for all p in the range $(1, \infty)$.

1.2.10. Further remarks on the equivalence of the LP and WM properties. In the formulation of the WM property of a decomposition $(\Delta_j)_{j \in J}$ (see 1.2.5), we specified not only that every bounded function ϕ on X which is constant on each of the sets Δ_j and 0 off the union of finitely many of them should be a multiplier of L^p for $1 < p < \infty$, but also that an inequality of the form

$$\|\phi\|_{p,p} \leqslant K_p \|\phi\|_\infty \tag{11}$$

should hold. The interesting outcome of the theorem we now present is that there is no need to specify the validity of the inequality (11); it is a consequence of the membership of the functions ϕ to $M_p(X)$.

Theorem. *Let $(\Delta_j)_{j \in J}$ be a decomposition of the LCA group X. For every subset I of J, write*

$$\sigma_I = \bigcup_{j \in I} \Delta_j.$$

The following statements are equivalent:

(i) *the decomposition (Δ_j) has the* LP *property;*

(ii) *$\xi_{\sigma_I} \in M_p(X)$ for every subset I of X and every index p in $(1, \infty)$;*

(iii) *for every p in $(1, \infty)$ there is a number K_p such that*

$$\|\xi_{\sigma_I}\|_{p,p} \leqslant K_p$$

for all finite subsets I of J.

Proof. It is a trivial consequence of 1.2.8(ii) and 1.2.7(i) that (i) implies (ii) and (i) implies (iii). A quick re-examination of 1.2.8(i) will show that it has already been proved there that (iii) implies (i). So it remains only to prove that (ii) implies (iii).

Introduce the compact space $A = \{0, 1\}^J$ with the metrisable product topology; for α in A, write

$$\phi_\alpha = \sum_{j \in J} \alpha(j)\xi_{\Delta_j}. \tag{12}$$

(The series on the right of (12) converges pointwise unconditionally.) Notice that

$$\phi_\alpha = \xi_{E_\alpha}$$

where

$$E_\alpha = \bigcup_{\alpha(j)=1} \Delta_j. \tag{13}$$

By hypothesis, $\phi_\alpha \in M_p(X)$ for every α in A and every p in $(1, \infty)$. Our aim is to prove that for each p in $(1, \infty)$, the function $v\colon \alpha \to \|\phi_\alpha\|_{p,p}$ is bounded on A. To this end we claim that the function v is lower semicontinuous on A, i.e. that for any c in $\mathbb{R}$,

$$\{\alpha \in A : v(\alpha) \leqslant c\}$$

is closed in A. Observe that, by virtue of 1.2.2(iii), to say that $v(\beta) = \|\phi_\beta\|_{p,p} \leqslant c$ is the same as saying that

$$\left| \int_X \phi_\beta(\chi)\hat{f}(\chi)\hat{g}(\chi)\, d\chi \right| \leqslant c\|f\|_p\|g\|_{p'} \tag{14}$$

for all integrable functions f and g on G having compactly supported Fourier transforms.

If (α_k) is a sequence of points in A for which $v(\alpha_k) \leqslant c$ and if $\alpha_k \to \alpha$ in A, then by the definition of the topology of A and (13), $\phi_{\alpha_k} \to \phi_\alpha$ pointwise. So by (14) and the dominated convergence theorem, $\|\phi_\alpha\| \leqslant c$ also. This establishes the lower semicontinuity of v.

By Baire's category theorem, there exists a nonvoid open subset of A on which v is bounded. In other words, there is a point α_0 of A, a finite subset F of J, and a number $d \geqslant 0$ such that

$$v(\alpha) \leqslant d \tag{15}$$

whenever $\alpha \in A$ and $\alpha = \alpha_0$ on F. Given any element β of A, write

$$\begin{aligned} \beta &= (\beta - \alpha_0)\xi_F + (\alpha_0\xi_F + \beta\xi_{J\setminus F}) \\ &= \sum_{j \in F} (\beta(j) - \alpha_0(j))\xi_{\{j\}} + \alpha \end{aligned}$$

where $\alpha \in A$ and $\alpha = \alpha_0$ on F. Hence

$$\phi_\beta = \sum_{j \in F} (\beta(j) - \alpha_0(j))\xi_{\Delta_j} + \phi_\alpha$$

and so (15) entails that

$$\begin{aligned}\|\phi_\beta\|_{p,p} &\leqslant \sum_{j\in F} \|\xi_{\Delta_j}\|_{p,p} + \|\phi_\alpha\|_{p,p} \\ &\leqslant \sum_{j\in F} \|\xi_{\Delta_j}\|_{p,p} + d = K_p.\end{aligned}$$

This establishes the condition (iii). □

1.2.11. For the LP property to hold, it is obviously necessary that every ξ_{Δ_j} belongs to $M_p(X)$ for every p in $(1, \infty)$ (though of course the LP property asserts more than this).

It is a tantalising problem to decide whether or not the characteristic function ξ_Δ of a given measurable subset Δ of X belongs to $M_p(X)$ for specified values of p different from 2. Even for such well known groups as $X = \mathbb{R}^n$ or $X = \mathbb{Z}$, and "simple" subsets Δ, very little is known. The famous M. Riesz theorem for $\mathbb{R}^n$ (see Chapter 6) implies that ξ_Δ belongs to $M_p(\mathbb{R}^n)$ for every p in $(1, \infty)$, when Δ is a half-space; and L. Schwartz [36] showed that the same is true when Δ is a projective polyhedron. On the other hand, there is the surprising result of Charles Fefferman [12] that when Δ is the unit ball in $\mathbb{R}^n$, $\xi_\Delta \in M_p(\mathbb{R}^n)$ only when *either* $p = 2$ *or* $n = 1$ *and* $p \in (1, \infty)$.

If X is discrete, $p \geqslant 2$, and Δ is a so-called $\Lambda(p)$ set then $\xi_\Delta \in M_p(X)$. This follows from the fact that when $p \geqslant 2$, $L^p \subseteq L^2$, and Δ is a $\Lambda(p)$ set if and only if it is of type $(2, p)$. See [9], 15.5, where the setting is $X = \mathbb{Z}$; the discussion and proofs there given are however valid with obvious changes for arbitrary discrete X.

Beside the M. Riesz multiplier theorem just mentioned, there is a vector-valued version of that theorem of which we shall make essential use in Chapter 7 when proving the classical Littlewood-Paley theorem. In the spirit of what we have done so far, we now explain what we mean by saying that a family (Δ_j) of measurable subsets of X (this family need not here be a decomposition) has the (*vector*) *Riesz property* or *the* R *property*.

1.2.12. The R property. *To every p in* $(1, \infty)$ *corresponds a number* $A_p > 0$ *such that*

$$\left\| \left(\sum_{j\in J} |S_j f_j|^2 \right)^{1/2} \right\|_p \leqslant A_p \left\| \left(\sum_{j\in J} |f_j|^2 \right)^{1/2} \right\|_p \tag{16}$$

for all families $(f_j)_{j\in J}$ *of elements of* $L^2 \cap L^p$. (*The assertion* (16) *is vacuous in case the right hand side of* (16) *is infinite.*)

1.2.13. Plainly, if J is a singleton and $\Delta_j = \Delta$ for the single element j of J, the family (Δ_j) has the LP property or the R property if and only if ξ_Δ belongs to $M_p(X)$ for every p in $(1, \infty)$. In general, if (Δ_j) is a family having the R property, then every ξ_{Δ_j} belongs to $M_p(X)$ for each p in $(1, \infty)$.

In case $G = \mathbb{T}$, $\mathbb{R}$ and $\mathbb{Z}$ we shall in Chapter 6 establish the R property for families of intervals in the character group $X = \mathbb{Z}$, $\mathbb{R}$ and $\mathbb{T}$ respectively. Analogous results for $G = \mathbb{T}^n$, $\mathbb{R}^n$ and $\mathbb{Z}^n$ will then follow from 1.3.5 below. From this, the fact that certain severely restricted families of intervals in $\mathbb{Z}$, $\mathbb{R}$ and $\mathbb{T}$ possess the LP property will be derived in Chapter 7; and similar results for the multidimensional case then follow from 1.3.4.

1.3. Extension of the LP and R Properties to Product Groups

It is our aim in this section to prove two "product" theorems which establish the LP and R properties for the product group $G = G_1 \times G_2$ when they are given for the factor groups. In doing this it will help notationally to avail ourselves of the fact that every operator S_Δ can be regarded as acting, either on $\mathscr{L}^2$, or on the associated quotient space L^2. (This comes about because the definition of T_ϕ in 1.2.2 is such that if $f \in \mathscr{L}^2$, we may understand $T_\phi f$ to mean $T_\phi(\mathrm{cl}\, f)$, where $\mathrm{cl}\, f$ (an element of L^2) is the class of f.) The ideas of the proofs of the product theorems are simple, but their execution requires care.

To explain what is involved, let G_1 and G_2 be Hausdorff LCA groups with character groups X_1 and X_2. Write $G = G_1 \times G_2$ and $X = X_1 \times X_2$, and identify X with the character group of G in the standard fashion. For $i = 1, 2$, let Δ_i be a measurable subset of X_i and let Δ denote $\Delta_1 \times \Delta_2$, which is a measurable subset of X. Since $\xi_\Delta = \xi_{\Delta_1 \times X_2} \xi_{X_1 \times \Delta_2}$, it is immediate that

$$S_\Delta = S_{\Delta_1 \times X_2} S_{X_1 \times \Delta_2}. \tag{1}$$

The main preliminary result, Lemma 1.3.2, makes precise the very natural but vague feeling that applying $S_{\Delta_1 \times X_2}$ to a function f on $G_1 \times G_2$ "comes to the same thing" as first fixing x_2, applying S_{Δ_1} to the function of x_1 thus obtained, and then restoring to x_2 its role as a variable. A similar assertion connects S_{Δ_2} and $S_{X_1 \times \Delta_2}$.

We now set down a result useful in the proof of 1.3.2.

1.3.1. Lemma. *Let $\mathscr{S}$ denote the subset of $\mathscr{L}^2(G)$ formed of the functions*

$$u_1 \otimes u_2 \colon (x_1, x_2) \to u_1(x_1) u_2(x_2),$$

where $u_i \in C_c(G_i)$ and $\hat{u}_i \in L^1(X_i)$ for $i = 1, 2$. Then the linear span of $\mathscr{S}$ is dense in $\mathscr{L}^2(G)$.

Proof. Since $C_c(G)$ is dense in $\mathscr{L}^2(G)$, it will suffice to show that the linear span of $\mathscr{S}$ has an $\mathscr{L}^2(G)$-closure containing $C_c(G)$. To this end, suppose that $f \in C_c(G)$ and $\varepsilon > 0$ is given. Choose compact sets $K_i \subseteq G_i$ so that f vanishes off $K = K_1 \times K_2$, and then choose relatively compact open subsets U_i of G_i containing K_i and a positive number δ such that $\delta \lambda_1(U_1)^{1/2} \lambda_2(U_2)^{1/2} \leqslant \varepsilon$, where λ_i denotes Haar measure on G_i. Reference to [2], Ch. IV, p. 89, Lemma 1 shows that there exists a finite sum $\sum_{j=1}^n f_{1,j} \otimes f_{2,j}$, wherein $f_{i,j} \in C_c(G_i)$ and vanishes off U_i, such that

$$\left| f - \sum_{j=1}^n f_{1,j} \otimes f_{2,j} \right| \leqslant \delta \xi_{U_1 \times U_2}$$

at all points of G and therefore also

$$\left\| f - \sum_{j=1}^{n} f_{1,j} \otimes f_{2,j} \right\|_2 \leqslant \delta\lambda_1(U_1)^{1/2}\lambda_2(U_2)^{1/2} \leqslant \varepsilon.$$

This being so, it will now be sufficient to show that any function $f_1 \otimes f_2$, where $f_i \in C_c(G_i)$ for $i = 1, 2$, is approximable arbitrarily closely in $\mathscr{L}^2(G)$ by functions $u_1 \otimes u_2$ in $\mathscr{S}$. Moreover, since

$$\begin{aligned}\|f_1 \otimes f_2 - u_1 \otimes u_2\|_2 &= \|(f_1 - u_1) \otimes u_2 + u_1 \otimes (f_2 - u_2)\|_2 \\ &\leqslant \|f_1 - u_1\|_2\|u_2\|_2 + \|u_1\|_2\|f_2 - u_2\|_2,\end{aligned}$$

it will be enough to show that $\|f_i - u_i\|_2$ can be made as small as we please. To this end, suppose f_i vanishes off a compact set E_i and suppose that $\varepsilon > 0$ is preassigned. Choose a relatively compact open neighborhood V_i of zero in G_i and $\delta > 0$ so that $\delta\lambda_i(E_i + V_i)^{1/2} \leqslant \varepsilon$. Then choose a positive definite function k_i in $C_c(G_i)$ which vanishes off V_i and is such that

$$|k_i * f_i - f_i| \leqslant \delta$$

at all points of G_i. Put $u_i = k_i * f_i$; then $u_i \in C_c(G_i)$, u_i vanishes off $E_i + V_i$, and $\hat{u}_i = \hat{k}_i\hat{f}_i \in L^1(X_i)$ since $\hat{k}_i \in L^1(X_i)$. Finally, therefore,

$$|u_i - f_i| \leqslant \delta\xi_{E_i+V_i}$$

at all points of G_i and so

$$\|u_i - f_i\|_2 \leqslant \delta\lambda_i(E_i + V_i)^{1/2} \leqslant \varepsilon,$$

and the proof is complete. □

We have now assembled the tools necessary to establish the connection between S_{Δ_1} and $S_{\Delta_1 \times X_2}$ and S_{Δ_2} and $S_{X_1 \times \Delta_2}$, alluded to above. In doing this, we use the following notation.

If f is a function on G and $x_2 \in G_2$, we write f_{*x_2} for the function $x_1 \to f(x_1, x_2)$ on G_1; and, if $x_1 \in G_1$, we write f_{x_1*} for the function $x_2 \to f(x_1, x_2)$ on G_2. Also, if g denotes an element of any $\mathscr{L}^2$ space, cl g will denote the class of g, an element of the corresponding L^2 space.

1.3.2. Lemma. (i) *Suppose $f \in \mathscr{L}^2(G)$ and let g be an element of $S_{\Delta_1 \times X_2}f$. Then for almost all x_2 in G_2, f_{*x_2} and g_{*x_2} belong to $\mathscr{L}^2(G_1)$ and*

$$\text{cl } g_{*x_2} = S_{\Delta_1}f_{*x_2}.$$

(ii) *Let f be an element of $\mathscr{L}^2(G)$ and h an element of $S_{X_1 \times \Delta_2}f$. Then for almost all x_1 in G_1, f_{x_1*} and h_{x_1*} belong to $\mathscr{L}^2(G_2)$ and*

$$\text{cl } h_{x_1*} = S_{\Delta_2}f_{x_1*}.$$

Proof. It will clearly be enough to give the proof of (i).

We first observe that, if (i) holds for *one* g in $S_{\Delta_1 \times X_2} f$, then it holds for *every* such g. In fact, Fubini's theorem implies that $\mathrm{cl}\, g_{*x_2} = \mathrm{cl}\, g'_{*x_2}$ for almost all x_2, whenever g and g' belong to the same class (element of $L^2(G)$).

Consider now the special case in which $f = u_1 \otimes u_2 \in \mathscr{S}$. It is then clear that $S_{\Delta_1} f_{*x_2}$ is the class of k_{*x_2}, where

$$k: (x_1, x_2) \to \int \hat{u}_1(\chi_1)\xi_{\Delta_1}(\chi_1)\chi_1(x_2)\, d\chi_1 \,.\, u_2(x_2).$$

On the other hand, $S_{\Delta_1 \times X_2} f$ plainly contains (and is therefore the class of) the function g, where

$$g: (x_1, x_2) \to \iint \hat{u}_1(\chi_1)\hat{u}_2(\chi_2)\xi_{\Delta_1 \times X_2}(\chi_1, \chi_2)\chi_1(x_1)\chi_2(x_2)\, d\chi_1 d\chi_2.$$

Since $\xi_{\Delta_1}(\chi_1) = \xi_{\Delta_1 \times X_2}(\chi_1, \chi_2)$ everywhere on X, the Fubini theorem and the Fourier inversion formula ensure that $g = k$. Consequently, $S_{\Delta_1} f_{*x_2} = \mathrm{cl}\, k_{*x_2} = \mathrm{cl}\, g_{*x_2}$ for every x_2 in G_2, and (i) is thus satisfied in this case.

It is now clear that (i) continues to hold for every f in the linear span of $\mathscr{S}$, and we now call upon 1.3.1 to complete the proof in the following way.

Given f in $\mathscr{L}^2(G)$, extract a sequence $(f^{(n)})$ from the linear span of $\mathscr{S}$ which converges in $\mathscr{L}^2(G)$ to f. Choose $g^{(n)}$ in $S_{\Delta_1 \times X_2} f^{(n)}$. By the continuity of $S_{\Delta_1 \times X_2}$, $g^{(n)} \to g$ in $\mathscr{L}^2(G)$. Also, by what we have already proved,

$$\mathrm{cl}\, g^{(n)}_{*x_2} = S_{\Delta_1} f^{(n)}_{*x_2} \quad \text{if} \quad x_2 \in G_2 \backslash N_n, \tag{2}$$

where N_n is a negligible subset of G_2. By Fubini's theorem, $f_{*x_2} \in \mathscr{L}^2(G_1)$ for almost every $x_2 \in G_2$, and

$$\int \left\{ \int |f_{*x_2} - f^{(n)}_{*x_2}|^2\, dx_1 \right\} dx_2 = \iint |f - f^{(n)}|^2\, dx_1\, dx_2 \to 0 \quad \text{as} \quad n \to \infty.$$

The same is true with g in place of f throughout. It follows that there are integers $n_1 < n_2 < \cdots$ and a negligible subset N_0 of G_2 such that

$$f^{(n_k)}_{*x_2} \to f_{*x_2},\; g^{(n_k)}_{*x_2} \to g_{*x_2} \quad \text{in} \quad \mathscr{L}^2(G_1), \quad \text{if} \quad x_2 \in G_2 \backslash N_0. \tag{3}$$

The continuity of S_{Δ_1} combines with the first clause of (3) to show that

$$S_{\Delta_1} f^{(n_k)}_{*x_2} \to S_{\Delta_1} f_{*x_2} \quad \text{in} \quad L^2(G_1), \quad \text{if} \quad x_2 \in G_2 \backslash N_0. \tag{4}$$

The second clause of (3) signifies that

$$\mathrm{cl}\, g^{(n_k)}_{*x_2} \to \mathrm{cl}\, g_{*x_2} \quad \text{in} \quad L^2(G_1), \quad \text{if} \quad x_2 \in G_2 \backslash N_0. \tag{5}$$

Put $N = N_0 \cup \bigcup_{n=1}^{\infty} N_n$; then N is a negligible subset of G_2 and, by comparison of

(2), (4) and (5), we infer that

$$\mathrm{cl}\, g_{*x_2} = S_{\Delta_1} f_{*x_2} \quad \text{if} \quad x_2 \in G_2 \backslash N,$$

which proves (i). □

We are now in a position to prove the first of the "product theorems". Throughout the proofs to follow, integrals over G are denoted by

$$\int \cdots dx \quad \text{or} \quad \iint \cdots dx_1\, dx_2,$$

while iterated integrals are written

$$\int\left\{\int \cdots dx_1\right\} dx_2 \quad \text{and} \quad \int\left\{\int \cdots dx_2\right\} dx_1.$$

In the course of the proof, we shall need to refer to the following simple result.

1.3.3. Lemma. *Suppose that the decomposition* (Δ_j) *of* X *possesses the* LP *property. Let* $j \to j^*$ *be an injection of* J *into* $\{0, 1, \ldots\}$. *Let* F *be a finite subset of* J *and define the function* ϕ_t, *for each* t *in* $[0, 1]$, *by the formula*

$$\phi_t = \sum_{j \in F} r_{j^*}(t) S_{\Delta_j};$$

cf. 1.2.8(i). *Then*

$$\| T_{\phi_t} f \|_p \leqslant B_p B_{p'} \| f \|_p$$

if $p \in (1, \infty)$, $f \in L^2 \cap L^p$, *and* $t \in [0, 1]$.

Proof. This is a restatement of a particular case of inequality (8) of 1.2.8(ii).

1.3.4. Theorem (**LP property for product decompositions**). *Let* X_1 *and* X_2 *be LCA groups with decompositions* $(\Delta_j^{(1)})_{j \in J_1}$ *and* $(\Delta_k^{(2)})_{k \in J_2}$ *respectively, both having the* LP *property. Then the decomposition* $(\Delta_j^{(1)} \times \Delta_k^{(2)})_{j \in J_1, k \in J_2}$ *of* $X = X_1 \times X_2$ *also possesses the* LP *property.*

Proof. Let $j \to j^*$ and $k \to k^*$ be injections of J_1 and J_2 respectively into $\{0, 1, \ldots\}$. In order to prove the theorem, it will be enough to prove the existence of a constant C_p such that

$$\int_{G_1 \times G_2} \left(\sum_{j \in F_1, k \in F_2} |S_{j,k} f|^2 \right)^{p/2} dx \leqslant C_p^p \| f \|_p^p \tag{6}$$

for all finite subsets F_1 of J_1 and F_2 of J_2 and all functions f in $L^2 \cap L^p$. (Here p is as usual a fixed but arbitrary index in the range $(1, \infty)$.) We have written $S_{j,k}$ instead of $S_{\Delta_j^{(1)} \times \Delta_k^{(2)}}$. But in order to prove (6), it is sufficient to prove the existence of a constant D_p such that

$$\iint \Big| \sum_{j \in F_1, k \in F_2} r_{j*}(s) r_{k*}(t) S_{j,k} f(x) \Big|^p dx \leqslant D_p^p \|f\|_p^p \tag{7}$$

for all points (s, t) in $[0, 1] \times [0, 1]$. For if (7) holds, the Fubini-Tonelli theorem can be used in conjunction with the fact that the set of Rademacher functions in two dimensions forms a $\Lambda(p)$ set ([38], Appendix D) to show that for some constant $\alpha_p > 0$,

$$\begin{aligned} \alpha_p^p \int \Big(\sum_{j \in F_1, k \in F_2} |S_{j,k}|^2 \Big)^{p/2} dx &\leqslant \int \Big\{ \int_{[0,1]\times[0,1]} \Big| \sum_{j \in F_1, k \in F_2} r_{j*}(s) r_{k*}(t) S_{j,k} f(x) \Big|^p ds\, dt \Big\} dx \\ &= \int_{[0,1]\times[0,1]} \Big\{ \int_G \cdots dx \Big\} ds\, dt \\ &\leqslant D_p^p \|f\|_p^p. \end{aligned}$$

If we recall (1) and let h denote a function of the class

$$\sum_{k \in F_2} r_{k*}(t) S_{X_1 \times \Delta_k^{(2)}} f, \tag{8}$$

the left side of (7) can be rewritten as

$$\iint \Big| \sum_{j \in F_1} r_{j*}(s) S_{\Delta_j^{(1)} \times X_2} h(x_1, x_2) \Big|^p dx_1\, dx_2.$$

By Lemma 1.3.2, the integrand is the class of a function g in $\mathscr{L}^2(G)$ which is such that for almost every x_2 in G_2,

$$\mathrm{cl}\, g_{*x_2} = \sum_{j \in F_1} r_{j*}(s) S_{\Delta_j^{(1)}} h_{*x_2}.$$

From this and the Fubini-Tonelli theorem (see for example [10], Theorem 4.17.8, noting that g vanishes off a σ-finite set) it follows that

$$\begin{aligned} S = \iint |g|^p dx_1\, dx_2 &= \int \Big\{ \int |g_{*x_2}|^p dx_1 \Big\} dx_2 \\ &= \int \Big\{ \int \Big| \sum_{j \in F_1} r_{j*}(s) S_{\Delta_j^{(1)}} h_{*x_2} \Big|^p dx_1 \Big\} dx_2. \end{aligned}$$

By Lemma 1.3.3 and the hypotheses concerning the family $(\Delta_j^{(1)})$, it now appears that

$$\begin{aligned} S &\leqslant \int \Big\{ B_{1,p}^p B_{1,p'}^p \int |h_{*x_2}|^p dx_1 \Big\} dx_2 \\ &= B_{1,p}^p B_{1,p'}^p \iint |h|^p dx_1\, dx_2, \end{aligned} \tag{9}$$

the last step by the Fubini-Tonelli theorem. (Here $B_{1,p}$ and $B_{1,p'}$ are naturally the constants appearing in inequality (1) of 1.2.4 for the decomposition $(\Delta_j^{(1)})$.) Now the class of h is (see (8)) the class of the function

$$\sum_{k \in F_2} r_{k*}(t) h_k$$

where h_k is a function of the class $S_{X_1 \times \Delta_k^{(2)}} f$. By the Fubini-Tonelli theorem, (9) leads to the estimate

$$\begin{aligned} S &\leqslant B_{1,p}^p B_{1,p'}^p \int\int \left| \sum_{k \in F_2} r_{k*}(t) h_k(x_1, x_2) \right|^p dx_1\, dx_2 \\ &= B_{1,p}^p B_{1,p'}^p \int \left\{ \int \left| \sum_{k \in F_2} r_{k*}(t) (h_k)_{x_1 *} \right|^p dx_2 \right\} dx_1. \end{aligned}$$

Since $h_k \in S_{X_1 \times \Delta_k^{(2)}} f$, Lemma 1.3.2(ii) says that, for almost every x_1 in G_1,

$$\mathrm{cl}\,(h_k)_{x_1 *} = S_{\Delta_k^{(2)}} f_{x_1 *}.$$

Hence

$$S \leqslant B_{1,p}^p B_{1,p'}^p \int \left\{ \int \left| \sum_{k \in F_2} r_{k*}(t) S_{\Delta_k^{(2)}} f_{x_1 *} \right|^p dx_2 \right\} dx_1$$

which, by Lemma 1.3.3 and the LP property for $(\Delta_k^{(2)})$, shows that

$$S \leqslant B_{1,p}^p B_{1,p'}^p \int \left\{ B_{2,p}^p B_{2,p'}^p \int |f_{x_1 *}|^p\, dx_2 \right\} dx_1.$$

Thus, by the Fubini-Tonelli theorem once again,

$$S \leqslant B_{1,p}^p B_{1,p'}^p B_{2,p}^p B_{2,p'}^p \|f\|_p^p = D_p^p \|f\|_p,$$

where D_p is independent of s, t, F_1 and F_2. This establishes (7) and, consequently, the theorem.

Remarks. (i) One of the essential ideas of the proof of Theorem 1.3.4, viz. the systematic use of Lemma 1.3.2, can be modified to show that, if $p \in [1, \infty]$ and $\phi \in M_p(X_1)$, then Φ, the natural extension of ϕ to $X_1 \times X_2$ defined by the rule

$$\Phi(\chi_1, \chi_2) = \phi(\chi_1) \quad \text{for} \quad (\chi_1, \chi_2) \text{ in } X_1 \times X_2,$$

belongs to $M_p(X_1 \times X_2)$ and

$$\|\Phi\|_{p,p} \leqslant \|\phi\|_{p,p}.$$

Notice that this is just Corollary B.2.2. The modification consists in proving that, if T_Φ and T_ϕ are the operators defined in 1.2.2, $f \in \mathscr{L}^2(G)$, and $g \in T_\Phi f$, then

$$\mathrm{cl}\,(g_{*x_2}) = T_\phi(f_{*x_2})$$

for almost all x_2 in G_2. The interested reader should fill in the details.

(ii) An alternative approach to Theorem 1.3.4 can be made via the homomorphism theorem for multipliers (Appendix B). Here is a sketch of the argument.

The essential step in the proof of Theorem 1.3.4 is to show that if $(s, t) \in [0, 1] \times [0, 1]$, the function

$$\Gamma_{s,t} = \sum_{j \in F_1, k \in F_2} r_{j*}(s) r_{k*}(t) \xi_{j,k}$$

belongs to $M_p(X_1 \times X_2)$ and that its multiplier norm is majorised by a number independent of s, t, F_1 and F_2: this is inequality (7). The symbol $\xi_{j,k}$ is a shorthand expression for $\xi_{\Delta_j^{(1)} \times \Delta_k^{(2)}}$.

Observe that

$$\Gamma_{s,t} = \Phi_s^{(1)} \Phi_t^{(2)} \tag{10}$$

where $\Phi_s^{(1)}$ is the function, constant on cosets of X_2, defined by the formula

$$\Phi_s^{(1)}(\chi_1, \chi_2) = \phi_s^{(1)}(\chi_1), \tag{11}$$

where

$$\phi_s^{(1)}(\chi_1) = \sum_{j \in F_1} r_{j*}(s) \xi_j^{(1)}(\chi_1) \tag{12}$$

and

$$\xi_j = \xi_{\Delta_j^{(1)}}; \tag{13}$$

and where $\Phi_t^{(2)}$ is obtained by replacing the subscript or superscript 1 by 2 at all places in (11)–(13) save in (χ_1, χ_2), and s by t in (11) and (12).

Now if the family $(\Delta_j^{(1)})$ has the LP property, it also has the WM property, and so there is a number $C^{(1)}(p)$, independent of s and F_1, such that

$$\|\phi_s^{(1)}\|_{p,p} \leqslant C^{(1)}(p).$$

Theorem B.2.2 implies then that

$$\|\Phi_s^{(1)}\|_{p,p} \leqslant C^{(1)}(p).$$

Similarly,

$$\|\Phi_t^{(2)}\|_{p,p} \leqslant C^{(2)}(p).$$

By virtue of (10), $\Gamma_{s,t} \in M_p(X_1 \times X_2)$ and

$$\|\Gamma_{s,t}\|_{p,p} \leqslant C^{(1)}(p)C^{(2)}(p),$$

which is independent of s, t, F_1 and F_2. $\square$

We pass on now to discuss the R property for product groups.

1.3.5. Theorem. (R property for product diagonal families). *Let* $(\Delta_j^{(1)})_{j\in J}$ *and* $(\Delta_j^{(2)})_{j\in J}$ *be families of measurable sets* (not *necessarily decompositions*) *in* X_1 *and* X_2 *respectively, each having the* R *property. Then the "diagonal product" family* $(\Delta_{j,j}) = (\Delta_j^{(1)} \times \Delta_j^{(2)})_{j\in J}$ *has the* R *property.*

Proof. This is similar to that of Theorem 1.3.4. Using (1), we have for f_j in $L^2 \cap L^p(G) = L^2 \cap L^p(G_1 \times G_2)$,

$$\begin{aligned} S' &= \left\| \left(\sum_{j\in J} |S_{\Delta_j{(1)} \times \Delta_j{(2)}} f|^2 \right)^{1/2} \right\|_p^p \\ &= \iint \left(\sum_{j\in J} |S_{\Delta_j{(1)} \times X_2} S_{X_1 \times \Delta_j{(2)}} f_j|^2 \right)^{p/2} dx_1\, dx_2. \end{aligned}$$

Let g_j be a function of the class $S_{X_1 \times \Delta_j{(2)}} f$. Then

$$S' = \iint \left(\sum_{j\in J} |S_{\Delta_j{(1)} \times X_2} g_j|^2 \right)^{p/2} dx_1\, dx_2.$$

By Lemma 1.3.2(i), $S_{\Delta_j{(1)} \times X_2} g_j$ is the class of a function h_j in $\mathscr{L}^2(G)$ such that, for almost every x_2 in G_2,

$$\mathrm{cl}\,(h_j)_{*x_2} = S_{\Delta_j{(1)}}(g_j)_{*x_2}.$$

Hence (Fubini-Tonelli theorem)

$$\begin{aligned} S' &= \iint \left(\sum_{j\in J} |h_j|^2 \right)^{p/2} dx_1\, dx_2 \\ &= \int \left\{ \int \left(\sum_{j\in J} |(h_j)_{*x_2}|^2 \right)^{p/2} dx_1 \right\} dx_2 \\ &= \int \left\{ \int \left(\sum_{j\in J} |S_{\Delta_j{(1)}}(g_j)_{*x_2}|^2 \right)^{p/2} dx_1 \right\} dx_2. \end{aligned}$$

Since $(\Delta_j^{(1)})$ is assumed to have the R property, it now appears that

$$\begin{aligned} S' &\leqslant \int \left\{ A_{1,p}^p \int \left(\sum_{j\in J} |(g_j)_{*x_2}|^2 \right)^{p/2} dx_1 \right\} dx_2 \\ &= A_{1,p}^p \int \left\{ \int \left(\sum_{j\in J} |(g_j)_{x_1 *}|^2 \right)^{p/2} dx_2 \right\} dx_1, \end{aligned}$$

the last step by the Fubini-Tonelli theorem. Since $g_j \in S_{X_1 \times \Delta_j^{(2)}} f_j$, 1.3.2(ii) shows that, for almost all x_1 in G_1,

$$\mathrm{cl}\,(g_j)_{x_1*} = S_{\Delta_j^{(2)}}(f_j)_{x_1*},$$

and so the last inequality can be written

$$S' \leqslant A_{1,p}^p \int \left\{ \int \left(\sum_{j \in J} |S_{\Delta_j^{(2)}}(f_j)_{x_1*}|^2 \right)^{p/2} dx_2 \right\} dx_1.$$

By applying the R property for the family $(\Delta_j^{(2)})$, it follows that

$$S' \leqslant A_{1,p}^p \int \left\{ A_{2,p}^p \int \left(\sum_{j \in J} |(f_j)_{x_1*}|^2 \right)^{p/2} dx_2 \right\} dx_1.$$

Yet another Fubini-Tonelli manipulation shows that

$$S' \leqslant A_{1,p}^p A_{2,p}^p \iint \left(\sum_{j \in J} |f_j(x_1, x_2)|^2 \right)^{p/2} dx_1\, dx_2;$$

the diagonal product decomposition therefore has the R property. □

1.4. Intersections of Decompositions Having the LP Property

The results of Section 1.3 already furnish a method for constructing decompositions with the LP property from given decompositions having that property. Here is another useful such technique.

1.4.1. Theorem. *Suppose that $(\Delta_i)_{i \in I}$ and $(\Delta'_j)_{j \in J}$ are decompositions of the LCA group X, both of which have the* LP *property. Then the same is true of the decomposition $(\Delta_i \cap \Delta'_j)_{(i,j) \in I \times J}$.*

Proof. Let $i \to i^*$ and $j \to j^*$ be injections of the sets I and J respectively into the set of nonnegative integers, and denote the Rademacher functions on [0, 1] by $r_0, r_1, \ldots$. Suppose that $1 < p < \infty$. Since $(\Delta_i)_{i \in I}$ has the LP property, the inequality 1.2(8) shows that there is a number C_p such that

$$\left\| \sum_{i \in F_1} r_{i*}(t) S_{\Delta_i} f \right\|_p \leqslant C_p \|f\|_p \tag{1}$$

for all finite subsets F_1 of I, all t in [0, 1], and all f in $L^2 \cap L^p(G)$. Similarly

$$\left\| \sum_{j \in F_2} r_{j*}(s) S_{\Delta'_j} f \right\|_p \leqslant C'_p \|f\|_p \tag{2}$$

for all finite subsets F_2 of J and all s in [0, 1]. Since

$$\sum_{i\in F_1}\sum_{j\in F_2} r_{i*}(t)r_{j*}(s)S_{\Delta_i\cap\Delta_j'}f = \sum_{j\in F_2} r_{j*}(s)S_{\Delta_{j'}}\left(\sum_{i\in F_1} r_{i*}(t)S_{\Delta_i}f\right),$$

it follows from (1) and (2) that

$$\int_G\left|\sum_{i\in F_1}\sum_{j\in F_2} r_{i*}(t)r_{j*}(s)S_{\Delta_i\cap\Delta_j'}f(x)\right|^p dx \leqslant C_p^p C_p'^p\|f\|_p^p. \tag{3}$$

Integrate (3) over the set [0, 1] × [0, 1], reverse the orders of integration on the left side of the resulting inequality, and use the fact that the set of Rademacher functions in two dimensions forms a $\Lambda(p)$ set, just as in the proof of Theorem 1.3.4, to conclude that

$$\int_G\left(\sum_{i\in F_1}\sum_{j\in F_2}|S_{\Delta_i\cap\Delta_j'}f|^2\right)^{p/2} dx \leqslant \alpha_p^{-p}C_p^p C_p'^p\|f\|_p^p. \tag{4}$$

Now let the sets F_1 and F_2 expand to fill out all of I and J respectively, and use the monotone convergence theorem to deduce from (4) that

$$\left\|\left(\sum_{(i,j)\in I\times J}|S_{\Delta_i\cap\Delta_j'}f|^2\right)^{1/2}\right\|_p \leqslant \alpha_p^{-1}C_p C_p'\|f\|_p. \quad \square$$

Chapter 2. Convolution Operators (Scalar-Valued Case)

This chapter is of quite general character: in it we make almost no direct reference to the Littlewood-Paley, Marcinkiewicz, and M. Riesz theorems, which are our main goals. The material developed here is, however, central to the proofs of all three types of result, and can be used to prove a great deal more besides.

The ultimate concern of the present chapter is with bounds for the norms of convolution operators

$$L_k : f \to k * f$$

mapping subsets of $L^p(G) = L^p$ into L^p, with special interest in the case where $p \in (1, \infty)$. These bounds are first established for the nonsingular case (i.e., $k \in L^1$), the results we subsequently need for the singular case being obtained thence by approximation arguments; further elaboration on this point appears in 2.4.1 and Corollary 2.4.5.

The methods are based on the use of covering families, covering lemmas and decomposition theorems of a type first given by Calderón and Zygmund [5] and since developed in very general situations. Of this extensive development, we shall use the minimum adequate for our purposes.

2.1. Covering Families .

2.1.1. We shall say that $(U_\alpha)_{\alpha \in \mathbb{Z}}$ is a *covering family in (or for) the* LCA *group G* if and only if the following conditions are satisfied:

(i) $(U_\alpha)_{\alpha \in \mathbb{Z}}$ is a decreasing base of neighborhoods of 0 in G formed of relatively compact measurable sets and $\bigcup_{\alpha \in \mathbb{Z}} U_\alpha = G$;

(ii) there exists a function $\theta: \mathbb{Z} \to \mathbb{Z}$ and a positive number A such that

$$\theta(\alpha) < \alpha,$$

$$U_\alpha - U_\alpha \subseteq U_{\theta(\alpha)},$$

and

$$m(U_{\theta(\alpha)}) \leqslant Am(U_\alpha)$$

for every α in $\mathbb{Z}$.

Note that (2) implies that $U_{\theta(\alpha)} \supseteq U_\alpha$, so that $A \geqslant 1$. Note also that, by replacing θ by

$$\theta_0 \colon \alpha \to \max \{\beta \in \mathbb{Z} \colon \beta < \alpha \quad \text{and} \quad U_\alpha - U_\alpha \subseteq U_\beta\},$$

we may always assume that θ is increasing.

2.1.2. Throughout this chapter, we shall assume that G is an LCA group and that $(U_\alpha)_{\alpha \in \mathbb{Z}}$ is a covering family in G. The group G is therefore necessarily first countable and σ-compact.

2.1.3. Examples.

(i) $G = \mathbb{R}$ and U_α is the open (or the closed) ball centre 0 and radius $2^{-\alpha}$, $\theta(\alpha) = \alpha - 1$, and $A = 2$. Similar definitions apply when $G = \mathbb{R}^n$ for any positive integer n, with $A = 2^n$.

(ii) $G = \mathbb{T}$; U_α is the set of numbers $\exp(\pi i t)$ with t real and $|t| \leqslant 2^{-\alpha}$, so that $U_\alpha = \mathbb{T}$ if $\alpha \leqslant 0$; $\theta(\alpha) = \alpha - 1$, $A = 2$.

(iii) G is an LCA group with a decreasing sequence $(U_\alpha)_{\alpha \in \mathbb{Z}}$ of compact open subgroups such that

$$\bigcup_\alpha U_\alpha = G, \qquad \bigcap_\alpha U_\alpha = \{0\}.$$

The U_α then form a base at 0. (If not, there would be an open neighbourhood V of 0 such that $U_\alpha \backslash V$ is nonvoid for every α; for $\alpha \geqslant 0$, $U_\alpha \backslash V$ is closed in U_0 and so, since U_0 is compact, $\bigcap_\alpha U_\alpha \backslash V$ is nonvoid, which contradicts the assumption that $\bigcap_\alpha U_\alpha = \{0\}$.) 2.1.1(ii) will be satisfied if and only if

$$A = \sup_{\alpha \in \mathbb{Z}} \{\text{index of } U_\alpha \text{ in } U_{\alpha-1}\}$$

is finite, in which case we can take $\theta(\alpha) = \alpha - 1$.

(iv) $G = \mathbb{Z}$; $U_\alpha = \{0\}$ if $\alpha \in \mathbb{Z}$ and $\alpha > 0$; $U_\alpha = \{n \in \mathbb{Z} \colon |n| < 2^{-\alpha}\}$ if $\alpha \in \mathbb{Z}$ and $\alpha \leqslant 0$; $\theta(\alpha) = \alpha - 1$, and $A = 3$.

We shall need the following lemma.

2.1.4. Lemma. *Suppose $f \in \mathscr{L}^1$ and*

$$F_\alpha(x) = m(U_\alpha)^{-1} \int_{U_\alpha} f(x + y)\, dy.$$

Then $\lim_{\alpha \to \infty} F_\alpha = f$ *in* $\mathscr{L}^1$.

Proof. We have that

$$f(x) - F_\alpha(x) = m(U_\alpha)^{-1} \int_{U_\alpha} (f(x) - f(x + y))\, dy$$

and so, by the Fubini-Tonelli theorem,

$$\|f - F_\alpha\|_1 \leqslant m(U_\alpha)^{-1} \int_{U_\alpha} \left\{ \int |f(x) - f(x + y)|\, dx \right\} dy.$$

Given $\varepsilon > 0$, there is a neighbourhood V of 0 in G such that

$$\int |f(x) - f(x + y)| \, dx \leqslant \varepsilon$$

for every y in V. By 2.1.1(i), $U_{\alpha_0} \subseteq V$ for some α_0. Then 2.1.1(i) ensures that $\|f - F_\alpha\|_1 \leqslant \varepsilon$ for every $\alpha \geqslant \alpha_0$. □

2.1.5. General comments. Families similar to our covering families $(U_\alpha)_{\alpha \in \mathbb{Z}}$ have been used by various writers. See for example [32] pp. 30–34 and [33]. Compare also the D'-sequences in Section 44 of [20].

2.2. The Covering Lemma

2.2.1. Lemma. *Let E be a subset of G and α: $E \to \mathbb{Z}$ a mapping such that*

(a) *$\alpha(E)$ is bounded below;*

(b) *for every α_0 in $\mathbb{Z}$, the set $\{x \in E: \alpha(x) \leqslant \alpha_0\}$ is relatively compact in G.*

The conclusion is that there is a finite or infinite sequence $(x_n)_{n \in Q}$ (Q is either $\mathbb{N} = \{1, 2, \ldots\}$ or $\{1, 2, \ldots, m\}$ for some m in $\mathbb{N}$) of points of E such that, writing $\alpha_n = \alpha(x_n)$, the sequence $(\alpha_n)_{n \in Q}$ is increasing and

(i) *the sets $x_n + U_{\alpha_n}$ are pairwise disjoint;*

(ii) $E \subseteq \bigcup_{n \in Q} (x_n + U_{\alpha_n} - U_{\alpha_n})$.

Proof. If there is a finite sequence $x_1, \ldots, x_k$ of points of E such that the $x_j + U_{\alpha(x_j)}$ are pairwise disjoint and

$$E \subseteq \bigcup_{j=1}^{k} (x_j + U_{\alpha(x_j)} - U_{\alpha(x_j)}),$$

we may always rename the x_j in such a way that the $\alpha(x_j)$ increase with j and there is nothing further to prove. In the contrary case, proceed as follows.

Begin by defining $\alpha_1 = \min\{\alpha(x): x \in E\}$, which is legitimate by (a), and choose x_1 in E such that $\alpha_1 = \alpha(x_1)$.

Suppose k is a positive integer and that the points $x_1, \ldots, x_k$ of E have been defined in such a way that

(A) the sets $x_j + U_{\alpha_j}$ ($j = 1, \ldots, k$) are pairwise disjoint, α_j denoting $\alpha(x_j)$;

(B) $\alpha_j = \min\{\alpha(x): x \in A_{j-1}\}$ for each j in $\{1, 2, \ldots, k\}$ where

$$A_r = E \Big\backslash \Big(\bigcup_{1 \leqslant j \leqslant r} x_j + U_{\alpha_j} - U_{\alpha_j} \Big)$$

for each r in $\{0, 1, 2, \ldots, k\}$. (A union over a void index set is interpreted to be $\varnothing$.)

Note that (A) and (B) are satisfied when $k = 1$.

Continue the process by defining x_{k+1} as follows: by hypothesis, A_k is nonvoid; since $A_k \subseteq E$, α is bounded below on A_k; put $\alpha_{k+1} = \min\{\alpha(x): x \in A_k\}$ and choose

x_{k+1} in A_k so that $\alpha_{k+1} = \alpha(x_{k+1})$. The first clause of (B) is then fulfilled when k is replaced by $k + 1$; and we define A_{k+1} in accordance with the second clause of (B). We claim that (A) holds with $k + 1$ in place of k, to verify which it suffices to prove that

(C) $(x_j + U_{\alpha_j}) \cap (x_{k+1} + U_{\alpha_{k+1}}) = \varnothing$ when $j \in \{1, 2, \ldots, k\}$.

However, if (C) were not true, there would exist j in $\{1, 2, \ldots, k\}$, u in U_{α_j} and v in $U_{\alpha_{k+1}}$ such that $x_j + u = x_{k+1} + v$. Then, since it is clear that $A_k \subseteq A_j$ and so $\alpha_{k+1} \geqslant \alpha_j$, it would follow that $x_{k+1} \in x_j + U_{\alpha_j} - U_{\alpha_j}$, which would contradict the fact that $x_{k+1} \in A_k$. This contradiction establishes (C).

We have thus defined the points x_k ($k \in \mathbb{N}$) by recurrence, and (A) and (B) are true for every k. Thus (i) is satisfied. At the same time we have arranged that $x_{k+1} \in A_k$ for $k = 0, 1, 2, \ldots$; and, since plainly $A_0 \supseteq A_1 \supseteq A_2 \supseteq \ldots$, (B) shows that $\alpha_1 \leqslant \alpha_2 \leqslant \ldots$.

It remains only to prove (ii), i.e., that $\bigcap_{n\in\mathbb{N}} A_n = \varnothing$. Were this not the case, there would exist an element x belonging to every A_n. Then $\alpha(x_n) = \alpha_n \leqslant \alpha(x)$ for every n in $\mathbb{N}$. By (b), therefore, the set $\{x_n : n \in \mathbb{N}\}$ would be relatively compact in G. Since $U_{\alpha_n} \subseteq U_{\alpha_1}$ is relatively compact, it follows that the set

$$F = \bigcup_{n\in\mathbb{N}} (x_n + U_{\alpha_n}) \subseteq \{x_n\} + U_{\alpha_1}$$

would be relatively compact and so $m(F)$ would be finite. But this is false because, by (i) and the fact that $U_{\alpha_n} \supseteq U_{\alpha(x)}$,

$$m(F) = \sum_{n\in\mathbb{N}} m(x_n + U_{\alpha_n}) = \sum_{n\in\mathbb{N}} m(U_{\alpha_n}) \geqslant \sum_{n\in\mathbb{N}} m(U_{\alpha(x)})$$

is infinite, by 2.1.1(i). This contradiction yields (ii). □

Remarks. The proof of the lemma depends only on the assumption that (U_α) is a decreasing family of relatively compact sets of positive measure.

The reader should take note of the way the compactness assumption (b) is used in the proof of Lemma 2.2.1 and compare it with the corresponding use of the *geometric* character of the balls B_j in the proof of Stein's lemma ([38], p. 9).

In applications, one may think of the function α in Lemma 2.2.1 as some sort of "choice function" which will vary from one application to another; condition (a) says that the $U_{\alpha(x)}$, $x \in E$, are not too large.

The lemma is similar to the type of covering lemma customarily used in proving Lebesgue's differentiation theorem for integrals; cf. [34], p. 154 and [38], Chapter 1. Indeed, it is possible to prove a differentiation theorem and theorems about the "maximal function" in the setting of a group G with a covering family (U_α), as follows. Write for f in $\mathscr{L}^1$:

$$Mf(x) = \sup_{\alpha\in\mathbb{Z}} m(U_\alpha)^{-1} \int_{x+U_\alpha} |f| \, dm;$$

define E to be the set of points x for which $Mf(x) > c$ where $c > 0$ is such that

$m(G)c > \|f\|_1$; and define $\alpha(x)$ for x in E to be the smallest integer α for which

$$m(U_\alpha)^{-1} \int_{x+U_\alpha} |f| \, dm > c.$$

Then, using the covering lemma and arguing almost exactly as in the proof of Theorem 2.3.2 down to line 9 on page 38, one finds that $E \subseteq \bigcup_{n \in Q} (x_n + U_{\alpha_n} - U_{\alpha_n})$ where the $x_n + U_{\alpha_n}$ are pairwise disjoint and $\int_{x_n + U_{\alpha_n}} |f| \, dm > cm(U_{\alpha_n})$. From this it follows that

(i) $m(\{x \in G: Mf(x) > c\}) \leqslant A\|f\|_1 c^{-1}$

provided $m(G)c > \|f\|_1$; in case $\|f\|_1 \geqslant m(G)c$, (i) is still true since the left side is trivially at most $m(G)$ and $A \geqslant 1$. If $M'f = Mf + |f|$, then $\{x: M'f(x) > c\} \subseteq \{x: Mf(x) > c/2\} \cup \{x: |f(x)| > c/2\}$ and so

(i′) $m(\{x \in G: M'f(x) > c\}) \leqslant A'\|f\|_1 c^{-1}$.

On the other hand, the hypotheses in 2.1.1 arrange that the formula

(ii) $\lim_{\alpha \to \infty} m(U_\alpha)^{-1} \int_{U_\alpha} | f(x+y) - f(x)| \, dm(y) = 0$

holds for every x in G and every f in C_c. If $f \in \mathscr{L}^1$, take a sequence (f_n) of elements of C_c such that $\|f - f_n\|_1 \to 0$ as $n \to \infty$. Write

$$I_\alpha f(x) = m(U_\alpha)^{-1} \int_{U_\alpha} |f(x + y) - f(x)| \, dm(y)$$

and notice that $I_\alpha f \leqslant M'f$ and also

$$I_\alpha f(x) \leqslant I_\alpha f_n(x) + I_\alpha(f - f_n)(x) \leqslant I_\alpha f_n(x) + M'(f - f_n)(x).$$

From this and the C_c-case of (ii), we infer that

$$\limsup_{\alpha \to \infty} I_\alpha f(x) \leqslant M'(f - f_n)(x)$$

and so, by (i′), that

$$m(\{x \in G: \limsup_{\alpha \to \infty} I_\alpha f(x) > c\}) \leqslant A'\|f - f_n\|_1/c$$

for every $c > 0$ and every n. Letting $n \to \infty$, we conclude that the left side is zero for every $c > 0$, and hence that

$$\lim_{\alpha \to \infty} I_\alpha f(x) = 0$$

for almost all x in G. This is a strong form of the differentiation theorem.

The "maximal function" Mf is a direct analogue of that introduced by Hardy and Littlewood for the case $G = \mathbb{R}$, and the inequality (i) is basic for the proof of other inequalities satisfied by Mf and associated with their names. For instance, it is easily deduced from (i) that

$$\|Mf\|_p \leqslant A_p \|f\|_p$$

for p in $(1, \infty)$. (This can be derived from (i) via a special case of the Marcinkiewicz interpolation theorem; alternatively and more directly, see [38], p. 7.)

2.3. The Decomposition Theorem

Our aim in this section is a decomposition theorem for functions similar to one proved originally for $G = \mathbb{R}^n$ by Calderón and Zygmund and used by them as the basis for the discussion of certain singular integrals.

From this point on we assume 2.1.1(i) and 2.1.1(ii).

We start with an auxiliary lemma.

2.3.1. Lemma. *Let Q be either $\mathbb{N}$ or $\{1, \ldots, m\}$ for some m in $\mathbb{N}$. Let $(X_n)_{n\in Q}$ and $(Y_n)_{n\in Q}$ be two sequences of measurable subsets of G, the first sequence being disjoint. Suppose also that $X_n \subseteq Y_n$ for every n in Q. Then there exists a* disjoint *sequence $(V_n)_{n\in Q}$ of measurable subsets of G such that*

(i) $X_n \subseteq V_n \subseteq Y_n$ *for every n in Q;*

(ii) $\bigcup_{n\in Q} V_n = \bigcup_{n\in Q} Y_n$.

Proof. We may obviously suppose that $Q = \mathbb{N}$; if $Q = \{1, 2, \ldots, m\}$ for some m in $\mathbb{N}$, simply define $X_n = Y_n = \varnothing$ for $n > m$ and work with these extended sequences.

This being so, define

$$V_n = Y_n \Big\backslash \Big(\bigcup_{1 \leqslant j < n} V_j \cup \bigcup_{j > n} X_j \Big).$$

Clearly, $V_n \subseteq Y_n$. In order to show that $X_n \subseteq V_n$, and so complete the proof of (i), it suffices, by examination of the definition of V_n and the fact that $X_n \subseteq Y_n$, to prove that

$$X_n \cap V_j = \varnothing \quad \text{if} \quad j < n$$

and

$$X_n \cap X_j = \varnothing \quad \text{if} \quad j > n.$$

The second of these statements is part of the hypothesis, while the first is clear from the definition of V_j.

It remains to prove (ii). Suppose $x \in \bigcup Y_n$. If $x \in \bigcup X_n$, there is nothing to prove, since $\bigcup X_n \subseteq \bigcup V_n$ by (i). Otherwise, $x \in \bigcup Y_n$ and $x \in G \backslash \bigcup X_j \subseteq G \backslash \bigcup_{j>n} X_j$ for all n in Q. Let n be the first positive integer such that $x \in Y_n$. Then $x \in G \backslash Y_j$ for $j < n$, which implies by (i) that $x \in G \backslash V_j$ for $j < n$. Hence

$$x \in Y_n \Big\backslash \Big(\bigcup_{1 \leqslant j < n} V_j \cup \bigcup_{j > n} X_j \Big) = V_n.$$

This establishes (ii) and completes the proof. □

We turn now to the principal result of this section.

2.3.2. Theorem. *Let f be a nonnegligible integrable function, and let c be a positive number such that $m(G)c > \|f\|_1$ (if $m(G) = \infty$, c can be an arbitrary positive number). Then one can write*

$$f = g + h, \tag{1}$$

where the measurable functions g and h satisfy the following conditions:

(a) $\|g\|_\infty \leqslant Ac;$ (2)

(b) $h = \sum_{n \in Q} h_n$ (3)

where $Q = \{1, \ldots, m\}$ for some m in $\mathbb{N}$ or $Q = \mathbb{N}$, and

$$\sum_{n \in Q} \|h_n\|_1 \leqslant 2\|f\|_1, \tag{4}$$

the series (3) *being absolutely convergent in the pointwise sense when $Q = \mathbb{N}$;*

(c) *to each n in Q, there is a point x_n of G, an integer β_n, and a measurable set V_n such that*

$$V_n \subseteq x_n + U_{\beta_n}, \tag{5}$$

$$h_n = 0 \quad \textit{off} \quad V_n, \tag{6}$$

$$\int h_n = 0, \tag{7}$$

(weak L^1 inequality)

$$\sum_{n \in Q} m(U_{\beta_n}) \leqslant \frac{A}{c} \|f\|_1, \tag{8}$$

and

$$V_n \cap V_m = \varnothing \quad \textit{if} \quad m \neq n; \tag{9}$$

(d) *if $p \in [1, \infty)$, then $g \in \mathscr{L}^p$ and*

$$\|g\|_p^p \leqslant (Ac)^{p-1} \|f\|_1; \tag{10}$$

furthermore, if $f \in \mathscr{L}^p$, then h and h_n belong to $\mathscr{L}^p$ for every n in Q;

(e) *if f vanishes off a compact subset K of G, there is a compact set K_1 such that g, h and all h_n vanish off K_1.*

Proof. We aim to apply Lemma 2.2.1, taking

$$E = \left\{x \in G: \int_{x+U_\alpha} |f|\, dm \geqslant cm(U_\alpha) \text{ for some } \alpha \text{ in } \mathbb{Z}\right\}.$$

If $x \in E$, the nonvoid set

$$\left\{\alpha \in \mathbb{Z}: \int_{x+U_\alpha} |f|\, dm \geqslant cm(U_\alpha)\right\}$$

is necessarily bounded below. Otherwise, in fact, we should have that

$$\|f\|_1 \geqslant \int_{x+U_{\alpha_k}} |f|\, dm \geqslant cm(U_{\alpha_k})$$

for certain α_k in $\mathbb{Z}$ such that $\alpha_k \to -\infty$, in which case 2.1.1(i) shows that $\|f\|_1 \geqslant cm(G)$, contrary to hypothesis. We may therefore define the mapping $\alpha: E \to \mathbb{Z}$ by setting

$$\alpha(x) = \min\left\{\alpha \in \mathbb{Z}: \int_{x+U_\alpha} |f|\, dm \geqslant cm(U_\alpha)\right\};$$

$\alpha(x)$ is thus the first index α in $\mathbb{Z}$ for which

$$\int_{x+U_\alpha} |f|\, dm \geqslant cm(U_\alpha).$$

We claim that 2.2.1(a) is satisfied, i.e., that $\alpha(E)$ is bounded below. For otherwise there would exist points x_k in E such that $\alpha_k = \alpha(x_k) \to -\infty$; since, for every k,

$$\|f\|_1 \geqslant \int_{x_k+U_{\alpha_k}} |f|\, dm \geqslant cm(U_{\alpha_k}),$$

we should again be led to a contradiction of the hypothesis that $\|f\|_1 < cm(G)$.

We claim further that 2.2.1(b) is satisfied. For if $\alpha_0 \in \mathbb{Z}$ and $\{x \in E: \alpha(x) \leqslant \alpha_0\}$ were not relatively compact in G, we could choose points x_n in E such that $x_n \to \infty$ and $\alpha_n = \alpha(x_n) \leqslant \alpha_0$. By the definition of α_n,

$$\int_{x_n+U_{\alpha_n}} |f|\, dm \geqslant cm(U_{\alpha_n})$$

for every n. Since $\alpha(E)$ is bounded below, we should have $\alpha_n \geqslant \beta$ for some β in $\mathbb{Z}$, and so $U_{\alpha_n} \subseteq U_\beta$, which is relatively compact in G. It follows that a subsequence (x_{n_k}) could be chosen so that the sets $x_{n_k} + U_{\alpha_{n_k}}$ are pairwise disjoint. Then, by adding the inequalities

$$\int_{x_{n_k}+U_{\alpha_{n_k}}} |f|\,dm \geqslant cm(U_{\alpha_{n_k}}) \geqslant cm(U_{\alpha_0}),$$

we should deduce that

$$\|f\|_1 \geqslant \sum_{k=1}^{\infty} cm(U_{\alpha_0}) = \infty,$$

which is absurd.

We may therefore appeal to Lemma 2.2.1 to conclude the existence of a finite or infinite sequence $(x_n)_{n\in Q}$ of points of E such that, if $\alpha_n = \alpha(x_n)$, the α_n increase with n, the $x_n + U_{\alpha_n}$ are pairwise disjoint, and

$$E \subseteq \bigcup_{n\in Q} (x_n + U_{\alpha_n} - U_{\alpha_n}).$$

We now define $\beta_n = \theta(\alpha_n)$ and inflate the neighbourhood $x_n + U_{\alpha_n} - U_{\alpha_n}$ of x_n into $x_n + U_{\beta_n}$. Then

$$E \subseteq \bigcup_{n\in Q} (x_n + U_{\beta_n}) = S.$$

The sets $x_n + U_{\alpha_n}$, while pairwise disjoint, do not in general cover E, while the sets $x_n + U_{\beta_n}$ cover E but are not pairwise disjoint. However, Lemma 2.3.1 assures us of the existence of pairwise disjoint measurable subsets V_n ($n \in Q$) of G such that

$$\begin{gathered} x_n + U_{\alpha_n} \subseteq V_n \subseteq x_n + U_{\beta_n} \text{ for each } n \text{ in } Q \\ \bigcup_{n\in Q} V_n = S = \bigcup_{n\in Q} (x_n + U_{\beta_n}). \end{gathered} \tag{11}$$

Plainly, (11) implies (5).

The disjointness of the V_n, which is vital, makes it possible to define

$$g(x) = \begin{cases} m(V_n)^{-1} \displaystyle\int_{V_n} f\,dm & \text{for } x \text{ in } V_n \text{ and } n \text{ in } Q \\ f(x) & \text{for } x \text{ in } G\backslash S, \end{cases} \tag{12}$$

$$h(x) = f(x) - g(x), \tag{13}$$

$$h_n(x) = h(x)\xi_{V_n}(x) \text{ for each } n \text{ in } Q, \tag{14}$$

where as usual ξ_A denotes the characteristic function of A relative to G. Certainly (1) is satisfied.

Proof of (a). This proceeds for the two sets S and $G\backslash S$ separately (recall (11)). If $x \in S$, then $x \in V_n \subseteq x_n + U_{\beta_n}$ for some n. By the definition of g ((12)),

$$\begin{aligned} |g(x)| &\leqslant m(V_n)^{-1} \int_{V_n} |f|\, dm \\ &\leqslant \frac{m(U_{\beta_n})}{m(V_n)} \frac{1}{m(U_{\beta_n})} \int_{x_n + U_{\beta_n}} |f|\, dm. \end{aligned} \tag{15}$$

However, $\beta_n = \theta(\alpha_n) < \alpha_n$ and so, by the definition of α,

$$m(U_{\beta_n})^{-1} \int_{x_n + U_{\beta_n}} |f|\, dm < c \leqslant m(U_{\alpha_n})^{-1} \int_{x_n + U_{\alpha_n}} |f|\, dm. \tag{16}$$

The inequalities (15) and (16) and 2.1.1(ii) together show that

$$\begin{aligned} |g(x)| &\leqslant \frac{m(U_{\beta_n})}{m(U_{\alpha_n})} \frac{1}{m(U_{\beta_n})} \int_{x + U_{\beta_n}} |f|\, dm \\ &\leqslant Ac. \end{aligned} \tag{17}$$

In the case that $x \in G\backslash S$, it is certainly the case that $x \notin E$, and therefore, by the definition of E,

$$m(U_\alpha)^{-1} \int_{x + U_\alpha} |f|\, dm < c \tag{18}$$

for all α in $\mathbb{Z}$. Now Lemma 2.1.4 shows that as $\alpha \to \infty$,

$$m(U_\alpha)^{-1} \int_{x + U_\alpha} |f|\, dm \to |f| \text{ in } \mathscr{L}^1,$$

so that a subsequence of the left side converges pointwise a.e. to $|f|$. Hence (18) implies that

$$|f(x)| \leqslant c \quad \text{for almost all } x \text{ in } G\backslash S,$$

and so, by (17) and (12),

$$|g(x)| \leqslant \max(c, Ac) = Ac \tag{19}$$

almost everywhere on G. This completes the proof of (a).

Proof of (b). The disjointness of the sets V_n ensures the validity of (3), the series converging pointwise absolutely. The definitions of h_n (see (14)) and g (see

(12)) show that

$$\begin{aligned}\|h_n\|_1 &= \int_{V_n} |f - g| \, dm \\ &= \int_{V_n} |f - m(V_n)^{-1} \int_{V_n} f \, dm| \, dm \\ &\leqslant 2 \int_{V_n} |f| \, dm. \end{aligned} \tag{20}$$

By using the disjointness of the V_n, we deduce from (20) that

$$\sum_{n \in Q} \|h_n\|_1 \leqslant 2 \sum_{n \in Q} \int_{V_n} |f| \, dm \leqslant 2\|f\|_1.$$

This completes the proof of (b).

Proof of (c). The statements (5), (6), (7) and (9) are immediate from the definitions and construction of V_n and h_n.

The crucial inequality (8) is verified by noting that, since $\theta(\alpha_n) = \beta_n$,

$$\sum_{n \in Q} m(U_{\beta_n}) \leqslant A \sum_{n \in Q} m(U_{\alpha_n}) \tag{21}$$

(see 2.1.1(2)); and by the definition of α_n,

$$m(U_{\alpha_n}) \leqslant \frac{1}{c} \int_{x_n + U_{\alpha_n}} |f| \, dm \leqslant \frac{1}{c} \int_{V_n} |f| \, dm \tag{22}$$

since $x_n + U_{\alpha_n} \subseteq V_n$. It follows from (21) and (22) and the disjointness of the V_n that

$$\sum_{n \in Q} m(U_{\beta_n}) \leqslant \frac{A}{c} \sum_{n \in Q} \int_{V_n} |f| \, dm \leqslant \frac{A}{c} \|f\|_1.$$

Proof of (d). It follows from (19) that, if $p \in [1, \infty)$,

$$|g(x)|^p \leqslant (Ac)^{p-1} |g(x)| \quad \text{a.e..}$$

So

$$\begin{aligned}\|g\|_p^p &\leqslant (Ac)^{p-1} \int_G |g| \, dm \\ &= (Ac)^{p-1} \left\{ \sum_{n \in Q} \int_{V_n} |g| \, dm + \int_{G \setminus S} |g| \, dm \right\} \\ &\leqslant (Ac)^{p-1} \left\{ \sum_{n \in Q} \int_{V_n} |f| \, dm + \int_{G \setminus S} |f| \, dm \right\}\end{aligned}$$

by (12). Hence

$$\|g\|_p^p \leqslant (Ac)^{p-1}\|f\|_1.$$

This is (10).

If $f \in \mathscr{L}^p$, then it is clear from (10) and (1) that both h and g belong to $\mathscr{L}^p$. The definition of the h_n (see (14)) makes it plain that $|h_n(x)| \leqslant |h(x)|$, and hence $h_n \in \mathscr{L}^p$ for every n in Q.

Proof of (e). Suppose that f vanishes off a compact set K. Since $\sum_{n \in Q} m(U_{\beta_n}) < \infty$, $\beta_n \to \infty$ in case $Q = \mathbb{N}$. So in all cases $\beta_n \geqslant \beta$ for some β in $\mathbb{Z}$, and every n in Q. By (16), $x_n + U_{\alpha_n}$ meets K for every n in Q, hence

$$x_n \in K - U_{\alpha_n} \subseteq K - U_{\alpha_1}$$

since, by Lemma 2.2.1, the sequence (α_j) is increasing. Thus, for every n in Q,

$$x_n + U_{\beta_n} \subseteq K - U_{\alpha_1} + U_{\beta_n} \subseteq K - U_{\alpha_1} + U_\beta \subseteq K^*,$$

say, where K^* is the compact closure of $K - U_{\alpha_1} + U_\beta$. Hence

$$S = \bigcup_{n \in Q} (x_n + U_{\beta_n}) \subseteq K^*.$$

Then (12) shows that g vanishes off the compact set $K_1 = K \cup K^*$; and (13) and (14) indicate that h and all h_n also vanish off K_1. The assertion (e) is therefore fully verified. □

Note. The approach to Theorem 2.3.2 is an adaptation of one due to Rivière [33].

2.4. Bounds for Convolution Operators

2.4.1. Convolution operators. The simplest so-called *convolution operators* are those of the form

$$L_k : f \to k * f, \tag{1}$$

where k is a given integrable function and is usually known as the *kernel of the operator* L_k. It is well known that, if $p \in [1, \infty]$, then L_k maps L^p continuously into itself and that the corresponding operator norm $\|L_k\|_{p,p}$ (see Appendix A) satisfies the inequality

$$\|L_k\|_{p,p} \leqslant \|k\|_1. \tag{2}$$

It is easy to verify that

$$(L_k f)^\wedge = \hat{k}.\hat{f}, \tag{3}$$

at least when $f \in L^2 \cap L^p$. Thus L_k is none other than an extension of the operator $T_{\hat{k}}$ as defined in 1.2.2. From this, together with the facts noted in 1.2.2(ii), it follows that

$$\|L_k\|_{2,2} = \|\hat{k}\|_\infty. \tag{4}$$

Another way of expressing what has just been written is to say that, when $k \in L^1$, $\hat{k} \in M_p(X)$ for every p in $[1, \infty]$; that L_k is (an extension of) $T_{\hat{k}}$; and that the multiplier norm $\|\hat{k}\|_{p,p}$ is at most $\|k\|_1$. More generally, it is a simple and well known fact that if μ is a bounded measure on G, then $\phi = \hat{\mu}$ belongs to $M_p(X)$ and $\|\hat{\mu}\|_{p,p} \leqslant \|\mu\|$ whenever $p \in [1, \infty]$. See 1.2.2(ii).

The results of 1.2.8 suggest the following approach to proving Littlewood-Paley-type theorems: show, by use of the Marcinkiewicz interpolation theorem, that functions ϕ in $\mathscr{L}^\infty(X)$, constant on elements of a decomposition (Δ_j), which are *not* equal l.a.e. to Fourier-Stieltjes transforms (i.e. which are *not* in $M_1(X)$), belong to $M_p(X)$ for each p in $(1, \infty)$ and have norm $\|\phi\|_{p,p}$ not greater than a constant multiple of $\|\phi\|_\infty$. Indeed this approach to the Littlewood-Paley theorem works well for certain disconnected groups: cf. Chapter 4. However, for groups such as $\mathbb{Z}$, $\mathbb{R}$ and $\mathbb{T}$ with their dyadic decompositions, it encounters difficulties. Cf. 7.1.3. The more circuitous path that is adopted in these cases still involves showing, among other things, that certain nontrivial functions are in $M_p(X)$.

Now it is possible to show that, whenever $\phi \in \mathscr{L}^\infty(X)$, T_ϕ is a generalised convolution operator. More precisely, there is a pseudomeasure k on G such that $\hat{k} = \phi$ and

$$T_\phi f = k * f$$

at least for every f in $L^1 \cap L^\infty$. In general, this k is neither an integrable function nor a bounded measure, and it is natural to speak of it as a *singular kernel.* One could therefore say that our ultimate concern lies with bounds for the operator norms of such singular kernels.

Nontheless, rather than become involved with detailed technical considerations of general singular kernels (which would add nothing to the essence of the matter, anyway), we shall seek to show that the particular functions ϕ of concern to us are multipliers, and to bound their norms, by trying to represent ϕ as a limit, in a suitable sense, of functions $\hat{k}_\nu$, where $k_\nu \in L^1(G)$, so that $T_{\hat{k}_\nu} = L_{k_\nu}$ is a convolution operator in the simplest (nonsingular) sense.

Here is a lemma which explains the general procedure.

2.4.2. Lemma. *Suppose that $\phi \in \mathscr{L}^\infty(X)$, that $p \in (1, \infty)$, that B is a positive number, and that (k_ν) is a sequence of integrable functions on G such that*

(a) $\|L_{k_\nu}\|_{p,p} = \|\hat{k}_\nu\|_{p,p} \leqslant B$;

(b) $\hat{k}_\nu \to \phi$ l.a.e..

Then $\phi \in M_p(X)$ and $\|\phi\|_{p,p} \leqslant B$. Roughly speaking, the norm-closed balls in $M_p(X)$ are in fact closed in a stronger sense.

Proof. By (a),

$$\|k_\nu * f\|_p \leqslant B\|f\|_p$$

for every ν and every f in L^p. So Hölder's inequality shows that

$$\left|\int k_\nu * f(x)\overline{g(x)}\,dx\right| \leqslant B\|f\|_p\|g\|_{p'}. \tag{5}$$

However, by the Parseval formula,

$$\int k_\nu * f(x)\overline{g(x)}\,dx = \int \hat{k}_\nu(\chi)\hat{f}(\chi)\overline{\hat{g}(\chi)}\,d\chi, \tag{6}$$

at least for f and g in L^2. By (a) and the results mentioned in 1.2.2(ii), $\|\hat{k}_\nu\|_\infty \leqslant B$ and so

$$|\hat{k}_\nu(\chi)\hat{f}(\chi)\overline{\hat{g}(\chi)}| \leqslant B|\hat{f}(\chi)\overline{\hat{g}(\chi)}| \quad \text{a.e..}$$

Thus we may appeal to Lebesgue's dominated convergence theorem to deduce from (5) and (6) that

$$\left|\int T_\phi f(x)g(x)\,dx\right| = \left|\int \phi(\chi)\hat{f}(\chi)\overline{\hat{g}(\chi)}\,d\chi\right| \leqslant B\|f\|_p\|g\|_{p'}$$

for f in $L^2 \cap L^p$ and g in $L^2 \cap L^{p'}$ whence (by the converse to Hölder's inequality) $\|T_\phi f\|_p \leqslant B\|f\|_p$ for f in $L^2 \cap L^p$. □

In putting Lemma 2.4.2 to work, the difficult task is to establish the uniform bound estimate (a) for some natural sequence (k_ν) which has property (b). The most simple-minded way of establishing (a) would be to establish the stronger estimate

$$\|k_\nu\|_1 \leqslant B.$$

But unless ϕ is equal l.a.e. to a Fourier-Stieltjes transform, any sequence (k_ν) of integrable functions satisfying (b) is necessarily unbounded in $L^1(G)$; and the particular functions ϕ of interest to us are certainly not equal l.a.e. to Fourier-Stieltjes transforms.

So a more subtle approach is necessary. The one we employ involves using the Marcinkiewicz interpolation theorem: the basic outline of the argument can be summarised in the recipe "weak (1, 1) estimate + strong (2, 2) estimate implies strong (p, p) estimate". To be more specific, given ϕ in $\mathscr{L}^\infty(X)$, we seek to produce a sequence (k_ν) of integrable functions such that

(i) $\|\hat{k}_\nu\|_\infty \leqslant B$;

(ii) weak (1, 1) norm of $L_{k_\nu} \leqslant B'$;
(iii) $\hat{k}_\nu \to \phi$ l.a.e.;
in (i) and (ii), B and B' denote numbers *independent of* ν.

Consider now the estimation of the weak (1, 1) norm of L_k, where k is an integrable kernel on a group G with a covering family (U_α). It seems to be desirable to say a few words about the basic ideas of the Calderón-Zygmund technique of estimating weak (1, 1) norms before launching into a full execution of the ideas. The first basic idea is contained in the following simple lemma.

2.4.3. Lemma. *Suppose that $\beta \in \mathbb{Z}$, $a \in G$, k is an integrable function on G, and that u is an integrable function on G which vanishes off $a + U_\beta$, and has zero integral. Then*

$$\int_{G\setminus(a+U_{\theta(\beta)})} |k * u(x)|\, dx \leqslant \|u\|_1 \sup_{y\in U_\beta} \int_{G\setminus U_{\theta(\beta)}} |k(x-y) - k(x)|\, dx.$$

Proof. Since u vanishes off $a + U_\beta$ and has zero integral, we have, for almost x in G:

$$k * u(x) = \int_{a+U_\beta} (k(x-y) - k(x-a))u(y)\, dy$$

and so, by the Fubini-Tonelli theorem,

$$\int_{G\setminus(a+U_{\theta(\beta)})} |k * u|\, dm \leqslant \int_{a+U_\beta} |u(y)| \left\{ \int_{G\setminus(a+U_{\theta(\beta)})} |k(x-y) - k(x-a)|\, dx \right\} dy$$

$$= \int_{a+U_\beta} |u(y)| \left\{ \int_{G\setminus U_{\theta(\beta)}} |k(z+a-y) - k(z)|\, dz \right\} dy. \tag{7}$$

If $y \in a + U_\beta$, $y - a \in U_\beta$, and so

$$\int_{G\setminus U_{\theta(\beta)}} |k(z+a-y) - k(z)|\, dz \leqslant \sup_{y\in U_\beta} \int_{G\setminus U_{\theta(\beta)}} |k(x-y) - k(x)|\, dx. \tag{8}$$

The lemma results by combining (7) and (8). □

It is an immediate consequence of Lemma 2.4.3 that

$$m(\{x \in G\setminus(a + U_{\theta(\beta)}) : |k * u(x)| > s\}) \leqslant J_\beta(k)\|u\|_1/s \tag{9}$$

where

$$J_\beta(k) = \sup_{y\in U_\beta} \int_{G\setminus U_{\theta(\beta)}} |k(x-y) - k(x)|\, dx$$

is akin to a modulus of continuity of k. In the proof of Theorem 2.4.4 the inequality (9) comes to be applied to each of the functions h_n and the corresponding $U_{\theta(\beta_n)}$ of the decomposition theorem 2.3.2. We are thus led to consider the numbers $J_{\beta_n}(k)$ with n varying. Accordingly, it is natural to introduce the supremum

$$J(k) = \sup_{\beta} J_\beta(k). \tag{10}$$

Theorem 2.4.4 gives a bound for $\|L_k\|_{p,p}$ in terms of $J(k)$ and $\|\hat{k}\|_\infty$.

2.4.4. Theorem. *Suppose $k \in L^1$. Then*
(i) *the weak* (1, 1) *norm of L_k on L^1 is at most*

$$B = A^2 + 4A\|\hat{k}\|_\infty^2 + 4J(k);$$

(ii) *for p in* $(1, \infty)$,

$$\|L_k\|_{p,p} \leqslant A^{2/p^*} B_p \max(J(k), \|\hat{k}\|_\infty),$$

where $p^* = \min(p, p')$ *and B_p depends solely on p.*

Proof. (i) Write temporarily f^* for $k * f$ and

$$\lambda(t) = m(\{x : |f^*(x)| > t\})$$

for $t > 0$. We have to show that

$$\lambda(t) \leqslant Bt^{-1}\|f\|_1 \tag{11}$$

for $t > 0$ and f in L^1. For this it is plainly necessary and sufficient to show that

$$\lambda(t) \leqslant Bt^{-1} \tag{12}$$

for $t > 0$ and all f in L^1 such that $\|f\|_1 = 1$.

If G is compact and $t \leqslant 1$, we have trivially that

$$\lambda(t) \leqslant m(G) = 1 \leqslant t^{-1}$$

and so, since $B \geqslant 1$, (12) holds. Hence in this case it suffices to prove (12) under the assumption that $tm(G) > \|f\|_1$.

This last assumption is automatically fulfilled when G is noncompact (in which case $m(G) = \infty$). Hence in either case we may and will suppose that $tm(G) > 1 = \|f\|_1$.

This being so, we may apply the decomposition theorem 2.3.2 with t in place of c. By 2.3(1) and the linearity of the operator L_k, $f^* = g^* + h^*$, hence $|f^*| \leqslant |g^*| + |h^*|$ and therefore

$$\begin{aligned}\lambda(t) &\leqslant m(\{x : |g^*(x)| > t/2\}) + m(\{x : |h^*(x)| > t/2\}) \\ &= (I) + (II),\end{aligned} \tag{13}$$

say. To estimate (I), note that by (4) and 2.3(10),

$$\|k * g\|_2 \leqslant \|\hat{k}\|_\infty \|g\|_2 \leqslant \|\hat{k}\|_\infty A^{1/2} t^{1/2} \|f\|_1 = \|\hat{k}\|_\infty A^{1/2} t^{1/2}$$

and hence also

$$t/2 . m(\{x\colon |g^*(x)| > t/2\})^{1/2} \leqslant \|g^*\|_2 \leqslant \|\hat{k}\|_\infty A^{1/2} t^{1/2}.$$

Therefore

$$(I) \leqslant 4t^{-1} A \|\hat{k}\|_\infty^2. \tag{14}$$

Turning to (II), we observe first that from 2.3(3) and 2.3(4) it follows easily that

$$h^* = k * h = \sum_{n \in Q} k * h_n = \sum_{n \in Q} h_n^*,$$

the series converging in L^1. Hence

$$|h^*| \leqslant \sum_{n \in Q} |h_n^*| \text{ a.e..} \tag{15}$$

Put $F = \bigcup_{n \in Q} (x_n + U_{\theta(\beta_n)})$, the points x_n and the indices β_n being as in 2.3.2, part (c). Then 2.3(8) and 2.1.1(ii) imply that

$$m(F) \leqslant A \sum_{n \in Q} m(U_{\beta_n}) \leqslant A^2 t^{-1} \|f\|_1 = A^2 t^{-1}. \tag{16}$$

According to Lemma 2.4.3,

$$\int_{G \setminus (x_n + U_{\theta(\beta_n)})} |h_n^*| \, dm \leqslant J(k) \|h_n\|_1$$

and so, by (15),

$$\begin{aligned} \int_{G \setminus F} |h^*| \, dm &\leqslant \sum_{n \in Q} \int_{G \setminus F} |h_n^*| \, dm \\ &\leqslant \sum_{n \in Q} \int_{G \setminus (x_n + U_{\theta(\beta_n)})} |h_n^*| \, dm \\ &\leqslant \sum_{n \in Q} J(k) \|h_n\|_1 \leqslant 2J(k), \end{aligned}$$

the last step by 2.3(4) and the assumption that $\|f\|_1 = 1$. This implies that

$$m(\{x \in G \setminus F\colon |h^*(x)| > t/2\}) \leqslant 4J(k) t^{-1}. \tag{17}$$

Combining (16) and (17), we get the estimate

$$(II) \leqslant m(F) + 4J(k) t^{-1} \leqslant (A^2 + 4J(k)) t^{-1}. \tag{18}$$

So, by (13), (14) and (18),

$$\lambda(t) \leqslant (4\|\hat{k}\|_\infty^2 A + A^2 + 4J(k))t^{-1},$$

which is (12). This proves (i).

(ii) Suppose first that $p \in (1, 2]$. Then (4), (i) and the Marcinkiewicz interpolation theorem (see Appendix A.2) applied with (say) $D = L^1 \cap L^\infty$ show that if $p \in (1, 2)$

$$\|L_k\|_{p,p}^p \leqslant \{2B(p-1)^{-1} + (2\|\hat{k}\|_\infty)^2(2-p)^{-1}\}p.$$

For $p = 2$, (4) itself applies. Since $A \geqslant 1$, simple calculations show that as a result, if $p \in (1, 2]$ and

$$\max(J(k), \|\hat{k}\|_\infty) \leqslant 1,$$

then

$$\|L_k\|_{p,p} \leqslant A^{2/p} B_p$$

where $B_p = 1$ if $p = 2$ and

$$B_p = \{p(32 - 14p)(p-1)^{-1}(2-p)^{-1}\}^{1/p}$$

if $p \in (1, 2)$. Hence, by homogeneity of the norm,

$$\|L_k\|_{p,p} \leqslant A^{2/p} B_p . \max(J(k), \|\hat{k}\|_\infty) \tag{19}$$

for all p in $(1, 2]$.

To handle the case where $p \in [2, \infty)$, and so complete the proof, one may now use the fact (stated in 1.2.2(i)) that $\|\phi\|_{p',p'} = \|\phi\|_{p,p}$ for every ϕ in $M_p(X)$. Alternatively one may begin by verifying that

$$L_k f * g = f * L_k g$$

at all points of G if (say) f and g belong to $L^1 \cap L^\infty$. So, evaluating at 0, we have for f and g in $L^1 \cap L^\infty$, the inequality

$$\begin{aligned} \left| \int L_k f(x) . g(-x)\, dx \right| &= \left| \int f(x) . L_k g(-x)\, dx \right| \\ &\leqslant \|f\|_p \|L_k g\|_{p'} \\ &\leqslant \|f\|_p . A^{2/p'} B_{p'} \max(J(k), \|\hat{k}\|_\infty) \|g\|_{p'}, \end{aligned}$$

the last step by (19) since $p' \in (1, 2]$. The converse of Hölder's inequality then shows that for f in $L^1 \cap L^\infty$ one has the estimate

$$\|L_k f\|_p \leqslant A^{2/p'} B_{p'} . \max(J(k), \|\hat{k}\|_\infty) . \|f\|_p .$$

If therefore we define B_p when $p \in [2, \infty)$ to be $B_{p'}$ and write p^* for $\min(p, p')$, we shall have for p in $(1, \infty)$ and f in $L^1 \cap L^\infty$ the inequality

$$\|L_k f\|_p \leqslant A^{2/p^*} B_p . \max(J(k), \|\hat{k}\|_\infty) . \|f\|_p,$$

which (see Appendix A.1) is equivalent to (ii). □

2.4.5. Corollary. *Suppose $\phi \in \mathscr{L}^\infty(X)$ and that there exists a sequence (k_ν) of integrable functions on G such that*

$$\kappa = \sup_\nu \max(J(k_\nu), \|\hat{k}_\nu\|_\infty) < \infty \tag{20}$$

and

$$\phi = \lim_{\nu \to \infty} \hat{k}_\nu \quad \textit{l.a.e. on } X. \tag{21}$$

Let T_ϕ be as defined in 1.2.2. *Then*

(i) *T_ϕ is of weak type* (1, 1) *on $L^1 \cap L^2$;*

(ii) *if $p \in (1, \infty)$,*

$$\|T_\phi f\|_p \leqslant A^{2/p^*} B_p \kappa \|f\|_p \tag{22}$$

for all f in $L^2 \cap L^p$; in other words, $\phi \in M_p(X)$ and

$$\|\phi\|_{p,p} \leqslant A^{2/p^*} B_p \kappa.$$

Proof. Part (i) of Theorem 2.4.4 combines with (20) to show that for f in $\mathscr{L}^1$ and $t > 0$,

$$m(\{x: |k_\nu * f(x)| > t\}) \leqslant B t^{-1} \|f\|_1 \tag{23}$$

for every ν, where B is independent of ν, t and f. If $f \in L^1 \cap L^2$, it follows from (20) and (21) that $F_\nu = k_\nu * f \to F = T_\phi f$ in L^2, and so some subsequence (F_{ν_j}) converges pointwise a.e. to F. Then

$$\{x: |F(x)| > t\} \subseteq \bigcup_i \bigcap_{j \geqslant i} \{x: |F_{\nu_j}(x)| > t\}$$

and so, by (23),

$$\begin{aligned} m(\{x: |F(x)| > t\}) &\leqslant \lim_i m\Big(\bigcap_{j \geqslant i} \{x: |F_{\nu_j}(x)| > t\}\Big) \\ &\leqslant \lim_i \inf_{j \geqslant i} m(\{x: |F_{\nu_j}(x)| > t\}) \\ &\leqslant \liminf_j B t^{-1} \|f\|_1 \\ &= B t^{-1} \|f\|_1, \end{aligned}$$

which proves (i).

Again, part (ii) of Theorem 2.4.4 combines with (20) and Lemma 2.4.2 to show that $\phi \in M_p(X)$ and $\|\phi\|_{p,p} \leqslant A^{2/p^*}B_p\kappa$.

(Alternatively,

$$\|k_\nu * f\|_p \leqslant A^{2/p^*}B_p\kappa\|f\|_p \tag{24}$$

for p in $(1, \infty)$, every ν and every f in $L^2 \cap L^p$. Taking a subsequence $(k_{\nu_j} * f)$ as before and applying Fatou's lemma to (24) leads to (22).)

2.4.6. *Remark.* Corollary 2.4.5 is a very powerful general result. By using it, quite large classes of functions belonging to M_p for every p in $(1, \infty)$ can be constructed. Cf. [38] and [21]. Nevertheless, it is a remarkable fact that there are functions ϕ belonging to $M_p(X)$ for every p in $(1, \infty)$ such that T_ϕ is *not* of weak type (1, 1): counterexamples will be given in 7.5 below. It follows that the process used above for establishing that ϕ is an element of $M_p(X)$ for every p in $(1, \infty)$ works only in rather special cases and falls far short of revealing the whole truth.

2.4.7. General comments. The decomposition theorem 2.3.2 provides the foundation for results considerably more general than Theorem 2.4.4 and Corollary 2.4.5 (which have been selected because they are just adequate for our purposes); for examples see [33], Theorem (3.3) and Remark (3.1), and [21], Chapter II (for the case $G = \mathbb{R}^n$).

Chapter 3. Convolution Operators (Vector-Valued Case)

3.1. Introduction

In Chapters 6 and 7 we shall need analogues of the results in Chapter 2 for operators of the type

$$L_K f(x) = \int K(x - y) f(y)\, dy,$$

where now f is a function on G taking values in a Hilbert space $\mathscr{H}_1$ (i.e., a vector-valued function) and K is a function on G taking values in $B(\mathscr{H}_1, \mathscr{H}_2)$, the space of bounded linear mappings of $\mathscr{H}_1$ into a second Hilbert space $\mathscr{H}_2$ (i.e., an operator-valued kernel).

For our purposes it suffices to handle the case in which every Hilbert space $\mathscr{H}$ involved is of the form $\mathbb{C}^I$, where I is some finite index set, a specialisation which allows us to shorten considerably the preliminaries.

Broadly speaking, the results of Chapter 2 carry over with no more than the obviously necessary changes, absolute values of complex numbers being replaced by appropriate vector or operator norms. We indicate briefly some necessary ideas and the method of approach.

3.2. Vector-Valued Functions

Throughout this section, $\mathscr{H}$ denotes the Hilbert space $\mathbb{C}^I$ with its canonical scalar product and norm

$$(a, b) = \sum_{i \in I} a_i \bar{b}_i, \qquad |a| = \left(\sum_{i \in I} |a_i|^2 \right)^{1/2}.$$

3.2.1. An $\mathscr{H}$-valued function f on G can be thought of as an I-tuple $(f_i)_{i \in I}$ of complex-valued functions on G. The function f is termed measurable if and only if every f_i is measurable; it is equivalent to demand that $x \to (f(x), a)$ be measurable for every a in $\mathscr{H}$. If f is measurable, so too is the function

$$|f| : x \to |f(x)| = \left(\sum_{i \in I} |f_i(x)|^2 \right)^{1/2}.$$

If $p \in [1, \infty)$, $L^p(G, \mathscr{H})$ denotes the set of measurable $\mathscr{H}$-valued functions f on G such that

$$\|f\|_p = \left(\int |f(x)|^p \, dx \right)^{1/p} < \infty;$$

$L^\infty(G, \mathscr{H})$ is the set of measurable $\mathscr{H}$-valued functions f on G such that

$$\|f\|_\infty = \text{loc ess sup } |f(x)| < \infty.$$

It is clear that $f \in L^p(G, \mathscr{H})$ if and only if $f_i \in L^p(G)$ for every i in I.

$L^p(G, \mathscr{H})$ is evidently a linear space, even a Banach space with the norm $\|.\|_p$ provided that (as in the case of complex-valued functions) we pass from functions to equivalence classes modulo equality a.e. (or l.a.e. in case $p = \infty$).

3.2.2. If $f \in L^1(G, \mathscr{H})$, $\int f(x)\, dx$ or $\int f\, dm$ is the unique element u of $\mathscr{H}$ such that

$$(u, a) = \int (f(x), a)\, dx$$

for every a in $\mathscr{H}$; this amounts to saying that

$$u_i = \int f_i(x)\, dx$$

for every i in I. It is clear that then

$$\left| \int f(x)\, dx \right| \leqslant \int |f(x)|\, dx.$$

3.2.3. If $f: G \to \mathscr{H}$ is measurable, and if (E_n) is an increasing sequence of compact subsets of G whose union is G, the functions $f_n: G \to \mathscr{H}$ defined by the formula

$$f_n(x) = \begin{cases} f(x) \text{ if } x \in E_n \text{ and } |f(x)| \leqslant n \\ 0 \quad \text{otherwise,} \end{cases}$$

belong to $L^1(G, \mathscr{H}) \cap L^\infty(G, \mathscr{H})$ and

$$\lim_{n \to \infty} \|f - f_n\|_p = 0$$

if $p \in [1, \infty)$ and $f \in L^p(G, \mathscr{H})$, while

$$\|f_n\|_\infty \leqslant \|f\|_\infty$$

if $f \in L^\infty(G, \mathscr{H})$. Moreover, $f_n \to f$ pointwise for every f.

3.2.4. There is no difficulty in verifying the validity of the analogue of Lemma 2.1.4.

3.2.5. A tedious inspection of the proof of the decomposition theorem 2.3.2 shows that the result carries over to functions f in $L^1(G, \mathscr{H})$.

3.3. Operator-Valued Kernels

3.3.1. The kernels are now functions K on G with values in $B(\mathscr{H}_1, \mathscr{H}_2)$, where $\mathscr{H}_1 = \mathbb{C}^I$ and $\mathscr{H}_2 = \mathbb{C}^J$. Such a kernel may be thought of as an $I \times J$-matrix (K_{ij}) of complex-valued kernels. K is termed *measurable* if and only if the function $x \to (K(x)a, b)$ is measurable for every a in $\mathscr{H}_1$ and every b in $\mathscr{H}_2$; this is equivalent to demanding that every K_{ij} be measurable. Since

$$|K(x)| = \max\left\{\sum_j \left|\sum_i K_{ij}(x)a_i\right|^2\right\}^{1/2},$$

the maximum being taken with respect to those a in $\mathscr{H}_1$ satisfying $|a| \leqslant 1$, it is clear that the function

$$|K| : x \to |K(x)|$$

is measurable whenever K is measurable. Similarly, the function

$$x \to K(x)f(x)$$

is measurable whenever K and f are measurable (the latter taking its values in $\mathscr{H}_1$).

The spaces $L^p(G, B(\mathscr{H}_1, \mathscr{H}_2))$ and the norms $\|.\|_p$ are defined in the expected way; cf. 3.2.1 above.

3.3.2. There are two special types of kernel which will come into use in Chapters 6 and 7.

(i) $I = J$ and

$$K(x) : (a_i)_{i \in I} \to (k_i(x)a_i)_{i \in I},$$

where the k_i are complex-valued kernels; then

$$|K(x)| = \max_{i \in I} |k_i(x)|.$$

A special subcase of this one is that in which $k_i = k$ for every $i \in I$, so that $K(x) = k(x)\mathbf{1}$ ($\mathbf{1}$ the identity operator on $\mathscr{H} = \mathbb{C}^I$), and $|K(x)| = |k(x)|$.

(ii) I a singleton and

$$K(x) : a \in \mathbb{C} \to (k_j(x)a)_{j \in J},$$

where the k_j are complex-valued kernels; here

$$K(x) = \left\{ \sum_{j \in J} |k_j(x)|^2 \right\}^{1/2}.$$

3.3.3. If $f\colon G \to \mathscr{H}_1$ and $K\colon G \to B(\mathscr{H}_1, \mathscr{H}_2)$ are measurable, and if

$$\int |K(x)f(x)|\, dx < \infty,$$

then 3.2.2 supplies the definition of

$$v = \int K(x)f(x)\, dx$$

as an element of $\mathscr{H}_2$; in fact,

$$v_j = \sum_i \int K_{ij}(x) f_i(x)\, dx$$

for every j in J.

3.4. Fourier Transforms

3.4.1. If $f \in L^1(G, \mathscr{H})$, its *Fourier transform* is the $\mathscr{H}$-valued function $\hat{f}$ on X defined by the formula

$$\hat{f}(\chi) = \int \overline{\chi(x)} f(x)\, dx, \tag{1}$$

the integral being defined as in 3.2.2. If $f = (f_i)_{i \in I}$, then $\hat{f} = (\hat{f}_i)_{i \in I}$. It is easy to verify that $\hat{f}$ is continuous from X into $\mathscr{H}$ and that $\hat{f} \in L^\infty(X, \mathscr{H})$.

By the Parseval formula for complex-valued functions,

$$\int |f_i(x)|^2\, dx = \int |\hat{f}_i(\chi)|^2\, d\chi$$

for every i in I, provided $f \in L^1(G, \mathscr{H}) \cap L^2(G, \mathscr{H})$. Summing over i, we get the Parseval formula

$$\int |f(x)|^2\, dx = \int |\hat{f}(\chi)|^2\, d\chi \tag{2}$$

valid for f in $L^1(G, \mathscr{H}) \cap L^2(G, \mathscr{H})$.

3.4.2. The Plancherel theorem for complex-valued functions shows that the Fourier transformation can be extended into an isometry of $L^2(G, \mathscr{H})$ onto $L^2(X, \mathscr{H})$, with (2) continuing to hold for every f in $L^2(G, \mathscr{H})$.

3.4.3. In an exactly similar way we can define $\hat{K}$ for K in $L^1(G, B(\mathscr{H}_1, \mathscr{H}_2))$, $\hat{K}$ being a continuous function from X into $B(\mathscr{H}_1, \mathscr{H}_2)$. Extensions to other kernels K are possible but we shall not need them.

3.5. Convolution Operators

3.5.1. Suppose that $K: G \to B(\mathscr{H}_1, \mathscr{H}_2)$ and $f: G \to \mathscr{H}_1$ are measurable functions. For every x in G such that $|K| * |f|(x) < \infty$, $K * f(x)$ is defined as an element of $\mathscr{H}_2$ (see 3.3.3) by the formula

$$K * f(x) = \int K(x - y) f(y) \, dy; \tag{1}$$

then, by 3.2.2,

$$|K * f(x)| \leqslant |K| * |f|(x) \tag{2}$$

whenever the left side is defined.

In particular, if $K \in L^1(G, B)$ and $f \in L^p(G, \mathscr{H}_1)$, $K * f(x)$ is defined for almost every x in G and (2) holds a.e.. It is simple to check that $K * f$, thus defined a.e., is measurable; and then (2) goes to show that $K * f \in L^p(G, \mathscr{H}_2)$ and

$$\|K * f\|_p \leqslant \|K\|_1 \|f\|_p. \tag{3}$$

3.5.2. It is possible to verify that

$$(K * f)^\wedge = \hat{K}\hat{f} \tag{4}$$

whenever $K \in L^1(G, B)$ and $f \in L^1(G, \mathscr{H}_1) \cup L^2(G, \mathscr{H}_2)$.

3.5.3. Just as in 2.4.1 for the case of complex-valued functions, we can now consider the *convolution operator*

$$L_K : f \to K * f \tag{5}$$

associated with a kernel K in $L^1(G, B)$. By (3), L_K maps $L^p(G, \mathscr{H}_1)$ continuously into $L^p(G, \mathscr{H}_2)$ and

$$\|L_K\|_{p,p} \leqslant \|K\|_1. \tag{6}$$

For $p = 2$ it is easy to infer from 3.4(2) and (4) that

$$\|L_K\|_{2,2} \leqslant \|\hat{K}\|_\infty. \tag{7}$$

Compare (6) and (7) with 2.4(2) and 2.4(4) respectively.

If ϕ is an element of $L^\infty(X, B)$, T_ϕ can be defined as a continuous linear operator from $L^2(G, \mathscr{H}_1)$ into $L^2(G, \mathscr{H}_2)$ by the formula

$$(T_\phi f)^\wedge = \phi\hat{f}, \tag{8}$$

the justification being provided by 3.4.2; cf. equation 2.4(3).

Equations (4) and (8) combine to show that $L_K = T_\phi$ where $\phi = \hat{K} \in L^\infty(X, B)$.

3.6. Bounds for Convolution Operators

3.6.1. *We now assume G has a covering family $(U_\alpha)_{\alpha\in\mathbb{Z}}$ as in 2.1.1.*

The foregoing preliminaries, together with the substance of Appendix A.3, make it possible (though somewhat tedious) to verify that the results of Chapter 2 carry over to vector-valued functions and operator-valued kernels, the only change being that absolute values are to be replaced by the appropriate vector or operator norms. The principal end results of this extension are the analogues of Theorem 2.4.4 and Corollary 2.4.5, which take the following forms, where A is as in 2.1.1(ii), $J(K)$ is defined exactly as in 2.4(10), p^* denotes $\min(p, p')$ as in Theorem 2.4.4, and B_p is also as in Theorem 2.4.4.

3.6.2. Theorem. *Suppose $K \in L^1(G, B(\mathscr{H}_1, \mathscr{H}_2))$ and $p \in (1, \infty)$. Then*

(i) *the weak* (1, 1) *norm of L_K on $L^1(G, \mathscr{H}_1)$ is at most*

$$B = A^2 + 4A\|\hat{K}\|_\infty^2 + 4J(K);$$

and

(ii) $$\|L_K\|_{p,p} \leqslant A^{2/p^*} B_p \max(J(K), \|\hat{K}\|_\infty). \tag{1}$$

3.6.3. Corollary. *Let (K_ν) be a sequence of elements of $L^1(G, B(\mathscr{H}_1, \mathscr{H}_2))$ such that*

$$\kappa = \sup_\nu \max(J(K_\nu), \|\hat{K}_\nu\|_\infty) < \infty.$$

Suppose also that $\lim \hat{K}_\nu(\chi) = \phi(\chi)$ *in $B(\mathscr{H}_1, \mathscr{H}_2)$ for l.a.e. χ in X. Let T_ϕ be defined as in* 3.5(8). *Then*

(i) *T_ϕ is of weak type* (1, 1) *on* $L^1(G, \mathscr{H}_1) \cap L^2(G, \mathscr{H}_1)$;

(ii) *if $p \in (1, \infty)$,*

$$\|T_\phi f\|_p \leqslant A^{2/p^*} B_p \kappa \|f\|_p \tag{2}$$

for all f in $L^2(G, \mathscr{H}_1) \cap L^p(G, \mathscr{H}_1)$.

3.6.4. Corollary. *Suppose that* $k_\nu \in L^1(G)$ *for* $\nu = 1, 2, \ldots, \phi \in \mathscr{L}^\infty(X)$,

$$\kappa = \sup_\nu \max(J(k_\nu), \|\hat{k}_\nu\|_\infty) < \infty$$

and

$$\lim_{\nu\to\infty} \hat{k}_\nu(\chi) = \phi(\chi) \quad \textit{l.a.e.}\,.$$

Let J *be any countable index set. Then*

$$\left\|\left(\sum_{j\in J} |T_\phi f_j|^2\right)^{1/2}\right\|_p \leqslant A^{2/p^*} B_p \kappa \left\|\left(\sum_{j\in J} |f_j|^2\right)^{1/2}\right\|_p$$

for p *in* $(1, \infty)$ *and every family* $(f_j)_{j\in J}$ *of elements of* $L^2 \cap L^p$.

Proof. Let I be any finite subset of J. In Corollary 3.6.3, take $\mathscr{H}_1 = \mathscr{H}_2 = \mathbb{C}^I$, $K_\nu(x) = k_\nu(x)\mathbf{1}$, and $\Phi_\nu(\chi) = \phi(\chi)\mathbf{1}$, where $\mathbf{1}$ is the identity endomorphism of $\mathbb{C}^I$. Then

$$\hat{K}_\nu = \hat{k}_\nu \mathbf{1}, \quad \|\hat{K}_\nu\|_\infty = \|\hat{k}_\nu\|_\infty,$$
$$J(K_\nu) = J(k_\nu),$$
$$\lim_{\nu\to\infty} \hat{K}_\nu(\chi) = \Phi(\chi) \text{ l.a.e.},$$

and

$$T_\Phi((f_j)_{j\in I}) = (T_\phi f_j)_{j\in I}.$$

Thus, by 3.6.3(ii),

$$\left\|\left(\sum_{j\in I} |T_\phi f_j|^2\right)^{1/2}\right\|_p \leqslant A^{2/p^*} B_p \kappa \left\|\left(\sum_{j\in I} |f_j|^2\right)^{1/2}\right\|_p \leqslant A^{2/p^*} B_p \kappa \left\|\left(\sum_{j\in J} |f_j|^2\right)^{1/2}\right\|_p,$$

and the conclusion follows (by the monotone convergence theorem) if we let I run through an increasing sequence of finite subsets of J with union equal to J. □

Chapter 4. The Littlewood-Paley Theorem for Certain Disconnected Groups

The first theorem of the type we are about to discuss was discovered by Paley [31]. It is a theorem about Walsh series or, equivalently, about Fourier series on the Cantor group $\mathbb{D}_2$. In Appendix C we give a brief discussion of the relationship between Fourier analysis on the Cantor group and the theory of Walsh series on [0, 1]. Here now are a description and a proof of Paley's theorem in the setting of harmonic analysis on $\mathbb{D}_2$.

Recall that the *Cantor group* is the group

$$\mathbb{D}_2 = \prod_{i=1}^{\infty} \mathbb{Z}(2)_i,$$

the direct product of countably many copies of the cyclic group $\mathbb{Z}(2)$. We consider $\mathbb{Z}(2)$ as the set $\{0, 1\}$ with the group operation addition modulo 2. Each factor $\mathbb{Z}(2)$ is given its discrete topology, and $\mathbb{D}_2$ is assumed to have the product topology, which makes it compact. $\mathbb{D}_2$ is a metrisable, compact Abelian group, which is totally disconnected and zero-dimensional. Set $G_0 = G = \mathbb{D}_2$ and, for $n \geqslant 1$,

$$G_n = \{(x_i)_{i=1}^{\infty} : x_i = 0 \text{ for } i \leqslant n\}.$$

Then the family $(G_n)_0^{\infty}$ of open compact subgroups forms a neighbourhood base at 0. The dual group $\hat{\mathbb{D}}_2$ of $\mathbb{D}_2$ is the weak direct product

$$\prod_{i=1}^{\infty}{}^{*} \mathbb{Z}(2)_i.$$

It is algebraically the subgroup of $\prod_{i=1}^{\infty} \mathbb{Z}(2)_i$ (the full direct product) generated by the elements $\rho_0, \rho_1, \ldots$, where

$$\rho_0 = (0, 0, \ldots),$$

and

$$\rho_i = (\delta_{ij})_{j=1}^{\infty}$$

for $i \geqslant 1$, δ_{ij} denoting Kronecker's delta.

With this background, we can state Paley's theorem quite simply.

Theorem (Paley). *Let X_n be the finite subgroup of $\hat{\mathbb{D}}_2$ generated by $\{\rho_0, \ldots, \rho_n\}$ $(n \geqslant 0)$. For $n \geqslant 1$, define Δ_n to be the "corona" $X_n \backslash X_{n-1}$, of cardinality 2^{n-1}; write $\Delta_0 = X_0 = \{0\}$. If $1 < p < \infty$, there is a pair of positive constants (A_p, B_p) such that*

$$A_p \|f\|_p \leqslant \left\| \left(\sum_0^\infty |S_{\Delta_n} f(x)|^2 \right)^{1/2} \right\|_p \leqslant B_p \|f\|_p$$

for all f in $L^p(\mathbb{D}_2)$.

In other words, the decomposition $(\Delta_j)_{j=0}^\infty$ of $\hat{\mathbb{D}}_2$ has the Littlewood-Paley property.

Proof. We prove the theorem by establishing the equivalent WM property.

To this end, let ϕ be a function constant on members of the decomposition and vanishing off the union of finitely many of them. With any such function ϕ we can associate an integer n such that ϕ vanishes off X_n and is constant on each of the coronas $X_n \backslash X_{n-1}, \ldots, X_1 \backslash X_0$.

For every integer $j \geqslant 0$,

$$\xi_{\Delta_{j+1}} = \xi_{X_{j+1}} - \xi_{X_j}.$$

Now

$$\xi_{X_j} = \hat{\xi}_{G_j} / m(G_j),$$

where m denotes the Haar measure on G. So it is clear that $\phi = \hat{k}$ where k is a function which is both constant on G_n and constant on each of the coronas $G \backslash G_1, \ldots, G_{n-1} \backslash G_n$.

Let $(U_\alpha)_{\alpha \in \mathbb{Z}}$ be the covering family, with associated function θ, defined as follows:

$$U_\alpha = G_0 = \mathbb{D}_2 \quad \text{for} \quad \alpha \leqslant 0;$$

$$U_\alpha = G_\alpha \quad \text{for} \quad \alpha > 0;$$

$$\theta(\alpha) = \alpha - 1.$$

The family (U_α) is a covering family in the sense of 2.1.1, with constant $A = 2$. See 2.1.3(iii).

In order to establish the theorem, it suffices, by Theorem 2.4.4(ii), to show that there is a constant C such that $J(k) \leqslant C$. In fact we show that $J(k) = 0$!

If $\alpha \in \mathbb{Z}$ and $\alpha \leqslant 1$, then $G \backslash G_{\alpha-1} = \varnothing$ and hence

$$\int_{G \backslash G_{\alpha-1}} |k(x-y) - k(x)| \, dx = 0$$

for all y in G_α. If $\alpha > 1$, $y \in G_\alpha$, and $x \in G \backslash G_{\alpha-1}$, then x and x-y both lie in the same one of the coronas $G_{\alpha-2}\backslash G_{\alpha-1}, \ldots, G\backslash G_1$. Since k is constant on each corona $G_{n-1}\backslash G_n$, it is plain that

$$\int_{G\backslash G_{\alpha-1}} |k(x-y) - k(x)|\, dx = 0$$

for all y in G_α. Hence $J(k) = 0$, and the proof is complete. □

Note. The proof just given works equally well for any group G of the form $\prod_1^\infty \mathbb{Z}(a_i)$ in which the integers a_i are *bounded*, and the subgroups X_j of $X = \prod_1^{*\infty} \mathbb{Z}(a_i)$ are defined in the natural way, viz.

$$X_j = \{\chi = (\chi_k)_1^\infty : \chi_k = 0 \quad \text{for} \quad k > j\}.$$

The conclusion is that the decomposition $(\Delta_j)_0^\infty$:

$$\Delta_0 = \{0\}$$
$$\Delta_j = X_j \backslash X_{j-1} \qquad (j \geqslant 1)$$

has the LP property. The boundedness condition on the a_i is necessary to ensure that

$$\sup_\alpha m(U_{\alpha-1})/m(U_\alpha) < \infty$$

if the U_α are defined in the natural way, viz.

$$U_\alpha = \{x = (x_j)_1^\infty : x_j = 0 \quad \text{for} \quad j \leqslant \alpha\}.$$

The purpose of this chapter is to exhibit and prove theorems of Littlewood-Paley type for certain kinds of disconnected groups, among them groups of the form $X = \prod_1^{*\infty} \mathbb{Z}(a_i)$ in which the a_i are *not necessarily bounded.*

4.1. The Littlewood-Paley Theorem for a Class of Totally Disconnected Groups

4.1.1. Groups having a suitable family of compact open subgroups. Throughout Section 4.1, we shall assume that X is an LCA group containing a two-way-infinite sequence $(X_n)_{-\infty}^\infty$ of closed subgroups having the following properties.

(i) X_n is open and compact;
(ii) $X_{n+1} \supsetneqq X_n$: i.e. the sequence is strictly increasing;
(iii) $\bigcup X_n = X$, and $\bigcap X_n = \{0\}$.

If X is such a group, we shall say that X is a *group with a suitable family of compact open subgroups.*

Observe that since each subgroup X_n is open and compact, the index k_n of X_n in X_{n+1} is finite. We make no assumption concerning boundedness of the k_n. Furthermore, the group X is noncompact and nondiscrete since the sequence (X_n) is strictly monotone. Clearly X is totally disconnected; it is simple to check that the family (X_n) is a neighbourhood base at 0.

4.1.2. Examples. (a) The group X is the product

$$X = \prod_{-\infty}^{0} \mathbb{Z}(a_i) \times \prod_{1}^{\infty}{}^{*} \mathbb{Z}(a_i)$$

of a compact direct product and a discrete weak direct product. The sequence $(a_i)_{-\infty}^{\infty}$ is a sequence of integers, each at least 2. By

$$\prod_{-\infty}^{0} \mathbb{Z}(a_i)$$

we mean the set of elements $(\ldots, x_{-2}, x_{-1}, x_0)$ in which the entry x_{-k} lies in the set $\{0, \ldots, a_{-k} - 1\}$. The group operations and topology are defined in the usual way.

If the elements $\boldsymbol{x}$ of X are thought of as two-way-infinite sequences $(x_i)_{-\infty}^{\infty}$, the subgroup X_n is defined to be

$$X_n = \{\boldsymbol{x} = (x_i) \colon x_i = 0 \quad \text{for} \quad i > n\}.$$

(b) The group X is the group $\Omega_{\boldsymbol{a}}$ of $\boldsymbol{a}$-adic numbers ([20], (10.2)), $\boldsymbol{a}$ being a fixed two-way-infinite sequence of positive integers, each at least 2. The subgroups

$$X_n = \{\boldsymbol{x} \colon x_j = 0 \quad \text{for} \quad j \leqslant -n\}$$

have all the desired properties. The group X_n is what Hewitt and Ross designate Λ_{-n}.

4.1.3. Embeddings of compact and discrete groups. The reader is no doubt puzzled that we began the chapter by discussing *discrete* groups of the type $\prod_1^{*\infty} \mathbb{Z}(a_i)$ and almost immediately afterwards turned to a study of the LP property for groups which are neither compact nor discrete. The explanation for this is that, if we begin with the group $\prod_1^{*\infty} \mathbb{Z}(a_i)$, we can think of it as a discrete subgroup of, for example, the group

$$X = \prod_{-\infty}^{0} \mathbb{Z}(a_{-i+1}) \times \prod_{1}^{\infty}{}^{*} \mathbb{Z}(a_i),$$

which has a suitable *two-way-infinite* (increasing) sequence of compact open subgroups. The compact group $\prod_1^{\infty} \mathbb{Z}(a_i)$ can moreover be identified with the open subgroup

$$\prod_{-\infty}^{0} \mathbb{Z}(a_{-i+1})$$

of X.

Our plan is to establish the LP property of the decomposition $(\Delta_n)_{n\in\mathbb{Z}} = (X_{n+1}\backslash X_n)_{n\in\mathbb{Z}}$ of X defined in 4.1.2(a) and then to *deduce* that the decomposition $(\Delta_n')_0^\infty$:

$$\Delta_0' = \{0\};$$

$$\Delta_n = \{x\colon x_n \neq 0 \quad\text{and}\quad x_m = 0 \quad\text{for}\quad m > n\} \quad (n \geqslant 1);$$

of $\prod_1^{*\infty}\mathbb{Z}(a_i)$ also has the LP property. Similarly, we shall deduce that the decomposition $(\Delta_n'')_1^\infty$:

$$\Delta_n'' = \{x\colon x_m = 0 \quad\text{for}\quad m < n,\ x_n \neq 0\}$$

of $\prod_1^\infty\mathbb{Z}(a_i)$ has the LP property. See 4.1.8.

While the approach of embedding $\prod_1^\infty\mathbb{Z}(a_i)$ and $\prod_1^{*\infty}\mathbb{Z}(a_i)$ into the group X, establishing the LP property for the decomposition $(\Delta_n)_{-\infty}^\infty$, and then deducing it for $(\prod_1^\infty\mathbb{Z}(a_i), (\Delta_n'))$ and $(\prod_1^{*\infty}\mathbb{Z}(a_i), (\Delta_n''))$, is unquestionably artificial, it has the advantage that the proofs of the convolution estimates need to be done only for the case of a group X with a suitable *two-way-infinite* increasing sequence of compact open subgroups. Were we to treat $\prod_1^{*\infty}\mathbb{Z}(a_i)$, for instance, directly, we would have to deal with a *one-way-infinite* sequence of subgroups. And we would then need a separate proof for groups like $\Omega_{\boldsymbol{a}}$.

The apology for the artificiality is therefore that it permits uniformity of proof.

4.1.4. Covering families. Given a group X with a suitable family of compact open subgroups, denote its dual by G. We now show how to construct a covering family for G, as defined in 2.1.1.

Denote by G_n the annihilator of X_n in G. Since X_n is open in X, the quotient group X/X_n is discrete; hence the dual group G_n is compact. Dually, since X_n is compact, G/G_n is discrete; hence G_n is open. The sequence $(G_n)_{-\infty}^\infty$ is *strictly decreasing*; furthermore,

$$\bigcup G_n = G \quad\text{since}\quad \bigcap X_n = X,$$

and

$$\bigcap G_n = \{0\} \quad\text{since}\quad \bigcup X_n = X.$$

In other words, G has the suitable family $(G_{-n})_{-\infty}^\infty$ of compact open subgroups.

For each n, the group G_{n-1}/G_n is finite, and is accordingly isomorphic to a finite product of cyclic groups, say s of them ([20], (A.27)). It is possible therefore to construct a chain of open subgroups

$$F_n^0 = G_n \subset F_n^1 \subset \cdots \subset F_n^s = G_{n-1}$$

such that each quotient group F_n^{r+1}/F_n^r is cyclic. An inductive argument, which it would be tedious to write out in detail, permits the construction of a strictly decreasing sequence $(H_n)_{-\infty}^{\infty}$ of open compact subgroups of G having the following properties:

(i) $H_0 = G_0$;

(ii) for each n, the group H_{n-1}/H_n is finite cyclic;

(iii) to each n there corresponds a unique integer m such that

$$G_m \subseteq H_n \subseteq G_{m-1},$$

and an integer m' such that

$$H_{n-1}\backslash H_n \subseteq G_{m'-1}\backslash G_{m'}.$$

In particular, G has the new suitable family $(H_n)_{-\infty}^{\infty}$ of compact open subgroups.

The idea for the construction of the covering family (U_α) is simple but its full execution is tedious. We give the construction in two versions, the first informal, the second a more formal inductive one.

Define $U_0 = H_0$. Consider the finite cyclic group H_{-1}/H_0, generated by $\dot{p}$, say, of order s_{-1}. Define $U_{-1} = U_{\theta(0)}$ to be the union of $H_0, p + H_0$, and $-p + H_0$. (There may be fewer than three sets in this collection.) Then certainly (cf. 2.1.1)

$$\theta(0) < 0,$$

$$U_0 - U_0 \subseteq U_{\theta(0)},$$

and

$$m(U_{\theta(0)}) \leqslant 3m(U_0).$$

If $U_{\theta(0)} = U_{-1} = H_{-1}$, which happens if $s_{-1} = 2$ or 3, pass to the finite cyclic group H_{-2}/H_1. If $s_{-1} > 3$, define $U_{-2} = U_{\theta(-1)}$ to be the union of the cosets

$$H_0, \pm p + H_0, \pm 2p + H_0.$$

(There may be fewer than five sets in this collection.) Evidently

$$\theta(-1) < -1,$$

$$U_{\theta(-1)} = U_{-1} - U_{-1},$$

and

$$m(U_{\theta(-1)}) \leqslant \frac{5}{3} m(U_1) < 3m(U_{-1}).$$

Continue the construction until the stage where $U_{-r-1} = U_{\theta(-r)} = H_{-2}$ is first reached, then start afresh, working with the cosets of H_{-2} in H_{-3}; and so on.

In more formal terms, we define $U_0 = H_0$ as before, and suppose that $\alpha \leqslant 0$

and that $U_0, \ldots, U_\alpha$ have been defined so that

$$H_r \subseteq U_\alpha \subsetneqq H_{r-1}$$

for some r in $\{0, -1, -2, \ldots\}$ and so that
either

(i) $U_\alpha = H_r$;

or

(ii) there is a generator $\dot{z}$ of H_{r-1}/H_r and a positive integer t for which the cosets H_r, $\pm z + H_r, \ldots, \pm tz + H_r$ are disjoint, and

$$U_\alpha = \bigcup_{|n| \leqslant t} (nz + H_r).$$

In case (i) select any generator $\dot{z}$ of H_{r-1}/H_r and define

$$U_{\alpha-1} = U_{\theta(\alpha)} = \bigcup_{|n| \leqslant 1} (nz + H_r).$$

In case (ii), define

$$U_{\alpha-1} = U_{\theta(\alpha)} = \bigcup_{|n| \leqslant 2t} (nz + H_r).$$

A similar construction prodeeds from H_0 downwards, and produces the sets U_α for $\alpha > 0$.

This construction has the consequence that for every integer α there is an integer r and an element z of $H_{r-1} \backslash H_r$ such that
either

$$U_\alpha = H_r \quad \text{and} \quad U_{\theta(\alpha)} = \bigcup_{|n| \leqslant 1} (nz + H_r) \subseteq H_{r-1};$$

or
there is a positive integer t for which the sets H_r, $z + H_r$, $-z + H_r, \ldots,$ $tz + H_r$, $-tz + H_r$ are pairwise disjoint,

$$U_\alpha = \bigcup_{|n| \leqslant t} (nz + H_r) \quad \text{and} \quad U_{\theta(\alpha)} = \bigcup_{|n| \leqslant 2t} (nz + H_r) \subseteq H_{r-1}.$$

This makes it clear that

$$U_\alpha - U_\alpha \subseteq U_{\theta(\alpha)}$$

and

$$\frac{m(U_{\theta(\alpha)})}{m(U_\alpha)} \leqslant \frac{3m(H_r)}{m(H_r)} = 3$$

in the first case, while

$$\frac{m(U_{\theta(\alpha)})}{m(U_\alpha)} \leqslant \frac{(4t + 1)m(H_r)}{(2t + 1)m(H_r)} \leqslant 2$$

in the second case. Thus we have a covering family in the sense of 2.1.1 with constant $A \leqslant 3$.

4.1.5. The weak Marcinkiewicz property. We aim now to show that the decomposition $(\Delta_n)_{-\infty}^{\infty} = (X_{n+1} \backslash X_n)$ of X has both the LP and WM properties. We establish the latter one.

4.1.6. Theorem. *Let X be an LCA group having a suitable family $(X_n)_{-\infty}^{\infty}$ of compact open subgroups, as in* 4.1.1. *Then the decomposition $(\Delta_n)_{-\infty}^{\infty}$:*

$$\Delta_n = X_{n+1} \backslash X_n$$

has the WM *property.*

Proof. We have to show that for each p in $(1, \infty)$ there is a constant C such that if ϕ is a bounded function which is constant on each corona and vanishes off the union of finitely many of them, then $\phi \in M_p(X)$ and

$$\|\phi\|_{p,p} \leqslant C\|\phi\|_\infty.$$

Assume that the Haar measures on G and X are, as usual, chosen so that the Fourier transformation is an isometry of $L^2(G)$ onto $L^2(X)$. Assume further that the measure of X_0 is 1. This entails that G_0, the annihilator of X_0, has measure 1 also.

If ϕ is as described in the first paragraph, then ϕ is a finite linear combination of the characteristic functions of coronas $X_{r+1} \backslash X_r$. However the characteristic function of X_r is the Fourier transform of c_r times the characteristic function of G_r where c_r is the measure of X_r. If we denote by k the function in $L^1(G)$ whose Fourier transform is ϕ, it follows that k is constant on the coronas $G_{j-1} \backslash G_j$. Therefore, by 4.1.4(iii), k is constant on each of the coronas $H_{s-1} \backslash H_s$ ($s \in \mathbb{Z}$). Theorem 2.4.4(ii) shows that, in order to complete the proof, it suffices to establish the existence of a constant B such that

$$\int_{G \backslash U_{\theta(\alpha)}} |k(x - y) - k(x)| \, dx \leqslant B \tag{1}$$

for y in U_α and all integers α. The dramatic aspect of the present set of circumstances is that the left side of (1) is always 0!

As we have seen in 4.1.4, there are two cases to consider:

(a) where $U_\alpha = H_r$ and $U_{\theta(\alpha)} = (z + H_r) \cup H_r \cup (-z + H_r) \subseteq H_{r-1}$;

and

(b) where $U_\alpha = \bigcup_{|n| \leqslant t} (nz + H_r)$ and

$$U_{\theta(\alpha)} = \bigcup_{|n| \leqslant 2t} (nz + H_r) \subseteq H_{r-1}.$$

Case (b). If $x \in G \backslash U_{\theta(\alpha)}$, and $y \in U_\alpha$, there are two possibilities. The first is that $x \in G \backslash H_{r-1}$, in which case $x - y \in G \backslash H_{r-1}$, and x and $x - y$ *lie in the same corona* $H_{s-1} \backslash H_s$, where s is the first integer for which $x \in H_{s-1}$. Therefore $k(x - y) - k(x) = 0$.

In the second case, $x \in H_{r-1} \backslash U_{\theta(\alpha)}$, which implies that $U_{\theta(\alpha)} \neq H_{r-1}$; then $x - y$ and x both lie in the corona $H_{r-1} \backslash H_r$. For otherwise $x - y$ would belong to H_r, which would imply that $x - y \in U_\alpha$; hence $x \in U_\alpha + U_\alpha \subseteq U_{\theta(\alpha)}$, a contradiction. So again $k(x - y) - k(x) = 0$.

Case (a) is treated in an analogous way. The details are simpler.

The proof of the theorem is now complete. □

The next stage of the development is to deduce the LP property for products of finite cyclic groups from Theorem 4.1.6. This is carried out in 4.1.8. But first, we establish a couple of simple results about L^p multipliers. These results are particular cases of those presented in Appendix B. However, they are so simple to prove from first principles that it seems a good idea to do so, and keep our present development self-contained.

4.1.7. Lemma. (i) *Let X be an infinite LCA group of the form $X = H \times Y$, where H is a compact group and Y is discrete. Denote by π the natural projection of X onto Y. Suppose that $1 \leqslant p \leqslant \infty$, that ϕ is a function on Y, and that $\phi \circ \pi$ belongs to $M_p(X)$. Then $\phi \in M_p(Y)$, and*

$$\|\phi\|_{p,p} \leqslant \|\phi \circ \pi\|_{p,p}.$$

(ii) *The notation is as in* (i). *If ψ is a function on H, and we denote by Ψ the function, supported in $H \times \{0\}$, such that*

$$\Psi(h, 0) = \psi(h),$$

and suppose that $\Psi \in M_p(X)$, where $1 \leqslant p \leqslant \infty$, then $\psi \in M_p(H)$, and

$$\|\psi\|_{p,p} \leqslant \|\Psi\|_{p,p}.$$

Proof. Since the values of $\|\phi\|_{p,p}$ and $\|\phi \circ \pi\|_{p,p}$ are not altered if the Haar measures on the respective groups are multiplied by positive scalars, we may assume that the Haar measure on X is the product of those on H and Y, that the Haar measure of H is 1, and that the Haar measure on Y is counting measure. It follows that the Haar measure on the dual group $\hat{H} \times \hat{Y}$ which makes the Fourier transformation an L^2-isometry is the product of those Haar measures on the factors which assign mass 1 to $\hat{Y}$ and mass 1 to $\{0\}$ in $\hat{H}$.

(i) By 1.2.2(iii), it suffices to prove that if f and g belong to $L^1(\hat{Y})$ and have compactly supported Fourier transforms, then

$$\left|\int_Y \phi(\chi_1)\hat{f}(\chi_1)\hat{g}(\chi_1)\, d\chi_1\right| \leqslant \|\phi\circ\pi\|_{p,p}\|f\|_p\|g\|_{p'}. \tag{2}$$

If we denote by F and G the functions on $\hat{H} \times \hat{Y}$ supported in $\{0\} \times Y$ such that

$$F(0, h) = f(h)$$

and

$$G(0, h) = g(h),$$

then $\hat{F} = \hat{f}\circ\pi$ and $\hat{G} = \hat{g}\circ\pi$. In view of the normalisations of the Haar measures we have assumed, the left side of (2) becomes

$$\left|\int_{H\times Y} \phi\circ\pi\hat{F}\hat{G}\, dm\right|,$$

which is bounded above by

$$\|\phi\circ\pi\|_{p,p}\|F\|_p\|G\|_{p'}.$$

Since it is clear that $\|F\|_p = \|f\|_p$ and $\|G\|_{p'} = \|g\|_{p'}$, this part of the proof is complete.

(ii) In this case it suffices, again by 1.2.2(iii), to show that if f and g belong to $L^1(\hat{H})$, then

$$\left|\int_H \psi(y)\hat{f}(y)\hat{g}(y)\, dy\right| \leqslant \|\Psi\|_{p,p}\|f\|_p\|g\|_{p'}. \tag{3}$$

However, if τ denotes the canonical projection of $\hat{H} \times \hat{Y}$ onto $\hat{H}$, then

$$\hat{f}(y) = (f\circ\tau)^\wedge(y, 0)$$

and

$$\hat{g}(y) = (g\circ\tau)^\wedge(y, 0).$$

The left side of (3) can be rewritten

$$\left|\int_{H\times Y} \Psi(f\circ\tau)^\wedge(g\circ\tau)^\wedge\, dm\right|.$$

Since $\|f\circ\tau\|_p = \|f\|_p$ and $\|g\circ\tau\|_{p'} = \|g\|_{p'}$, the inequality (3) is an immediate consequence of the inequality

$$\left| \int_{H \times Y} \Psi(f \circ \tau)^\wedge (g \circ \tau)^\wedge \, dm \right| \leqslant \|\Psi\|_{p,p} \|f \circ \tau\|_p \|g \circ \tau\|_{p'}. \quad \square$$

4.1.8. The Littlewood-Paley theorem for products of cyclic groups.

Theorem. (i) *Let* $\mathbf{a}' = (a_j)_1^\infty$ *be a sequence of integers each greater than* 1, *and* Y *the discrete group*

$$Y = \prod_1^\infty{}^* \mathbb{Z}(a_j).$$

Define the subgroups $(Y_n)_0^\infty$ *of* Y *by the rule*:

$$Y_n = \{y : y_m = 0 \quad for \quad m > n\},$$

and let $(\Delta_n')_0^\infty$ *be the decomposition of* Y *defined by the formulas*

$$\Delta_0' = \{0\}$$
$$\Delta_n' = Y_n \backslash Y_{n-1} \qquad (n \geqslant 1).$$

Then $(\Delta_n')_0^\infty$ *has the* LP *property.*

(ii) *Let* $\mathbf{a}'' = (a_j)_{-\infty}^0$ *be a sequence of integers each greater than* 1 *and* H *the compact group*

$$H = \prod_{-\infty}^0 \mathbb{Z}(a_j).$$

Let H_n *be the subgroup*

$$H_n = \{h : h_m = 0 \quad for \quad m > n\},$$

for $n = 0, 1, \ldots,$ *and define the decomposition* $(\Delta_n'')_{-\infty}^0$ *by the formulas*

$$\Delta_n'' = H_n \backslash H_{n-1}.$$

Then $(\Delta_n'')_0^\infty$ *has the* LP *property.*

Proof. We prove that each decomposition has the equivalent WM property.

(i) It is enough to show that if ϕ is a function on Y which is bounded, and constant on each set Δ_n', then $\phi \in M_p(Y)$ (cf. 1.2.7(ii)).

Consider the group

$$X = H \times Y = \prod_{-\infty}^0 \mathbb{Z}(a_i) \times \prod_1^\infty{}^* \mathbb{Z}(a_i),$$

which is a group with a suitable family of compact open subgroups (4.1.2(a)), viz.

$$X_n = H \times Y_n, \quad (n \geqslant 0)$$

and

$$X_n = H_n \times \{0\}, \quad (n < 0).$$

The corresponding decomposition $(\Delta_n)_{-\infty}^{\infty}$ of X is defined by the rule

$$\Delta_n = X_{n+1} \backslash X_n.$$

By Theorem 4.1.6, it has the LP and WM properties.

Now if π is the canonical projection of X onto Y, the function $\phi \circ \pi$ is a bounded function, constant on the coronas Δ_n of X, and hence, by what we have just proved, belongs to $M_p(X)$. By Lemma 4.1.7(i), $\phi \in M_p(Y)$.

The proof for case (ii) is equally simple, and involves the use of Lemma 4.1.7(ii). We leave the writing of the details to the reader. □

4.1.9. A characterisation of certain groups having an LP **decomposition.** The proofs just given in 4.1.8 need very little modification in order to establish the following more general results.

Theorem. (i) *Let Y be an infinite discrete Abelian group in which there is a sequence $(Y_n)_0^{\infty}$ of subgroups having the following properties*:

(a) *each group Y_n is finite*;

(b) $Y_n \subsetneqq Y_{n+1}$;

(c) $\bigcup Y_n = Y$.

Then the decomposition $(\Delta_n)_0^{\infty}$:

$$\Delta_0 = Y_0;$$
$$\Delta_n = Y_n \backslash Y_{n-1} \qquad (n \geqslant 1);$$

has the LP *property.*

(ii) *Let H be a compact Abelian group having a sequence $(H_n)_0^{\infty}$ of closed subgroups such that*

(a) *each H_n is open*;

(b) $H_n \supsetneqq H_{n+1}$;

and

(c) $\bigcap H_n = \{0\}$; $H_0 = H$.

Then the decomposition $(\Delta_n)_0^{\infty} = (H_n \backslash H_{n+1})_0^{\infty}$ has the LP *property.*

Remark. It is easy to show from first principles that the groups having the properties described in (i) are precisely the countably infinite Abelian torsion groups.

Reference to Theorems (24.15) and (24.26) of [20] shows that the groups described in part (ii) of the theorem are exactly the infinite, compact, Abelian, metrisable, totally disconnected groups.

4.2. The Littlewood-Paley Theorem for a More General Class of Disconnected Groups

The results of Section 4.1 have been established for groups X having a suitable family of compact open subgroups. It is our intention now to show that similar results hold for groups X having a family of open subgroups subject to much less stringent restrictions than those expressed in 4.1.1. However, the more general results will be seen to be corollaries of those in Section 4.1.

4.2.1. Groups having a suitable family of open subgroups. Let X be an LCA group. We say that *X has a suitable family of open subgroups* if there is a two-way-infinite sequence $(X_n)_{-\infty}^{\infty}$ of subgroups of X having the following properties:

(i) each X_n is open;

(ii) $X_n \subsetneqq X_{n+1}$ for all n;

and

(iii) for every n, the index of X_n in X_{n+1} is finite.

4.2.2. Examples. (a) $\boldsymbol{a} = (a_i)_{-\infty}^{\infty}$ is a sequence of integers, each greater than 1, and

$$X = \prod_{-\infty}^{\infty}{}^{*} \mathbb{Z}(a_i);$$

$(I_n)_{-\infty}^{\infty}$ is a sequence of subsets of $\mathbb{Z}$ such that

(i) $I_n \subsetneqq I_{n+1}$ for all n;

and

(ii) $I_{n+1} \backslash I_n$ is finite;

and

$$X_n = \{\boldsymbol{x} \in X : x_m = 0 \quad \text{for} \quad m \in \mathbb{Z} \backslash I_m\}.$$

(b)

$$X = \mathbb{Z} \times \prod_{1}^{\infty}{}^{*} \mathbb{Z}(a_j),$$

$(a_j)_1^{\infty}$ being a sequence of integers, each greater than 1. Consider each element of X as a sequence indexed from 0 to ∞. The subgroups X_n are then as follows. For $n \geqslant 0$,

$$X_n = \{(x_j) : x_j = 0 \quad \text{for} \quad j > n\},$$

while for $n < 0$,

$$X_n = \{(x_j) : x_j = 0 \text{ for all } j > 0, \text{ and } x_0 = 2^{-n}k, k \in \mathbb{Z}\}$$
$$= (2^{-n}\mathbb{Z}) \times \{0\}.$$

(c) $\boldsymbol{a} = (a_i)_{-\infty}^{\infty}$ is an arbitrary sequence of positive integers, each greater than 1, X is the group of $\boldsymbol{a}$-adic numbers with the discrete topology, and

$$X_n = \{(x_j): x_j = 0 \quad \text{for} \quad j < -n\}.$$

Remarks. There is an obvious source of confusion in the use of the phrase "suitable family of open subgroups" in 4.2.1. A reader already accustomed to the definition in 4.1.1 might mistakenly assume that a suitable family of open subgroups is to have intersection 0 and union X. Cf. condition 4.1.1(iii).

We now present two lemmas, necessary for the proof of the LP property in the new setting, the first of which shows that, for the purpose of that proof, there is no loss of generality in assuming that $\bigcup X_n = X$. (It is possible also to show, by arguing on the quotient group $X/\bigcap X_n$ and using the homomorphism theorem for multipliers (Appendix B), that there is no loss of generality in assuming that $\bigcap X_n = \{0\}$; but this assumption would not help us in establishing the LP property, so we omit it.)

4.2.3. Lemma. *Let X be an LCA group with a suitable family of open subgroups, as described in* 4.2.1. *Then the characteristic function of the set $X \backslash \bigcup X_n$ is the Fourier-Stieltjes transform of a measure of mass at most* 2, *and is therefore, for every p, a multiplier of norm at most* 2.

Proof. If $\bigcup X_n = X$, there is nothing to prove. In the contrary case, $X' = \bigcup X_n$ is a proper open and closed subgroup of X.

The characteristic function of X is the Fourier transform of the unit point mass ε_0 placed at 0 in G, the dual of X. Similarly, $\xi_{X'}$ is the Fourier transform of the measure m_{G_0} on G, where m_{G_0} is the measure, concentrated on G_0, equal on G_0 to the normalised Haar measure of G_0. (G_0 is compact since X/X' is discrete.) It is now apparent that

$$\xi_{X \backslash X'} = (\varepsilon_0 - m_{G_0})^\wedge,$$

so that the assertion of the lemma is clear.

4.2.4. Corollary. *Let X be as in* 4.2.3. *The decomposition $((X_{n+1} \backslash X_n)_{-\infty}^{\infty}, \bigcap X_n, X \backslash \bigcup X_n)$ of X has the* WM *property if and only if the decomposition $((X_{n+1} \backslash X_n)_{-\infty}^{\infty}, \bigcap X_n)$ of $X' = \bigcup X_n$ has the* WM *property.*

Proof. This follows from the observation, itself a consequence of 1.2.2(iii), that if ϕ is a function on X', and Φ is the function on X which vanishes off X' and agrees with ϕ on X', then

$$\|\phi\|_{M_p(X')} = \|\Phi\|_{M_p(X)}.$$

4.2.5. Lemma. *Let X be an LCA group having a suitable family $(X_n)_{-\infty}^{\infty}$ of open subgroups for which $\bigcup X_n = X$. In order to prove that the decomposition $((X_{n+1} \backslash X_n)_{-\infty}^{\infty}, \bigcap X_n)$ has the* WM *property, it suffices to prove that when*

$1 < p < \infty$, *there is a constant* C_p *such that*

$$\|\phi\|_{p,p} \leqslant C_p \|\phi\|_\infty$$

for all functions ϕ *which are constant on members of the decomposition and vanish off the union of finitely many of the sets* $X_{n+1} \backslash X_n$.

Proof. The closed subgroup $\bigcap X_n$ is either locally negligible or open. If it is locally negligible, the result is immediate. In the opposite case, the characteristic function of $\bigcap X_n$ is the Fourier-Stieltjes transform of the Haar measure on H, the compact annihilator of $\bigcap X_n$ in G.

4.2.6. Embeddings into the Bohr compactification. The key idea in establishing the LP and WM properties for the cases described in 4.2.1 is to replace the groups X and X_n by their Bohr compactifications and then so to change the topology on the Bohr compactification of X that use can be made of the simple form of the WM theorem established in 4.1.6.

Let X be an LCA group having a suitable family of open subgroups $(X_n)_{-\infty}^{\infty}$. Denote by Y the Bohr compactification of X. If G is the dual group of X, then Y is the dual group of G_d, the group G with the discrete topology. The natural injection i of X into Y is a continuous algebraic isomorphism ([20], (26.11)).

For each n, let $Y_n = \overline{i(X_n)}$, the closure of $i(X_n)$ in Y. The group Y_n is identifiable with the Bohr compactification of X_n. If G_n is the annihilator of X_n in G, then the dual group of X_{n+1}/X_n is the subgroup G_n/G_{n+1} of G/G_{n+1}. Consequently, the group G_n/G_{n+1} is finite (4.2.1(iii)). The dual group of Y_{n+1}/Y_n is the group $(G_n)_d/(G_{n+1})_d = (G_n/G_{n+1})_d$ so that Y_{n+1}/Y_n is a finite group. The device of replacing the groups X_n by their Bohr compactifications Y_n has therefore arranged that the groups Y_n are compact and each has finite index in its successor. Clearly, openness has been lost; and we cannot be sure that $\bigcup Y_n = Y$. The next step shows that by declaring the Y_n to be open, *the compactness is not lost.*

4.2.7. Altering the Bohr topology. In the notation of 4.2.6, let Y' be the subgroup of Y:

$$Y' = \bigcup Y_n.$$

Declare the sets Y_n to be open in Y'. More precisely, let $(Y_n)_{-\infty}^{\infty}$ be a neighbourhood base of open sets for a topological group topology on Y'. This is legitimate by appeal to (4.21) of [20]. The following lemma shows that, although the topology has been changed, the subgroups Y_n remain compact.

Lemma. *Let* Y, $(Y_n)_{-\infty}^{\infty}$ *and* Y' *be as above,* Y' *being given the topology of which* $(Y_n)_{-\infty}^{\infty}$ *is a neighbourhood base at* 0. *Then each subgroup* Y_n *is compact in* Y'.

Proof. Since, for every index j, the group Y_{j+1}/Y_j is finite, it follows that for every index $m < n$, Y_n/Y_m is finite. This shows that Y_n is precompact for the new structure on Y', and it thus suffices to show that Y_n is complete for this new structure.

Let (y_i) be a net extracted from Y_n which is Cauchy for the new structure. To every integer m there is an index i_m such that

$$y_i - y_j \in Y_m$$

for $i, j \geqslant i_m$. Now the net (y_j) has a limiting point y in Y_n for the initial structure (induced by that of Y)—remember that Y is compact and that Y_n is closed in Y. Hence $y_i - y$ is a limiting point of $(y_i - y_j)_j$ in the sense of the initial structure. So, since Y_m is closed in Y,

$$y_i - y \in Y_m$$

for $i \geqslant i_m$. Hence $y_i \to y$ for the new structure, showing that Y_n is complete for this new structure. □

4.2.8. The Littlewood-Paley theorem for groups having a suitable family of open subgroups.

Theorem. *Let X be an LCA group having a suitable family $(X_n)_{-\infty}^{\infty}$ of open subgroups. The decomposition $(\bigcap X_n, X\setminus\bigcup X_n, (X_{n+1}\setminus X_n)_{-\infty}^{\infty})$ of X has the* LP *and* WM *properties.*

Proof. We establish the WM property.

By Corollary 4.2.4, we may assume that $\bigcup X_n = X$. Let Y, the Bohr compactification of X, $(Y_n)_{-\infty}^{\infty}$ and Y' be as in 4.2.6. Denote by Y'' the closed subgroup

$$Y'' = \bigcap Y_n$$

of Y'. The group Y'/Y'' is an LCA group with a family $(Y_n/Y'')_{-\infty}^{\infty}$ of compact open subgroups that is suitable in the sense of 4.1.1. Therefore the decomposition $((Y_{n+1}\setminus Y_n)/Y'')_{-\infty}^{\infty}$ of Y'/Y'' has the WM property, by Theorem 4.1.6.

Suppose that ϕ is a function on X which is constant on members of the decomposition of X and vanishes except on finitely many of the sets $X_{n+1}\setminus X_n$. The function ϕ can be written

$$\phi = \sum c_n \xi_{X_{n+1}\setminus X_n}$$

where all but finitely many c_n are 0. Consider the function

$$\Phi = \sum c_n \xi_{(Y_{n+1}\setminus Y_n)/Y''}.$$

We claim that

$$\phi = \Phi \circ (\pi \circ i), \tag{1}$$

where π denotes the canonical projection of Y' onto Y'/Y''. To prove (1), it suffices to verify that

$$i(X_{n+1}\setminus X_n) \subseteq Y_{n+1}\setminus Y_n$$

for every n. Since it is clear that $i(x) \in Y_{n+1}(=\overline{i(X_{n+1})})$ for x in X_{n+1}, it remains to show that

$$i(x) \notin Y_n \quad \text{if} \quad x \in X_{n+1}\backslash X_n. \tag{2}$$

Recall that Y is the dual group of G_d, that X is the dual of G, and that the element $i(x)$ of Y is defined by the specification that

$$i(x)(\zeta) = x(\zeta) \tag{3}$$

for all ζ in G_d. Now $Y_n = \overline{i(X_n)}$ is the annihilator in Y of the annihilator G_n of X_n in G. So $i(x) \in Y_n$ if and only if

$$i(x)(\zeta) = 1$$

for all ζ in G_n; i.e., by (3), if and only if

$$x(\zeta) = 1$$

for all ζ in G_n. But this would mean that x belongs to the annihilator in X of G_n, which is X_n; a contradiction. So (1) is established.

Notice next that the function Φ is continuous on Y'/Y'', since each function $\xi_{(Y_{n+1}\backslash Y_n)/Y''}$ is continuous. For $(Y_{n+1}\backslash Y_n)/Y''$ is both open and closed in Y'/Y''.

Now since Y'/Y'' is a group with a suitable family of compact open subgroups, the decomposition $((Y_{n+1}\backslash Y_n)/Y'')$ has the WM property by Theorem 4.1.6. Hence there is for each p in $(1, \infty)$ a constant C_p such that $\Phi \in M_p$ and

$$\|\Phi\|_{p,p} \leqslant C_p\|\Phi\|_\infty. \tag{4}$$

The homomorphism $\pi \circ i$ is continuous and Φ is continuous. So, by the homomorphism theorem for multipliers (Theorem B.2.1), $\phi = \Phi \circ \pi \circ i$ is in $M_p(X)$, and

$$\|\phi\|_{p,p} \leqslant \|\Phi\|_{p,p}. \tag{5}$$

Since $\|\phi\|_\infty = \|\Phi\|_\infty$, we conclude, by combining (4) and (5), that

$$\|\phi\|_{p,p} \leqslant C_p\|\phi\|_\infty. \tag{6}$$

Hence by Lemma 4.2.5 the decomposition $(\bigcap X_n, (X_{n+1}\backslash X_n)_{-\infty}^{\infty})$ of X has the WM property. □

4.3 A Littlewood-Paley Theorem for Decompositions of $\mathbb{Z}$ Determined by a Decreasing Sequence of Subgroups

The traditional forms of the Littlewood-Paley theorem concern dyadic block, and more generally, Hadamard block decompositions of $\mathbb{Z}$. See Chapters 1, 7 and 8.

In the present section, we establish the LP theorem for decompositions of $\mathbb{Z}$ determined by decreasing sequences of subgroups. The result is an easy consequence of Theorem 4.2.8, but deserves to be singled out for special mention because of its novelty.

4.3.1. Theorem. *Let $(k_n)_1^\infty$ be a sequence of positive integers, each at least 2. Denote by $\mathbb{Z}^{(n)}$ the subgroup of $\mathbb{Z}$ consisting of all integral multiples of $k_1 \ldots k_n$, for $n = 1, 2, \ldots$. In shorthand,*

$$\mathbb{Z}^{(n)} = k_1 \cdots k_n \mathbb{Z}.$$

The decomposition $(\{0\}, \mathbb{Z}\backslash\mathbb{Z}^{(1)}, \mathbb{Z}^{(1)}\backslash\mathbb{Z}^{(2)}, \ldots)$ of $\mathbb{Z}$ has the LP *and* WM *properties.*

Proof. Consider the discrete group

$$X = \mathbb{Z} \times \prod_1^\infty{}^* \mathbb{Z}(k_n).$$

It will suffice to prove that the decomposition $(\Delta_j')_0^\infty = (\{0\} \times \{\mathbf{0}\}, (\mathbb{Z}\backslash\mathbb{Z}^{(1)}) \times \{\mathbf{0}\}, \ldots)$ of the subgroup $X_0 = \mathbb{Z} \times \{\mathbf{0}\}$ of X has the WM property. Define the subgroups X_n $(n \in \mathbb{Z})$ of X by the following rules.

$$X_n = \{\mathbf{x}: x_j = 0 \quad \text{for} \quad j > n\}, \quad \text{when } n \geqslant 0;$$

$$X_n = \{\mathbf{x} = (m, \mathbf{0}): m \in \mathbb{Z}^{(-n)}\}, \quad \text{when } n < 0.$$

Since $X_0 = \mathbb{Z} \times \{\mathbf{0}\}$ is an open subgroup of X, reference to 1.2.2(iii) shows that (Δ_j') has the WM property if the decomposition $(\{0\}, (X_{n+1}\backslash X_n)_{-\infty}^\infty)$ of X has the WM property. But this is a consequence of Theorem 4.2.8. □

Remarks. The technical device just used of embedding a group in a non-compact product group with a suitable two-way-infinite family of open subgroups can be employed in other circumstances to prove theorems of Littlewood-Paley type. For instance, it could be used to prove that if

$$Y = \prod_1^\infty{}^* \mathbb{Z}(a_i)$$

and

$$Y_n = \{\mathbf{y}: y_j = 0 \quad \text{for} \quad j = 2^0, \ldots, 2^n\}$$

then the decomposition $(Y\backslash Y_1, Y_1\backslash Y_2, \ldots)$ of Y has the LP property.

We re-emphasise that the device is forced upon us because we have insisted on proving the general theorem 4.2.8 only for groups having a suitable (*two-way-infinite strictly decreasing*) sequence of open subgroups.

Concluding notes. The main results of this chapter, viz. Theorems 4.1.6, 4.2.8, and 4.3.1 are due to Spector [37]. The general approach is also due to him. There is one significant difference in detail between Spector's approach and the one presented here: we have made a more specific splitting of the group G_n into blocks of cosets modulo G_{n+1} in 4.1.4 than is carried out in V-3 of [37]. This was necessary because we wished to fit the arguments into the framework of "covering families" explained in Chapter 2, and to make use of the general results developed there.

Chapter 5. Martingales and the Littlewood-Paley Theorem

We have seen that it is possible to prove a quite general Littlewood-Paley theorem (Theorem 4.2.8) for certain disconnected groups by combining the results of Chapter 2 with arguments about topological groups, notably concerning the Bohr compactification. (The Paley theorem on $\mathbb{D}_2$ was of course much simpler to establish.) We intend to show now that it is possible to adopt an alternative approach, namely to prove a Littlewood-Paley theorem for martingales and then deduce Theorem 4.2.8 from it. Indeed we shall show more: that Theorem 4.2.8 is valid even without the condition of finiteness of the indices of X_n in X_{n+1} ($n \in \mathbb{Z}$). This approach has a commendable directness and an elementary character. Moreover, it affords an introduction to the relation between martingales and Littlewood-Paley theory which is only just beginning to be systematically explored. See for instance [39].

Interestingly, the ideas we are now going to set down hark back in a number of ways to Paley's original paper [31]. Our presentation owes a great deal in many places to the lecture notes of Garsia [16] on the spaces H^1 and BMO for martingales. The thesis of Inglis [22] contains a proof of the Littlewood-Paley theorem for martingales, under conditions slightly more restrictive than ours. His and our presentations of the material in Sections 5.1–5.3 are however substantially the same.

The reader is referred also to the earlier papers of Burkholder [4] and Gundy [17] where somewhat different proofs of the Littlewood-Paley theorem for martingales are given.

5.1. Conditional Expectations

5.1.1. Let $(\Omega, \mathscr{F}, \mu)$ be a measure space and $(\mathscr{F}_j)_{-\infty}^{\infty}$ a sequence of sub-σ-algebras of $\mathscr{F}$ such that

(i) $\mathscr{F}_j \subseteq \mathscr{F}_{j+1}$ for all j;

(ii) (0–1 property) the sets in $\bigcap_{j\in\mathbb{Z}} \mathscr{F}_j$ which are of finite measure are either null or have null complements;

(iii) $\bigcup_{n\in\mathbb{Z}} \mathscr{F}_j$ generates $\mathscr{F}$;

(iv) if $n \in \mathbb{Z}$, $F \in \mathscr{F}$, and $\mu(F) < \infty$, there is a sequence $(U_j)_{j=1}^{\infty}$ of elements of

$\mathscr{F}_n$ such that

$$F \subseteq \bigcup_{j=1}^{\infty} U_j$$

and

$$\mu(U_j) < \infty$$

for every j.

We make the standing assumption throughout Sections 5.1–5.3 that the measure space $(\Omega, \mathscr{F}, \mu)$ has associated with it a sequence $(\mathscr{F}_j)_{-\infty}^{\infty}$ of σ-algebras satisfying all the conditions in 5.1.1.

Throughout this chapter we take the liberty of not distinguishing between a function and the corresponding class (modulo the space of null or locally null functions). Whereas in Section 1.3, for instance, the distinction between the two is vital, this is not the case here.

We show first that, under these assumptions, it is possible to define the *conditional expectation* of an element f of $L^p(\Omega, \mathscr{F}, \mu)$ with respect to $\mathscr{F}_n$, when $p < \infty$.

5.1.2. Lemma. *If $1 \leqslant p < \infty, f \in L^p(\Omega, \mathscr{F}, \mu)$, and $n \in \mathbb{Z}$, there is an essentially unique locally integrable function g_n, measurable for $\mathscr{F}_n$, and vanishing off a σ-finite set, such that*

$$\int_A g_n \, d\mu = \int_A f \, d\mu \tag{1}$$

for all sets A of finite measure in $\mathscr{F}_n$.

Proof. Since $f \in L^p$ and $p < \infty, f$ vanishes off a set of the form

$$B = \bigcup_{i=1}^{\infty} B_i$$

where $B_i \in \mathscr{F}$ and $\mu(B_i) < \infty$ for each i. By 5.1.1(iv),

$$B_i \subseteq \bigcup_{j=1}^{\infty} U_j^{(i)}.$$

We set $g_n = 0$ off $\bigcup_{i,j} U_j^{(i)}$. On each set $U_j^{(i)}$, we define g_n as follows. Observe that, since $U_j^{(i)} \in \mathscr{F}_n$, and $\mu(U_j^{(i)}) < \infty$, the Radon-Nikodym theorem ([34], 6.10) can be applied to the measures $\rho = f\mu | U_j^{(i)}$ and $\sigma = \mu | U_j^{(i)}$ on the measurable space $(U_j^{(i)}, \mathscr{F}_n | U_j^{(i)})$, ρ being clearly absolutely continuous with respect to σ. So we set $g_j^{(i)}$ equal to 0 off $U_j^{(i)}$ and equal to the Radon-Nikodym derivative of ρ with respect to σ on $U_j^{(i)}$. Then, if $A \in \mathscr{F}_n$ and $\mu(A) < \infty$, the definition of $g_j^{(i)}$ shows that

$$\int_A g_j^{(i)} \, d\mu = \int_A f \, d\mu. \tag{2}$$

It is a routine matter to check that if $U_j^{(i)} \cap U_m^{(k)} \neq \emptyset$, then

$$g_j^{(i)} = g_m^{(k)} \text{ a.e.}$$

on $U_j^{(i)} \cap U_m^{(k)}$. It therefore makes sense to define g_n on $\bigcup_{i,j} U_j^{(i)}$ by specifying that

$$g_n = g_j^{(i)} \quad \text{on} \quad U_j^{(i)}$$

for all i, j and $g_n = 0$ off $\bigcup_{i,j} U_j^{(i)}$. It follows directly from (2) that g_n has the property (1); and it is entirely routine to check that g_n is unique modulo the space of null, $\mathscr{F}_n$-measurable functions.

5.1.3. Definition (Conditional expectation). $(\Omega, \mathscr{F}, \mu)$ and $(\mathscr{F}_j)_{-\infty}^{\infty}$ being as in 5.1.1, the function (class) g_n referred to in 5.1.2 is called *the conditional expectation of f relative to, or given, $\mathscr{F}_n$*, and is denoted by $\mathsf{E}_n f$ or $\mathsf{E}(f|\mathscr{F}_n)$. Its characteristic properties are that

(i) it is $\mathscr{F}_n$-measurable and vanishes off a set σ-finite for $\mathscr{F}_n$;

(ii) $$\int_A \mathsf{E}_n f \, d\mu = \int_A f \, d\mu, \tag{3}$$

for all sets A in $\mathscr{F}_n$ which have finite measure.

5.1.4. Basic properties of the conditional expectation operator E_n.

Lemma. *Denote by E_n the operator $f \to \mathsf{E}_n f$ on $\bigcup_{1 \leqq p < \infty} L^p(\Omega)$ defined in* 5.1.3. *Then*

(i) *E_n is linear;*

(ii) *$\mathsf{E}_n f \geqslant 0$ if $f \geqslant 0$; $\overline{\mathsf{E}_n f} = \mathsf{E}_n \bar{f}$;*

(iii) *$|\mathsf{E}_n f| \leqslant \mathsf{E}_n(|f|)$;*

(iv) *if $\mu(\Omega) < \infty$, then $E_n 1 = 1$ for all n;*

(v) *$\|\mathsf{E}_n f\|_p \leqslant \|f\|_p$ if $f \in L^p(\Omega)$ and $1 \leqslant p < \infty$;*

(vi) *if $1 \leqslant p < \infty$, $f \in L^p(\Omega)$, $g \in L^{p'}(\Omega)$, and g is $\mathscr{F}_n$-measurable, then*

$$\mathsf{E}_n(fg) = g \mathsf{E}_n f; \tag{4}$$

(vii) (the "Parseval" relation) *if $1 < p < \infty$, $f \in L^p(\Omega)$, and $g \in L^{p'}(\Omega)$*, or *if $f \in L^1(\Omega)$ and $g \in \bigcup_{1 \leqq r < \infty} L^\infty \cap L^r(\Omega)$, then*

$$\int_\Omega \mathsf{E}_n f \cdot \bar{g} \, d\mu = \int_\Omega \mathsf{E}_n f \cdot \overline{\mathsf{E}_n g} \, d\mu$$
$$= \int_\Omega f \, \overline{\mathsf{E}_n g} \, d\mu; \tag{5}$$

(viii) (the martingale condition) *if $f \in L^p(\Omega)$, $1 \leqslant p < \infty$, and $m \leqslant n$, then*

$$\mathsf{E}_m \mathsf{E}_n f = \mathsf{E}_m f. \tag{6}$$

(ix) *if $f \in L^p(\Omega)$, $1 \leqslant p < \infty$ and f is measurable for $\mathscr{F}_n$, then $\mathsf{E}_n f = f$.*

Proof. The properties (i)–(iv) are evident.

(v) It is enough, by the converse of Hölder's inequality, to show that

$$\left| \int_\Omega \mathsf{E}_n f . g \, d\mu \right| \leqslant \|f\|_p \|g\|_{p'} \tag{7}$$

for all functions g simple and integrable for $\mathscr{F}_n$. But if

$$g = \sum a_i \xi_{A_i}$$

is such a function,

$$\begin{aligned} \int_\Omega \mathsf{E}_n f . g \, d\mu &= \sum a_i \int_{A_i} \mathsf{E}_n f \, d\mu \\ &= \sum a_i \int_{A_i} f \, d\mu \\ &= \int_\Omega f \sum a_i \xi_{A_i} \, d\mu \end{aligned}$$

by (3). The inequality (7) follows.

(vi) There is a sequence (g_j) of functions simple for $\mathscr{F}_n$, and integrable, such that

$$|g_j| \leqslant |g|$$

and

$$g_j \to g$$

pointwise. This remark, coupled with (v), proves that it is enough to establish (vi) for functions g which are simple and integrable. But if

$$g = \sum a_i \xi_{A_i},$$

$A \in \mathscr{F}_n$, and $\mu(A) < \infty$, then

$$\begin{aligned} \int_A \mathsf{E}_n(fg) \, d\mu = \int_A fg \, d\mu &= \sum a_i \int_{A \cap A_i} f \, d\mu \\ &= \sum a_i \int_{A \cap A_i} \mathsf{E}_n f \, d\mu \\ &= \sum a_i \int_A \mathsf{E}_n f . \xi_{A_i} \, d\mu \\ &= \int_A \mathsf{E}_n f . \sum a_i \xi_{A_i} \, d\mu \\ &= \int_A \mathsf{E}_n f . g \, d\mu. \end{aligned}$$

So (vi) is proved, by reference to 5.1.3.

(vii) Thanks to (3), (vi) and (ii),

$$\begin{aligned}\int_\Omega \mathsf{E}_n f.\bar{g}\, d\mu &= \int_\Omega \mathsf{E}_n(\mathsf{E}_n f.\bar{g})\, d\mu \\ &= \int_\Omega \mathsf{E}_n f.\, \mathsf{E}_n\bar{g}\, d\mu \\ &= \int_\Omega \mathsf{E}_n f.\, \overline{\mathsf{E}_n g}\, d\mu.\end{aligned}$$

The second equality in (5) is proved similarly.

(viii) If $A \in \mathscr{F}_m$ and $\mu(A) < \infty$, then $A \in \mathscr{F}_n$, since $\mathscr{F}_m \subseteq \mathscr{F}_n$. By the definition of the m-th and n-th conditional expectations,

$$\int_A \mathsf{E}_m\mathsf{E}_n f\, d\mu = \int_A \mathsf{E}_n f\, d\mu = \int_A f\, d\mu = \int_A \mathsf{E}_m f\, d\mu.$$

But $\mathsf{E}_m\mathsf{E}_n f$ is also $\mathscr{F}_m$-measurable. Therefore (see Definition 5.1.3),

$$\mathsf{E}_m\mathsf{E}_n f = \mathsf{E}_m f.$$

(ix) This is clear from the uniqueness part of 5.1.2 since f is assumed to be measurable for $\mathscr{F}_n$. □

5.2. Martingales and Martingale Difference Series

If $(\Omega, \mathscr{F}, \mu)$ is as in 5.1.1 and $f \in L^p(\Omega)$, where $1 \leqslant p < \infty$, then the sequence $(\mathsf{E}_j f)_{-\infty}^{\infty}$ of conditional expectations of f is defined. It is a sequence of L^p functions having the properties

(i) $\mathsf{E}_j f$ is measurable for $\mathscr{F}_j$;

and

(ii) $\mathsf{E}_m\mathsf{E}_n f = \mathsf{E}_m f$ when $m < n$.

5.2.1. Definition. We say that a sequence $(g_j)_{-\infty}^{\infty}$ of functions in $\bigcup_{1 \leqslant p < \infty} L^p(\Omega, \mathscr{F}, \mu)$ is a *martingale* if

(i) g_j is $\mathscr{F}_j$-measurable;

(ii) $\mathsf{E}_j g_k = g_j$ whenever $j < k$.

5.2.2. Definition. If $f \in L^p(\Omega)$, where $1 \leqslant p < \infty$, we call $(\mathsf{E}_j f)_{-\infty}^{\infty}$ *the martingale associated with* f, and

$$\mathsf{S}_j f = \mathsf{E}_{j+1} f - \mathsf{E}_j f \tag{1}$$

the *j-th martingale difference of* f.

5.2.3. Examples. (a) Let X be an LCA group having a suitable family $(X_j)_{-\infty}^{\infty}$ of compact open subgroups, as defined in 4.1.1, and let $\Omega = G$, the dual group of X. If G_j is the annihilator of X_j in G, then

$$G_{j+1} \subseteq G_j \tag{2}$$

since $X_j \subseteq X_{j+1}$ by assumption. Further,

$$\bigcup_{j\in\mathbb{Z}} G_j = G \tag{3}$$

and

$$\bigcap_{j\in\mathbb{Z}} G_j = \{0\}. \tag{4}$$

See 4.1.4. Define $\mathscr{F}$ to be the Borel σ-algebra on G, μ to be the Haar measure m on G, and let $\mathscr{F}_j$ be the σ-algebra generated by the collection of cosets of G_j in G. Since the index of X_j in X_{j+1} is finite for each j, by assumption, so too is the index of G_{j+1} in G_j and hence there are only countably many cosets of G_j in G. So $\mathscr{F}_j$ can be described as the collection of sets each of which is a union of cosets of G_j in G. It is immediate from (3) that $\bigcap_{j\in\mathbb{Z}} \mathscr{F}_j$ comprises exactly Ω and $\varnothing$. Condition 5.1.1(ii) is thus satisfied; so also is (i), because of (2). Because $(G_j)_{j\in\mathbb{Z}}$ is a base of open sets at 0 (see 4.1.1), every open set in G belongs to $\bigcup_{j\in\mathbb{Z}} \mathscr{F}_j$ and therefore 5.1.1(iii) holds. Finally, for each j, G is the union of the countably many cosets of G_j, and therefore condition 5.1.1(iv) is trivially satisfied.

Each measure space $(\Omega, \mathscr{F}_j, \mu)$ is atomic, with the various cosets of G_j as the atoms; these atoms are of positive and finite measure, since G_j is open and compact. The functions measurable for $\mathscr{F}_j$ are precisely those constant on each coset of G_j in G. As a consequence of these remarks and Definition 5.1.3, we see that if $f \in L^1(\Omega)$, $\mathsf{E}_j f$ is that function on G whose value on the coset $x + G_j$ is

$$\frac{1}{m(x + G_j)} \int_{x+G_j} f\,dm = \frac{1}{m(G_j)} \int_G f(x - y)\xi_{G_j}(y)\,dy.$$

In other words,

$$\mathsf{E}_j f = \frac{\xi_{G_j}}{m(G_j)} * f,$$

and so

$$(\mathsf{E}_j f)^\wedge = \xi_{X_j}\hat{f}.$$

That is, $\mathsf{E}_j f$ is the "Fourier partial sum of f over X_j".

(b) Let $\Omega = \mathbb{D}_2$, the Cantor group, and define $\mathscr{F}$ to be the Borel σ-algebra on $\mathbb{D}_2$. For $j \leqslant 0$, set $\mathscr{F}_j = \{\varnothing, \Omega\}$; if $j > 0$, let $\mathscr{F}_j$ be the finite σ-algebra generated by the cosets of the subgroup

$$\Omega_j = \{\mathbf{x} = (x_i)_1^\infty : x_i = 0, 1 \leqslant i \leqslant j\}.$$

The conditions in 5.1.1 are plainly satisfied. In this case, $\mathsf{E}_j f$ is, for $j \geqslant 0$, the partial sum of the Fourier series of f over the subgroup of $\hat{\mathbb{D}}_2$ generated by the characters $\rho_0, \ldots, \rho_j$ defined in the introduction to Chapter 4.

(c) Suppose X is an LCA group having a sequence $(X_j)_{-\infty}^\infty$ of subgroups such that

(i) X_j is open for every j;
(ii) $X_j \subseteq X_{j+1}$;
(iii) $\bigcup X_j = X$;
(iv) $\bigcap X_j = \{0\}$.

(There is no condition of finiteness of the index of each group in its successor.) Let G_j be the annihilator of X_j in G_j and define $\mathscr{F}$ to be the Borel σ-algebra on $\Omega = G$. For each j, let $\mathscr{F}_j$ be the collection of Borel sets in Ω that are "periodic modulo G_j"; i.e. set

$$\mathscr{F}_j = \{A \in \mathscr{F} : A + G_j = A\}.$$

We shall show in the proof of Theorems 5.4.1 and 5.4.2 that all the conditions in 5.1.1 are satisfied in this case and that if $f \in L^1 \cup L^2(\Omega)$, then

$$(\mathsf{E}_j f)^\wedge = \xi_{X_j} \hat{f}.$$

The examples just given indicate the analogy between the j-th conditional expectation of a function f and the j-th partial sum (integral) of its Fourier series (integral). We now further substantiate this analogy by proving theorems about the "maximal function" and the convergence of the martingale associated with f.

We begin by proving that $\mathsf{E}_j f \to f$ in L^p as $j \to +\infty$ when $f \in L^p(\Omega)$ and $1 \leqslant p < \infty$. The main ingredients of the proof are contained in the following two lemmas of a measure-theoretic character.

5.2.4. Lemma. *If X is a set, $A \subseteq X$, and $\mathscr{K}$ is a collection of subsets of X, then*

$$\sigma(\mathscr{K} | A) = \sigma(\mathscr{K}) | A.$$

Proof. The inclusion $\sigma(\mathscr{K} | A) \subseteq \sigma(\mathscr{K}) | A$ is clear since $\sigma(\mathscr{K}) | A$ is a σ-algebra containing all sets of the form $K \cap A$ where $K \in \mathscr{K}$.

On the other hand, if we write

$$\mathscr{T} = \{T \subseteq X : T \cap A \in \sigma(\mathscr{K} | A)\}$$

it is a routine exercise to check that $\mathscr{T}$ is a σ-algebra containing $\mathscr{K}$ and hence containing $\sigma(\mathscr{K})$. By the very definition of $\mathscr{T}$,

$$\mathscr{T}|A \subseteq \sigma(\mathscr{K}|A),$$

and so $\sigma(\mathscr{K})|A \subseteq \sigma(\mathscr{K}|A)$. $\square$

5.2.5. Lemma. *Given a set A in $\mathscr{F}$ of finite measure and $\varepsilon > 0$, there is a set B belonging to some $\mathscr{F}_n$ and also of finite measure such that*

$$\mu(A \triangle B) \leqslant \varepsilon.$$

Proof. By virtue of condition 5.1.1(iv) and the countable additivity of μ, we may assume without loss of generality that $A \subseteq U$ where

$$U \in \bigcup_{j\in\mathbb{Z}} \mathscr{F}_j \quad \text{and} \quad \mu(U) < \infty.$$

Since by assumption $\bigcup \mathscr{F}_j$ generates $\mathscr{F}$, it follows from Lemma 5.2.4 that $(\bigcup \mathscr{F}_j)|U$ generates $\mathscr{F}|U$. Since the argument to follow depends only on assumptions 5.1.1(i) and 5.1.1(iii), we may assume without loss of generality that $U = \Omega$ and that $\mu(\Omega) < \infty$.

Define

$$\mathscr{A} = \{A \in \mathscr{F} : \text{for every } \varepsilon > 0, \text{ there exists } B \text{ in } \bigcup_{j\in\mathbb{Z}} \mathscr{F}_j \text{ such that } \mu(A \triangle B) \leqslant \varepsilon\}.$$

It will suffice to prove that $\mathscr{A}$ is a σ-algebra.

First, if $A \in \mathscr{F}$ and $\varepsilon > 0$, choose B in some $\mathscr{F}_n$ so that $\mu(A \triangle B) \leqslant \varepsilon$. Since

$$\mu\{(\Omega\backslash A) \triangle (\Omega\backslash B)\} = \mu(A \triangle B),$$

$\Omega\backslash A \in \mathscr{F}$ and $\Omega\backslash B \in \mathscr{F}_n$, it is plain that $\mathscr{A}$ is stable under the formation of complements.

Next suppose A_1 and A_2 belong to $\mathscr{A}$. If $\varepsilon > 0$, we can choose B_1 and B_2 in the same $\mathscr{F}_n$ (since $(\mathscr{F}_j)_{-\infty}^{\infty}$ is increasing) such that

$$\mu(A_1 \triangle B_1) \leqslant \frac{\varepsilon}{2}$$

and

$$\mu(A_2 \triangle B_2) \leqslant \frac{\varepsilon}{2}.$$

Now let $A = A_1 \cap A_2$ and $B = B_1 \cap B_2$. Then

$$\begin{aligned}\mu(A \triangle B) &= \mu[(A_1 \cap A_2)\backslash(B_1 \cap B_2)] + \mu[(B_1 \cap B_2)\backslash(A_1 \cap A_2)]\\ &\leqslant \mu(A_1\backslash B_1) + \mu(A_2\backslash B_2) + \mu(B_1\backslash A_1) + \mu(B_2\backslash A_2)\\ &\leqslant \varepsilon.\end{aligned}$$

Therefore $A \in \mathscr{A}$, and $\mathscr{A}$ is stable under the formation of finite intersections. So also then is $\mathscr{A}$ stable under the formation of finite unions and is thus an algebra.

Finally, let $(A_j)_1^\infty$ be an increasing sequence of elements of $\mathscr{A}$, with union A. Given $\varepsilon > 0$, we choose N so that $\mu(A \backslash \bigcup_1^N A_j) \leqslant \varepsilon/2$. This is possible since, by assumption, $\mu(\Omega) < \infty$. Let

$$B = \bigcup_1^N A_j.$$

Then $B \in \mathscr{A}$, and so there is a set B_0 in $\bigcup_{j \in \mathbb{Z}} \mathscr{F}_j$ such that

$$\mu(B \triangle B_0) \leqslant \frac{\varepsilon}{2}.$$

We deduce then that $A \in \mathscr{A}$ since

$$\begin{aligned}\mu(A \triangle B_0) &\leqslant \mu(A \backslash B) + \mu(B \backslash B_0) + \mu(B_0 \backslash B)\\ &\leqslant \varepsilon. \quad \square\end{aligned}$$

5.2.6. Theorem. *If $(\Omega, \mathscr{F}, \mu)$ and $(\mathscr{F}_j)_{-\infty}^\infty$ are as in 5.1.1 and $f \in L^p(\Omega)$, where $1 \leqslant p < \infty$, then*

$$\mathsf{E}_j f \to f$$

in $L^p(\Omega)$ as $j \to +\infty$.

Proof. The statement is clearly true if f is a function which is simple for some $\mathscr{F}_n$ (dependent on f) and integrable, since then $\mathsf{E}_j f = f$ from a certain stage on. By virtue of Lemma 5.2.5, the space of such simple functions is dense in L^p. On the other hand, the operators E_j are uniformly bounded on L^p (Lemma 5.1.4(v)). So the result follows by a standard approximation argument ([10], 7.1.4). $\square$

We pass on now to the consideration of the convergence of the sequence $(\mathsf{E}_j f)_{-\infty}^\infty$ as $j \to -\infty$. Here we shall prove first a result, of independent interest, about the maximal function associated with the martingale $(\mathsf{E}_j f)_{-\infty}^\infty$.

5.2.7. Theorem. *If $f \in \bigcup_{1 \leqslant p < \infty} L^p(\Omega)$, define the maximal function $\mathsf{M}f$ to be*

$$\mathsf{M}f = \sup_{j \in \mathbb{Z}} \mathsf{E}_j |f|. \tag{5}$$

Then

(i) *the operator $f \to \mathsf{M}f$ is of weak type* (1, 1): *indeed,*

$$\mu(\{x \in \Omega : \mathsf{M}f(x) > \lambda\}) \leqslant \frac{2}{\lambda} \|f\|_1; \tag{6}$$

and

$$\mu(\{x \in \Omega : \mathsf{M}f(x) > \lambda\}) \leqslant \frac{4}{\lambda} \int_{\{x : |f(x)| > \lambda/2\}} |f| \, d\mu \tag{7}$$

for all f in $L^1(\Omega)$ and all $\lambda > 0$;

(ii) *the operator* $f \to \mathsf{M}f$ *is of strong type* (p, p) *when* $1 < p < \infty$; *in fact there is a number* C_p *dependent only on* p *such that*

$$\|\mathsf{M}f\|_p \leqslant C_p \|f\|_p \tag{8}$$

for all f *in* $L^p(\Omega)$.

Proof. (i) Let $\mathsf{M}_0 f = \sup_{j<0} \mathsf{E}_j |f|$ and $\mathsf{M}_1 f = \sup_{j\geqslant 0} \mathsf{E}_j |f|$. Then if $\lambda > 0$,

$$\{x: \mathsf{M}f(x) > \lambda\} = \{x: \mathsf{M}_0 f(x) > \lambda\} \cup \{x: \mathsf{M}_1 f(x) > \lambda\}.$$

Reasoning with $\mathsf{M}_0 f$ first, set

$$A_1 = \{x: \mathsf{E}_{-1}|f|(x) > \lambda\}$$

and for $n \geqslant 2$, set

$$A_n = \{x \in \Omega: \mathsf{E}_{-n}|f|(x) > \lambda,\ \mathsf{E}_{-j}|f|(x) \leqslant \lambda \ \text{ for } j = 1, \ldots, n-1\}.$$

Then

$$\{x: \mathsf{M}_0 f(x) > \lambda\} = \bigcup_{j=1}^{\infty} A_j. \tag{9}$$

Yet the sets A_j are pairwise disjoint, and clearly

$$\mu(A_j)\lambda \leqslant \int_{A_j} |f| \, d\mu. \tag{10}$$

Adding the inequalities (10), and using (9), we conclude that

$$\lambda\mu(\{x: \mathsf{M}_0 f(x) > \lambda\}) \leqslant \int_\Omega |f| \, d\mu = \|f\|_1.$$

Since a similar argument applies to $\mathsf{M}_1 f$, (6) is established.

To prove (7), let $f = f_1 + f_2$ where

$$f_1(x) = \begin{cases} f(x) & \text{if } |f(x)| > \dfrac{\lambda}{2} \\ 0 & \text{otherwise.} \end{cases}$$

Then $|f_2(x)| \leqslant \lambda/2$ and so

$$\mathsf{M}f_2 \leqslant \frac{\lambda}{2}.$$

Therefore

$$\mathsf{M}f \leqslant \mathsf{M}f_1 + \frac{\lambda}{2}$$

and

$$\{x: \mathsf{M}f(x) > \lambda\} \subseteq \left\{x: \mathsf{M}f_1(x) > \frac{\lambda}{2}\right\}.$$

By (6) then,

$$\begin{aligned}\mu(\{x: \mathsf{M}f(x) > \lambda\}) &\leqslant \mu\left(\left\{x: \mathsf{M}f_1(x) > \frac{\lambda}{2}\right\}\right) \\ &\leqslant \frac{4}{\lambda}\|f_1\|_1 = \frac{4}{\lambda}\int_{\{x:|f(x)|>\lambda/2\}} |f|\, d\mu.\end{aligned}$$

(ii) Recall that if $g \in L^p(\Omega)$, $g \geqslant 0$ and α is the distribution function of g, then

$$\|g\|_p^p = p\int_0^\infty t^{p-1}\,\alpha(t)\,dt. \tag{11}$$

Therefore, if we take $g = \mathsf{M}f$ and use (7) and (11), we deduce that

$$\begin{aligned}\|\mathsf{M}f\|_p^p &\leqslant 4p\int_0^\infty t^{p-1}\cdot\frac{4}{t}\int_{\{x:|f(x)|>t/2\}} |f|\, d\mu\, dt \\ &= 4p\int_0^\infty t^{p-2}\int_\Omega \xi_{\{x:|f(x)|>t/2\}}(y)|f(y)|\, d\mu(y)\, dt.\end{aligned} \tag{12}$$

It is simple to check that the integrand in the second integral in (12) is measurable on $\mathbb{R} \times \Omega$ for the product σ-algebra, and so the Fubini-Tonelli theorem can be applied to (12) to give the estimate

$$\begin{aligned}\|\mathsf{M}f\|_p^p &\leqslant 4p\int_\Omega\int_0^\infty t^{p-2}\xi_{\{x:|f(x)|>t/2\}}(y)|f(y)|\, dt\, d\mu(y) \\ &= 4p\int_\Omega\int_0^{2|f(y)|} t^{p-2}|f(y)|\, dt\, d\mu(y) \\ &= \frac{4p}{p-1}\int_\Omega 2^{p-1}|f(y)|^{p-1}|f(y)|\, d\mu(y) \\ &= \frac{4p.2^{p-1}}{p-1}\|f\|_p^p. \quad \square\end{aligned}$$

We pass now to a discussion of the behaviour of $\mathsf{E}_{-N}f$ as $N \to +\infty$ in the case that $\mu(\Omega) = +\infty$.

5.2.8. Theorem. *If* $1 < p < \infty$, $\mu(\Omega) = +\infty$, *and* $f \in L^p(\Omega)$, *then* $\mathsf{E}_j f \to 0$, *both pointwise a.e. and in norm, as* $j \to -\infty$.

Proof. There is no loss of generality in assuming that $f \geqslant 0$. With this assumption, define

$$\phi(x) = \limsup_{j\to-\infty} \mathsf{E}_j f(x) = \lim_{N\to-\infty} \sup_{j\leqslant N} \mathsf{E}_j f(x). \tag{13}$$

It is plain from (13) that

$$\phi(x) \leqslant \sup_{j\leqslant N} \mathsf{E}_j f(x) \leqslant \mathsf{M}f(x) \tag{14}$$

for every index N, and so $\phi \in L^p(\Omega)$, by Theorem 5.2.6. On the other hand, ϕ is measurable for the σ-algebra $\bigcap_{-\infty}^{\infty} \mathscr{F}_j$ and so, by 5.1.1(ii), ϕ is equal a.e. to a constant, which means that $\phi = 0$ a.e., since $\phi \in L^p(\Omega)$ and $\mu(\Omega) = +\infty$. This is to say that $\mathsf{E}_j f(x) \to 0$ a.e. as $j \to -\infty$.

At the same time, (14) taken in conjunction with 5.2.7(ii) and the dominated convergence theorem then shows that $\mathsf{E}_j f \to 0$ in $L^p(\Omega)$ as $j \to -\infty$. □

The discussion of the case where $\mu(\Omega) < \infty$ is considerably more complicated and necessitates the introduction of the notion of *up-crossing* due to Doob.

5.2.9. Definition. Let N be a positive integer, f an integrable function, and c and d real numbers such that $c < d$. For each point x in Ω, consider the finite sequence $\mathsf{E}_{-N}f(x), \ldots, \mathsf{E}_{-1}f(x)$ of real numbers. Define the indices $v_1(x), \ldots, v_N(x)$ inductively as follows.

First set

$$v_1(x) = \inf\{j \in [-N, -1] : \mathsf{E}_j f(x) \leqslant c\} \tag{15}$$

with the convention that the right side is to be interpreted as -1 if the set written there is empty. Then define the successive even and odd indices by specifying that

$$v_{2k}(x) = \inf\{j \in [-N, -1] : j \geqslant v_{2k-1}, \mathsf{E}_j f(x) \geqslant d\}$$

and (16)

$$v_{2k+1}(x) = \inf\{j \in [-N, -1] : j \geqslant v_{2k}, \mathsf{E}_j f(x) \leqslant c\},$$

whenever $k \geqslant 1$ and $2k$ (resp. $2k + 1$) $\leqslant N$. Once more, interpret the right side as -1 if the corresponding set is empty.

The sequence

$$\mathsf{E}_{v_1(x)} f(x), \ldots, \mathsf{E}_{v_N(x)} f(x) \tag{17}$$

is called the *up-crossing arrangement* of the given sequence relative to the interval $[c, d]$.

The number $u_N(x)$ of indices j in $[1, N)$ such that

$$\mathsf{E}_{v_j(x)} f(x) \leqslant c \quad \text{and} \quad \mathsf{E}_{v_{j+1}(x)} f(x) \geqslant \mathrm{d}$$

is called the number of *up-crossings* of $[c, d]$ by the sequence $\mathsf{E}_{-N}f(x), \ldots, \mathsf{E}_{-1}f(x)$.

Remark. Although it would be a tedious matter to write out all the details of a complete argument, a few moments' reflection should convince the reader that u_N is a measurable function of x for each positive integer N. Similarly, each of the functions $x \to \mathsf{E}_{v_j(x)} f(x)$ is measurable and integrable.

The important properties of the function u_N and of the up-crossing arrangement (17) are contained in the next lemma.

5.2.10. Lemma. *Let f, N, c and d be as in* 5.2.9, *and consider the up-crossing arrangement* (17). *Then*

(i) $\int_\Omega \mathsf{E}_{v_j(x)} f(x)\, d\mu(x) = \int_\Omega \mathsf{E}_{-1} f\, d\mu$
for $j = 1, \ldots, N$.

(ii) $\sum_{j \text{ odd},\, j \in [1, N)} \{\mathsf{E}_{v_{j+1}(x)} f(x) - \mathsf{E}_{v_j(x)} f(x)\} \geqslant (d - c) u_N(x)$.

Proof. (i) The definitions (15) and (16) show that if k is any integer in $[-N, -1]$, then

$$A_k = \{x \colon v_j(x) = k\} \in \mathscr{F}_{-1}.$$

Hence

$$\begin{aligned} \int_{A_k} \mathsf{E}_{v_j(x)} f(x)\, d\mu(x) &= \int_{A_k} \mathsf{E}_k f\, d\mu \\ &= \int_{A_k} \mathsf{E}_{-1} f\, d\mu \end{aligned} \tag{18}$$

by the martingale property (Definition 5.2.1). Summing (18) from $k = -N$ to $k = -1$, we conclude that

$$\begin{aligned} \int_\Omega \mathsf{E}_{v_j(x)} f(x)\, d\mu(x) &= \sum_{-N}^{-1} \int_{A_k} \mathsf{E}_{-1} f\, d\mu \\ &= \int_\Omega \mathsf{E}_{-1} f\, d\mu. \end{aligned}$$

(ii) This is quite obvious since by Definition 5.2.9, an index j for which

$$\mathsf{E}_{v_j(x)} f(x) \leqslant c \quad \text{and} \quad \mathsf{E}_{v_{j+1}(x)} f(x) \geqslant d$$

is necessarily odd. □

Here now is the remaining case of convergence of $\mathsf{E}_j f$ as $j \to -\infty$.

5.2.11. Theorem. *Assume that $\mu(\Omega) = 1$, $f \in L^p(\Omega)$, and $1 \leqslant p < \infty$. Then $\mathsf{E}_j f \to \int_\Omega f\, d\mu$, both pointwise a.e. and in norm, as $j \to -\infty$.*

Proof. We suppose without loss of generality that $f \geqslant 0$. Define

$$\phi(x) = \limsup_{j \to -\infty} \mathsf{E}_j f(x)$$

and

$$\psi(x) = \liminf_{j \to -\infty} \mathsf{E}_j f(x).$$

Since ϕ and ψ are both measurable for the σ-algebra $\bigcap_{-\infty}^{\infty} \mathscr{F}_j$, 5.1.1(ii) tells us that each is constant a.e.. Write $a = \psi(x)$, $b = \phi(x)$. We aim to prove that $a = b$. Assume the contrary: then there exist numbers c and d such that

$$0 \leqslant a < c < d < b.$$

In this case, we deduce from the definitions of ϕ and ψ and Definition 5.2.9 that $u_N(x) \to +\infty$ for almost every x as $N \to +\infty$, $u_N(x)$ being defined relative to the interval $[c, d]$. But again by Definition 5.2.9,

$$\begin{aligned} \mathsf{E}_{-1} f(x) &= \sum_{j=1}^{N-1} \{\mathsf{E}_{\nu_{j+1}(x)} f(x) - \mathsf{E}_{\nu_j(x)} f(x)\} + \mathsf{E}_{\nu_1(x)} f(x) \\ &\geqslant \sum_{j=1}^{N-1} \{\mathsf{E}_{\nu_{j+1}(x)} f(x) - \mathsf{E}_{\nu_j(x)} f(x)\}. \end{aligned}$$

Using Lemma 5.2.10(i), we deduce that

$$\int_\Omega \mathsf{E}_{-1} f(x)\, d\mu(x) \geqslant \sum_{j \text{ odd},\, j \in [1,N)} \int_\Omega \{\mathsf{E}_{\nu_{j+1}(x)} f(x) - \mathsf{E}_{\nu_j(x)} f(x)\}\, d\mu(x)$$

and finally, by Lemma 5.2.10(ii) that

$$\int_\Omega \mathsf{E}_{-1} f(x)\, d\mu(x) \geqslant (d - c) \int_\Omega u_N(x)\, d\mu(x). \tag{19}$$

But (19) entails that $u_N(x)$ is bounded as $N \to +\infty$ for almost every x, and this is a contradiction. So we conclude that $\mathsf{E}_j f(x) \to a$ a.e., as $j \to -\infty$, a being a constant.

However, $\mathsf{E}_j f(x) \leqslant \mathsf{M} f(x)$ for every x, and $\mathsf{M} f \in L^p(\Omega)$, by Theorem 5.2.7. Hence, by the dominated convergence theorem, $\mathsf{E}_j f \to a$ in norm. It then follows from 5.3.1(ii) and the hypothesis $\mu(\Omega) = 1$ that

$$\int_\Omega f\, d\mu = \int_\Omega \mathsf{E}_j f\, d\mu \to \int_\Omega a\, d\mu = a.$$

We have proved then that

$$\mathsf{E}_j f(x) \to a = \int_\Omega f\, d\mu,$$

both a.e. and in norm, as $j \to -\infty$. □

The examples given in 5.2.3, and Theorems 5.2.6, 5.2.8 and 5.2.11, make plain the analogy between ordinary trigonometric series and the martingale difference series which we now define.

5.2.12. Definition. (i) If $\mu(\Omega) = +\infty$, a *martingale difference series on* $(\Omega, \mathcal{F}, \mu)$ is a series of the form

$$\sum_{j\in\mathbb{Z}} \phi_j$$

where

(a) $\phi_j \in \bigcup_{1\leqslant p<\infty} L^p(\Omega)$ for each j;

and

(b) ϕ_j is measurable for $\mathcal{F}_{j+1}$ and

$$\mathsf{E}_j\phi_j = 0.$$

(ii) If $\mu(\Omega) < +\infty$, a *martingale difference series on* Ω is a series of the form

$$c + \sum_{j\in\mathbb{Z}} \phi_j,$$

in which c is a constant, and the functions ϕ_j satisfy (a) and (b) above.

5.2.13. Definition. (i) If $\mu(\Omega) = +\infty$, and $f \in L^p(\Omega)$, where $1 \leqslant p < \infty$, the *martingale difference series of* f is the series

$$\sum_{j\in\mathbb{Z}} \mathsf{S}_j f$$

where

$$\mathsf{S}_j f = \mathsf{E}_{j+1} f - \mathsf{E}_j f.$$

The *square function* $\mathsf{Q}f$ *of* f is the function

$$\mathsf{Q}f = \left(\sum_{j\in\mathbb{Z}} |\mathsf{S}_j f|^2\right)^{1/2}.$$

(ii) When $\mu(\Omega) < \infty$, we normalise μ to arrange that $\mu(\Omega) = 1$. If $f \in L^1(\Omega)$, the martingale difference series of f is the series

$$\int_\Omega f\,d\mu + \sum_{j\in\mathbb{Z}} \mathsf{S}_j f,$$

where $\mathsf{S}_j f = \mathsf{E}_{j+1} f - \mathsf{E}_j f$. The *square function of* f is the function

$$\mathsf{Q}f = \left(\left|\int_\Omega f\,d\mu\right|^2 + \sum_{j\in\mathbb{Z}} |\mathsf{S}_j f|^2\right)^{1/2}.$$

Taken together, Theorems 5.2.6, 5.2.8 and 5.2.11 imply that the partial sums

$$\sum_{M}^{N} \mathsf{S}_j f$$

of the martingale difference series of f converge in norm to f as $N \to +\infty$ and $M \to -\infty$ when $1 < p < \infty$ and $\mu(\Omega) = +\infty$. When $\mu(\Omega) = 1$, they converge to $f - \int_\Omega f\,d\mu$.

We aim now to prove the corresponding Littlewood-Paley theorem, namely that the norms $\|f\|_p$ and $\|\mathsf{Q}f\|_p$ are equivalent when $1 < p < \infty$.

5.3. The Littlewood-Paley Theorem

Just as in the case of Fourier series and integrals (see 1.2.6(ii)) we begin by showing that we need only prove a one-way inequality.

The conditions on $(\Omega, \mathscr{F}, \mu)$ stated in 5.1.2 are still in force. In addition, if $\mu(\Omega) < \infty$, we shall assume without loss of generality that $\mu(\Omega) = 1$.

5.3.1. Theorem. *Suppose $1 < p < \infty$. If there is a number B_p such that*

$$\|\mathsf{Q}f\|_p \leqslant B_p\|f\|_p, \tag{1}$$

for all f in $L^p(\Omega)$, then there is a number $A_p > 0$ such that

$$A_p\|f\|_p \leqslant \|\mathsf{Q}f\|_p \leqslant B_p\|f\|_p \tag{2}$$

for all f in $L^p(\Omega)$.

Proof. This is an exact analogue of the proof in 1.2.6(ii). □

5.3.2. Corollary. *If $\mu(\Omega) = 1$, then in proving (2) it is sufficient to prove the existence of a number B_p such that*

$$\left\|\left(\sum_{j\in\mathbb{Z}} |\mathsf{S}_j f|^2\right)^{1/2}\right\|_p \leqslant B_p\|f\|_p$$

for all f in $L^p(\Omega)$.

Proof. Since $|\int_\Omega f\,d\mu|^2 \leqslant \|f\|_1 \leqslant \|f\|_p$, and

$$\begin{aligned}\mathsf{Q}f &= \left(\left|\int_\Omega f\,d\mu\right|^2 + \sum_{j\in\mathbb{Z}} |\mathsf{S}_j f|^2\right)^{1/2}\\ &\leqslant \left|\int_\Omega f\,d\mu\right| + \left(\sum_{j\in\mathbb{Z}} |\mathsf{S}_j f|^2\right)^{1/2},\end{aligned}$$

the statement is clear. □

We are now in a position to outline the strategy of the proof of (2). It is to show that the subadditive operator Q is of weak type $(1, 1)$ and of strong type (q, q) when

$2 < q < \infty$. Once these statements are established, the Marcinkiewicz interpolation theorem (Theorem A.2.1) and some elementary density arguments will do the rest. It is probably not surprising that the strong continuity is the more difficult to establish. It will necessitate the introduction of some new concepts. So we begin with the weak (1, 1) continuity.

In view of Corollary 5.3.2, we may and will in these proofs commit the mild abuse of notation of interpreting $\mathsf{Q}f$ always to mean $(\sum |\mathsf{S}_j f|^2)^{1/2}$.

First a simple result about conditional expectations.

5.3.3. Lemma. *Suppose that* $f \in L^1(\Omega)$, $f \geqslant 0$, *and* $\lambda > 0$. *For each integer* k,

$$\mathsf{E}_{k-1}(\min\{\mathsf{E}_k f, \lambda\}) \leqslant \min\{\mathsf{E}_{k-1} f, \lambda\}. \tag{3}$$

(The inequality (3) expresses the fact that the sequence $(\min\{\mathsf{E}_k f, \lambda\})$ is a so-called *supermartingale*. See [6], Chapter 9.)

Proof. Since $\mathsf{E}_{k-1}\mathsf{E}_k f = \mathsf{E}_{k-1} f$, it will be enough to prove that

$$\mathsf{E}_{k-1}(\min\{g, \lambda\}) \leqslant \min\{\mathsf{E}_{k-1} g, \lambda\}$$

when $g \geqslant 0$, $g \in L^1(\Omega)$, and $\lambda > 0$. Suppose the contrary. Then there exists a set A in $\mathscr{F}_{k-1}$ of finite measure and numbers r and s such that $\mu(A) > 0$ and

$$\mathsf{E}_{k-1}(\min\{g, \lambda\}) \geqslant r > s \geqslant \min\{\mathsf{E}_{k-1} g, \lambda\}$$

on A. Let $A_0 = \{x \in A : \mathsf{E}_{k-1} g(x) > \lambda\}$, and $A_1 = A \backslash A_0$. Then either $\mu(A_0) > 0$ or $\mu(A_1) > 0$. In the first case, since $A_0 \in \mathscr{F}_{k-1}$, and has finite measure, 5.1.3(ii) shows that

$$\begin{aligned}\int_{A_0} \lambda \, d\mu \geqslant \int_{A_0} \min\{g, \lambda\} \, d\mu &= \int_{A_0} \mathsf{E}_{k-1}(\min\{g, \lambda\}) \, d\mu \\ &> \int_{A_0} \min\{\mathsf{E}_{k-1} g, \lambda\} \, d\mu \\ &= \lambda \, \mu(A_0) = \int_{A_0} \lambda \, d\mu.\end{aligned}$$

This is absurd.

In the second case, one has similarly that

$$\begin{aligned}\int_{A_1} g \, d\mu \geqslant \int_{A_1} \min\{g, \lambda\} \, d\mu &= \int_{A_1} \mathsf{E}_{k-1}(\min\{g, \lambda\}) \, d\mu \\ &> \int_{A_1} \min\{\mathsf{E}_{k-1} g, \lambda\} \, d\mu \\ &= \int_{A_1} \mathsf{E}_{k-1} g \, d\mu \\ &= \int_{A_1} g \, d\mu,\end{aligned}$$

and this also is absurd. $\square$

5.3.4. Theorem. *If $f \in L^1(\Omega)$ and $\lambda > 0$, there is a number C independent of f and λ such that*

$$\mu(\{x\colon \mathsf{Q}f(x) > \lambda\}) \leqslant \frac{C}{\lambda}\|f\|_1. \tag{4}$$

That is to say, the operator Q is of weak type (1, 1) on $L^1(\Omega)$.

Proof. As remarked earlier, we are going to assume that

$$\mathsf{Q}f = \left(\sum_{-\infty}^{\infty} |\mathsf{S}_j f|^2\right)^{1/2}$$

regardless of whether $\mu(\Omega)$ is finite or infinite.

We begin with the case $f \geqslant 0$. Since

$$\{x\colon \mathsf{Q}f(x) > \lambda\} = \bigcup_{N=1}^{\infty} \{x\colon \mathsf{Q}_N f(x) > \lambda\}$$

where

$$\mathsf{Q}_N f = \left(\sum_{-N}^{N-1} |\mathsf{S}_k f|^2\right)^{1/2},$$

it will suffice to prove that inequality (4) holds with $\mathsf{Q}_N f$ in place of $\mathsf{Q}f$ and a number C independent of N as well as of f and λ. For technical reasons which will appear in a moment, we consider, for each integer $N \geqslant 1$,

$$\mathsf{T}_N f = \left(\sum_{-N}^{N-1} |\mathsf{S}_k f|^2 + |\mathsf{E}_{-N} f|^2\right)^{1/2} \tag{5}$$

instead of $\mathsf{Q}_N f$.

Now if $(g_k)_{k\in\mathbb{Z}}$ is any sequence of real-valued functions on Ω, we have the elementary identity

$$g_N^2 = \sum_{-N}^{N-1} (g_{k+1} - g_k)^2 + 2\sum_{-N}^{N-1} g_k(g_{k+1} - g_k) + g_{-N}^2. \tag{6}$$

If $\boldsymbol{\Phi} = (\Phi_k)_{k\in\mathbb{Z}}$ is any sequence of real integrable functions on Ω, write

$$R_N\boldsymbol{\Phi} = \{\Phi_{-N}^2 + \sum_{-N}^{N-1} (\Phi_{k+1} - \Phi_k)^2\}^{1/2}.$$

By (6),

$$\int_\Omega R_N\boldsymbol{\Phi}^2\,d\mu = \int_\Omega \Phi_N^2\,d\mu + 2\int_\Omega \sum_{-N}^{N-1} \Phi_k(\Phi_k - \Phi_{k+1})\,d\mu. \tag{7}$$

We apply the identity (7) to the sequence (Φ_k)

$$\Phi_k = \min\{\mathsf{E}_k f, \lambda\}, \tag{8}$$

noting the fact that Φ_k is $\mathscr{F}_{k+1}$-measurable, that by Lemmas 5.1.4 and 5.3.3

$$\mathsf{E}_k(\Phi_k - \Phi_{k+1}) \geqslant 0,$$

and that by (8) $\Phi_k \leqslant \lambda$. We deduce that

$$\begin{aligned}\int_\Omega R_N \Phi^2\, d\mu &= \int_\Omega \Phi_N^2\, d\mu + 2\int_\Omega \sum_{-N}^{N-1} \Phi_k \mathsf{E}_k(\Phi_k - \Phi_{k+1})\, d\mu \\ &\leqslant \lambda \int_\Omega \Phi_N\, d\mu + 2\lambda \int_\Omega \sum_{-N}^{N-1} (\Phi_k - \Phi_{k+1})\, d\mu \\ &= 2\lambda \int_\Omega \Phi_{-N}\, d\mu - \lambda \int_\Omega \Phi_N\, d\mu \\ &\leqslant 2\lambda \int_\Omega \Phi_{-N}\, d\mu. \end{aligned} \tag{9}$$

Now if x is a point at which $\mathsf{M}f(x) \leqslant \lambda$, then by (8), $\Phi_k(x) = \mathsf{E}_k f(x)$ for all k. Therefore

$$\begin{aligned}\mu(\{x\colon \mathsf{T}_N f(x) > \lambda,\, \mathsf{M}f(x) \leqslant \lambda\}) &\leqslant \mu(\{x\colon R_N\Phi(x) > \lambda\}) \\ &\leqslant \frac{\|R_N \Phi\|_2^2}{\lambda^2} \\ &\leqslant \frac{2}{\lambda}\int_\Omega \Phi_{-N}\, d\mu \\ &\leqslant \frac{2}{\lambda}\int_\Omega \mathsf{E}_{-N} f\, d\mu \\ &= \frac{2}{\lambda}\|f\|_1, \end{aligned} \tag{10}$$

the third step by (9), and the last by 5.1.3(ii).

Thanks to (10) and Theorem 5.2.7(i),

$$\begin{aligned}\mu(\{x\colon \mathsf{T}_N f(x) > \lambda\}) &\leqslant \mu(\{x\colon \mathsf{M}f(x) > \lambda\}) + \mu(\{x\colon \mathsf{T}_N f(x) > \lambda,\, \mathsf{M}f(x) \leqslant \lambda\}) \\ &\leqslant \frac{2\|f\|_1}{\lambda} + \frac{2\|f\|_1}{\lambda} = \frac{4\|f\|_1}{\lambda}. \end{aligned}$$

Consequently

$$\mu(\{x\colon \mathsf{Q}_N f(x) > \lambda\}) \leqslant \frac{4}{\lambda}\|f\|_1,$$

and therefore

$$\mu(\{x\colon \mathsf{Q}f(x) > \lambda\}) \leqslant \frac{4}{\lambda}\|f\|_1.$$

This is under the assumption that $f \geqslant 0$. For a general real-valued function f, the usual device of writing $f = f_+ - f_-$ leads to (4) with $C = 8$. If $f = g + ih$ is complex-valued, we deduce that

$$\begin{aligned}\mu(\{x\colon \mathsf{Q}f(x) > \lambda\}) &= \mu(\{x\colon \mathsf{Q}f(x)^2 > \lambda^2\})\\ &\leqslant \mu(\{x\colon \mathsf{Q}g(x)^2 + \mathsf{Q}h(x)^2 > \lambda^2\})\\ &\leqslant \mu\left(\left\{x\colon \mathsf{Q}g(x)^2 > \frac{\lambda^2}{2}\right\}\right) + \mu\left(\left\{x\colon \mathsf{Q}h(x)^2 > \frac{\lambda^2}{2}\right\}\right)\\ &= \mu\left(\left\{x\colon \mathsf{Q}g(x) > \frac{\lambda}{\sqrt{2}}\right\}\right) + \mu\left(\left\{x\colon \mathsf{Q}h(x) > \frac{\lambda}{\sqrt{2}}\right\}\right)\\ &\leqslant \frac{8\sqrt{2}}{\lambda}(\|g\|_1 + \|h\|_1) \leqslant \frac{16\sqrt{2}}{\lambda}\|f\|_1. \quad \square\end{aligned}$$

To complete the main steps in the proof of the Littlewood-Paley theorem, we introduce a new operator.

5.3.5. Definition. If $f \in L^2(\Omega)$, we define the function $f^\#$ by the formula

$$f^\# = \sup_{n\in\mathbb{Z}} \{\mathsf{E}_n(|f - \mathsf{E}_{n-1}f|^2)\}^{1/2}. \tag{11}$$

Remarks. The introduction of the function $f^\#$ will quite probably strike the reader as unnatural and unmotivated. However, it must be pointed out that the function $f^\#$ plays a fundamental role in the theory of the spaces H^1 and BMO for martingales. The Littlewood-Paley theory we are in the midst of presenting properly amounts to an important chapter of that more general theory.

If $\mu(\Omega) = 1$, the space BMO(Ω) is usually defined as the set of functions f in $L^2(\Omega)$ such that $f^\# \in L^\infty(\Omega)$. The norm of an element f of BMO is taken to be $\|f^\#\|_\infty$. The space $H^1(\Omega)$ is just the set of integrable functions f for which $\mathsf{Q}f$ is integrable. The situation is more complicated when $\mu(\Omega) = +\infty$, but, in broad terms, the same principles apply.

The affinity of the space $H^1(\Omega)$ with Littlewood-Paley theory should be apparent from these remarks, for, roughly speaking, the Littlewood-Paley theorem amounts to the statement that when $1 < p < \infty$, $L^p(\Omega)$ is precisely the set of functions for which $\mathsf{Q}f$ belongs to $L^p(\Omega)$. The book of Garsia [16] presents a detailed and fascinating account of the H^1-BMO story for martingales.

Here now are the two important results we shall need concerning the mapping $f \to f^\#$.

5.3.6. Lemma. *Suppose that $2 < q < \infty$ and that $f \in L^1 \cap L^\infty(\Omega)$. Then*

$$\|f^\#\|_q \leqslant C_q\|f\|_q$$

where C_q is a number independent of f, but dependent upon q. That is to say, the mapping $f \to f^{\#}$ is of strong type (q, q) from $L^1 \cap L^{\infty}(\Omega)$ to $L^q(\Omega)$.

Proof. By Lemma 5.1.4(v) and (vi) and the converse of Hölder's inequality,

$$\|\mathsf{E}_n f\|_{\infty} \leqslant \|f\|_{\infty}$$

for all f in $L^1 \cap L^{\infty}(\Omega)$. Therefore

$$\begin{aligned}\mathsf{E}_n(|f - \mathsf{E}_{n-1}f|^2) &= \mathsf{E}_n(|f|^2 + |\mathsf{E}_{n-1}f|^2 - \bar{f}.\mathsf{E}_{n-1}f - f.\overline{\mathsf{E}_{n-1}f}) \\ &\leqslant \mathsf{E}_n(|f|^2) + \mathsf{E}_n[(\mathsf{E}_{n-1}|f|)^2] + 2\mathsf{E}_n[|f.\mathsf{E}_{n-1}f|] \\ &\leqslant \|f\|_{\infty}^2 + (\mathsf{E}_{n-1}|f|)^2 + 2\|f\|_{\infty}\mathsf{E}_n(|\mathsf{E}_{n-1}f|) \\ &\leqslant 4\|f\|_{\infty}^2,\end{aligned}$$

where we have used 5.1.4(ix) several times. It is now clear from the definition (11) of $f^{\#}$ that

$$\|f^{\#}\|_{\infty} \leqslant 2\|f\|_{\infty},$$

and that therefore the mapping $f \to f^{\#}$ is of strong type (∞, ∞). The theorem will result from an application of the Marcinkiewicz interpolation theorem (Theorem A.2.2) once we prove the mapping to be of weak type (2, 2).

To this end observe that if $\lambda > 0$ and x is a point for which $f^{\#}(x) > \lambda$, then for some integer n,

$$\mathsf{E}_n(|f - \mathsf{E}_{n-1}f|^2) > \lambda^2,$$

that is

$$\mathsf{E}_n(|f|^2 + |\mathsf{E}_{n-1}f|^2 - f.\overline{\mathsf{E}_{n-1}f} - \bar{f}.\mathsf{E}_{n-1}f)(x) > \lambda^2$$

or equivalently,

$$\mathsf{E}_n(|f|^2)(x) + |\mathsf{E}_{n-1}f(x)|^2 \dot{-} \mathsf{E}_n f(x).\overline{\mathsf{E}_{n-1}f(x)} - \overline{\mathsf{E}_n f(x)}.\mathsf{E}_{n-1}f(x) > \lambda^2.$$

It follows that

$$\mathsf{M}(|f|^2)(x) + 3(\mathsf{M}f)^2(x) > \lambda^2$$

and hence that

$$\begin{aligned}\mu(\{x : f^{\#}(x) > \lambda\}) &\leqslant \mu\left(\left\{x : \mathsf{M}(|f|^2)(x) > \frac{\lambda^2}{2}\right\}\right) + \mu\left(\left\{x : \mathsf{M}f(x) > \frac{\lambda}{\sqrt{6}}\right\}\right) \\ &\leqslant 2.2\frac{\||f|^2\|_1}{\lambda^2} + \frac{6C\|f\|_2^2}{\lambda^2} \\ &\leqslant D\frac{\|f\|_2^2}{\lambda^2},\end{aligned}$$

C and D being constants. The penultimate inequality uses the fact that M is of weak type (1, 1) and of strong (a fortiori weak) type (2, 2): see Theorem 5.2.7. This completes the proof. □

5.3.7. Lemma. *If $2 < q < \infty$, there is a number D_q dependent only on q such that*

$$\|\mathsf{Q}f\|_q \leqslant D_q \|f^{\#}\|_q \tag{12}$$

for all f in $L^1 \cap L^\infty(\Omega)$.

Proof. The proof depends in two places on the following Plancherel-type formula: if n and j are integers, $j < n$, and $f \in L^1 \cap L^\infty(\Omega)$, then

$$\mathsf{E}_{j+1}(|\mathsf{E}_n f - \mathsf{E}_j f|^2) = \sum_{k=j}^{n-1} \mathsf{E}_{j+1}(|\mathsf{S}_k f|^2). \tag{13}$$

This formula is valid because

$$\mathsf{E}_{j+1}(|\mathsf{E}_n f - \mathsf{E}_j f|^2) = \mathsf{E}_{j+1}\left(\sum_{k,m=j}^{n-1} \mathsf{S}_k f.\overline{\mathsf{S}_m f}\right)$$

and

$$\begin{aligned} \mathsf{E}_{j+1}(\overline{\mathsf{S}_k f}.\mathsf{S}_m f)^{-} &= \mathsf{E}_{j+1}(\mathsf{S}_k f.\overline{\mathsf{S}_m f}) \\ &= \mathsf{E}_{j+1}\{\mathsf{S}_k f.\overline{\mathsf{E}_{k+1}[\mathsf{E}_{m+1} f - \mathsf{E}_m f]}\} \\ &= 0 \end{aligned}$$

if $j \leqslant k < m$ since $\mathsf{S}_k f$ is measurable for $\mathscr{F}_{k+1}$, and

$$\mathsf{E}_{k+1}\mathsf{E}_{m+1} f = \mathsf{E}_{k+1} f = \mathsf{E}_{k+1}\mathsf{E}_m f.$$

Returning to the proof of (12), suppose that $f \in L^1 \cap L^\infty(\Omega)$ and that N is a positive integer. For each integer $n \geqslant -N$, define

$$\mathsf{Q}_{-N}^n f = \left(\sum_{k=-N}^{n-1} |\mathsf{S}_k f|^2\right)^{1/2}, \tag{14}$$

$\mathsf{Q}_{-N}^{-N} f$ being interpreted as 0. Write

$$\int_\Omega (\mathsf{Q}_{-N}^N f)^q \, d\mu = \sum_{k=-N}^{N-1} \int_\Omega \{(\mathsf{Q}_{-N}^{k+1} f)^q - (\mathsf{Q}_{-N}^k f)^q\} \, d\mu. \tag{15}$$

Now

$$\begin{aligned} (\mathsf{Q}_{-N}^{k+1} f)^q - (\mathsf{Q}_{-N}^k f)^q &= (\mathsf{Q}_{-N}^{k+1} f)^{2.q/2} - (\mathsf{Q}_{-N}^k f)^{2.q/2} \\ &\leqslant \frac{q}{2}\{(\mathsf{Q}_{-N}^{k+1} f)^2 - (\mathsf{Q}_{-N}^k f)^2\}\{(\mathsf{Q}_{-N}^{k+1} f)^2\}^{q/2-1} \end{aligned}$$

by the mean value theorem and the fact that $Q_{-N}^{k+1}f \geqslant Q_{-N}^{k}f$. Hence by (15),

$$\int_\Omega (Q_{-N}^N f)^q \, d\mu \leqslant \frac{q}{2} \sum_{k=-N}^{N-1} \int_\Omega (Q_{-N}^{k+1} f)^{q-2} \{(Q_{-N}^{k+1} f)^2 - (Q_{-N}^k f)^2\} \, d\mu. \tag{16}$$

For each k in $[-N, N-1]$, define the function θ_k by the formula

$$\theta_k = (Q_{-N}^{k+1} f)^{q-2} - (Q_{-N}^k f)^{q-2}. \tag{17}$$

Proceeding on from (16), we deduce that

$$\begin{aligned}
\int_\Omega (Q_{-N}^N f)^q \, d\mu &\leqslant \frac{q}{2} \sum_{k=-N}^{N-1} \sum_{j=-N}^{k} \int_\Omega \theta_j \{(Q_{-N}^{k+1} f)^2 - (Q_{-N}^k f)^2\} \, d\mu \\
&= \frac{q}{2} \sum_{j=-N}^{N-1} \sum_{k=j}^{N-1} \int_\Omega \theta_j \{(Q_{-N}^{k+1} f)^2 - (Q_{-N}^k f)^2\} \, d\mu \\
&= \frac{q}{2} \sum_{j=-N}^{N-1} \int_\Omega \theta_j \{(Q_{-N}^N f)^2 - (Q_{-N}^j f)^2\} \, d\mu \\
&= \frac{q}{2} \sum_{j=-N}^{N-1} \int_\Omega \theta_j \mathsf{E}_{j+1} \{(Q_{-N}^N f)^2 - (Q_{-N}^j f)^2\} \, d\mu
\end{aligned}$$

since it is clear from (14) and (17) that θ_j is measurable for $\mathscr{F}_{j+1}$. So by (14) and (13),

$$\begin{aligned}
\int_\Omega (Q_{-N}^N f)^q \, d\mu &\leqslant \frac{q}{2} \sum_{j=-N}^{N-1} \int_\Omega \theta_j \mathsf{E}_{j+1} (|\mathsf{E}_N f - \mathsf{E}_j f|^2) \, d\mu \\
&\leqslant \frac{q}{2} \sum_{j=-N}^{N-1} \int_\Omega \theta_j \{(\mathsf{E}_N f)^\#\}^2 \, d\mu \\
&= \frac{q}{2} \int_\Omega \{(\mathsf{E}_N f)^\#\}^2 \{Q_{-N}^N f\}^{q-2} \, d\mu,
\end{aligned} \tag{18}$$

where in the second step we have used the fact that $\mathsf{E}_j \mathsf{E}_N f = \mathsf{E}_j f$, and the last step follows from (17). Using Hölder's inequality for the indices $q/(q-2)$ and $q/2$ on the right side of (18) results in the inequality

$$\int_\Omega (Q_{-N}^N f)^q \, d\mu \leqslant \frac{q}{2} \left\{ \int_\Omega [Q_{-N}^N f]^q \right\}^{(q-2)/q} \left\{ \int_\Omega [(\mathsf{E}_N f)^\#]^q \, d\mu \right\}^{2/q}. \tag{19}$$

Yet

$$Q_{-N}^N f = \left(\sum_{-N}^{N-1} |\mathsf{S}_k f|^2 \right)^{1/2} \leqslant \sum_{-N}^{N-1} |\mathsf{S}_k f| \in L^q$$

since $\mathsf{E}_j f \in L^q$ for every j. So we can divide (19) through by $\{\int_\Omega (Q_{-N}^N f)^q \, d\mu\}^{1-q/2}$

and take square roots to conclude that

$$\|\mathsf{Q}_{-N}^{N} f\|_q \leqslant \sqrt{\frac{q}{2}}\,\|(\mathsf{E}_N f)^{\#}\|_q. \tag{20}$$

But, if $n - 1 \geqslant N$, $\mathsf{E}_{n-1}\mathsf{E}_N f = \mathsf{E}_N f$ and so

$$\mathsf{E}_n(|\mathsf{E}_N f - \mathsf{E}_{n-1}(\mathsf{E}_N f)|^2) = 0. \tag{21}$$

On the other hand, if $n - 1 < N$, $\mathsf{E}_{n-1}\mathsf{E}_N f = \mathsf{E}_{n-1} f$ and so (13) shows that

$$\begin{aligned}\mathsf{E}_n(|\mathsf{E}_N f - \mathsf{E}_{n-1}(\mathsf{E}_N f)|^2) = \mathsf{E}_n(|\mathsf{E}_N f - \mathsf{E}_{n-1} f|^2) &= \sum_{k=n-1}^{N-1} \mathsf{E}_n(|\mathsf{S}_k f|^2)\\ &\leqslant \sum_{k=n-1}^{\infty} \mathsf{E}_n(|\mathsf{S}_k f|^2)\\ &= \lim_{M\to\infty} \mathsf{E}_n(|\mathsf{E}_M f - \mathsf{E}_{n-1} f|^2). \end{aligned} \tag{22}$$

Since $\mathsf{E}_M f \to f$ in L^2 as $M \to +\infty$ (Theorem 5.2.6), we deduce from (22) and Lemma 5.1.4(v) that

$$\begin{aligned}\mathsf{E}_n(|\mathsf{E}_N f - \mathsf{E}_{n-1}(\mathsf{E}_N f)|^2) &\leqslant \mathsf{E}_n(|f - \mathsf{E}_{n-1} f|^2)\\ &\leqslant (f^{\#})^2\end{aligned} \tag{23}$$

if $n - 1 < N$. It follows from (21) and (23) that

$$(\mathsf{E}_N f)^{\#} \leqslant f^{\#}$$

and hence from (20) that

$$\|\mathsf{Q}_{-N}^{N} f\|_q \leqslant \sqrt{\frac{q}{2}}\,\|f^{\#}\|_q.$$

An application of the monotone convergence theorem now completes the proof. □

Finally, with all the ground-work prepared, we can quickly prove the main result.

5.3.8. Theorem (Littlewood-Paley). *If $1 < p < \infty$, there are numbers A_p and B_p such that*

$$A_p\|f\|_p \leqslant \|\mathsf{Q} f\|_p \leqslant B_p\|f\|_p \tag{24}$$

for all f in $L^p(\Omega)$.

Proof. By Theorem 5.3.1 and Corollary 5.3.2, it is enough to prove the right hand inequality in (24) with Qf interpreted to mean

$$Qf = \left(\sum_{k\in\mathbb{Z}} |S_k f|^2\right).$$

An elementary density argument just like the one in 1.2.6(i) shows that it will suffice to prove the inequality for all f in $L^1 \cap L^\infty(\Omega)$.

But, by Theorem 5.3.4, the mapping $f \to Qf$ is of weak type (1, 1) on $L^1 \cap L^\infty(\Omega)$; and by Lemmas 5.3.6 and 5.3.7, the same mapping is of strong type (q, q) whenever $q > 2$. By the Marcinkiewicz interpolation theorem (Theorem A.2.1) it is of strong type (p, p) whenever $1 < p < \infty$. □

5.4. Applications to Disconnected Groups

Our intention is now to read off from Theorem 5.3.8 several important instances of the Littlewood-Paley theorem for disconnected groups. Among these will be the promised new proof and generalisation of Theorem 4.2.8. We begin by verifying the conditions in 5.1.1 for the particular sequence of σ-algebras of interest to us.

5.4.1. Theorem. *Let X be an LCA group containing a sequence $(X_j)_{-\infty}^{\infty}$ of subgroups such that*

(a) $X_j \subseteq X_{j+1}$ *for all* j;

(b) *each* X_j *is open in* X;

(c) $\bigcap_{j\in\mathbb{Z}} X_j = \{0\}$;

(d) $\bigcup_{j\in\mathbb{Z}} X_j = X$.

Let G be the dual group of X, and G_j the annihilator of X_j in G. Denote by $\mathscr{F}$ the Borel σ-algebra on G, and $\mathscr{F}_j$ the σ-algebra comprising those Borel sets F such that $F + G_j = F$. Then the measure space $(\Omega = G, \mathscr{F}, \mu = m)$ and the sequence $(\mathscr{F}_j)_{-\infty}^{\infty}$ have all the properties stated in 5.1.1.

Proof. Property 5.1.1(i). This is clear since $G_{j+1} \subseteq G_j$ for each j.

Property 5.1.1(ii). If $F \in \bigcap_{j\in\mathbb{Z}} \mathscr{F}_j$, $\mu(F) < \infty$, and $j \in \mathbb{Z}$, then the characteristic function ξ_F is periodic modulo G_j; i.e.

$$\xi_F(x + y) = \xi_F(x)$$

for all y in G_j. Since G_j is the dual of the discrete group X/X_j, it is compact. It follows from [20], Theorems (28.54), (23.19) and (24.11) that

$$\hat{\xi}_F(\chi) = 0$$

when $\chi \in X\backslash X_j$. This being the case for every j, we conclude from (c) that $\hat{\xi}_F = 0$ on $X\backslash\{0\}$. So if X is nondiscrete, $\hat{\xi}_F = 0$ by continuity; on the other hand, if X is discrete,

$$\hat{\xi}_F = \hat{\xi}_F(0)\hat{\xi}_G$$

and hence F or $G \backslash F$ is null.

Property 5.1.1(iii). It is sufficient to prove that each nonempty open set U in G belongs to $\sigma(\bigcup_{j \in \mathbb{Z}} \mathscr{F}_j)$. To this end, notice first that G_j is compact, being the dual group of the discrete group X/X_j. Therefore if we let

$$V_j = \{x \in U: x + G_j \subseteq U\},$$

then V_j is open and $V_j \in \mathscr{F}_j$. If $x \in U$, the sets

$$(x + G_j) \cap (G \backslash U)$$

are compact and decrease with j. By assumption (d),

$$\bigcap_{j \in \mathbb{Z}} (x + G_j) = \{x\}.$$

Hence $(x + G_j) \cap (G \backslash U) = \varnothing$, and $x \in V_j$ for all sufficiently large j. Thus

$$\bigcup_{j \in \mathbb{Z}} V_j = U,$$

and this completes this part of the proof.

Property 5.1.1(iv). We may suppose without loss of generality that F is open and of finite measure. The function Φ on G/G_n defined by the formula

$$\Phi(\dot{x}) = \int_{G_n} \xi_F(x + y)\, dm_{G_n}(y)$$

is nonnegative, and positive precisely on $\pi(F)$, π denoting the canonical projection of G on G/G_n. By [20], Theorem (28.54) Φ is an integrable function. In addition to all this, it is lower semicontinuous, because we can choose a sequence (f_k) of nonnegative functions in $C_c(G)$ such that $f_k(x) \uparrow \xi_F(x)$ everywhere. But then the functions Φ_k, where

$$\Phi_k(\dot{x}) = \int_{G_n} f_k(x + y)\, dm_{G_n}(y)$$

are in $C_c(G/G_n)$ by [20], Theorem (15.21) and increase pointwise to Φ. So Φ is lower semicontinuous.

If now we let

$$W_j = \left\{\dot{x} \in G/G_n: \Phi(\dot{x}) > \frac{1}{j}\right\},$$

then W_j is open, and of finite measure since

$$\frac{1}{j} m_{G/G_n}(W_j) \leqslant \|\Phi\|_1.$$

We let $U_j = \pi^{-1}(W_j)$ and observe that U_j is clearly open and in $\mathscr{F}_j$. Each U_j is of finite measure since, by [20] Theorem (28.54),

$$\begin{aligned} m(U_j) &= \int_{G/G_n} \int_{G_n} \xi_{\pi^{-1}(W_j)}(x + y)\, dm_{G_n}(y)\, dm_{G/G_n}(\dot{x}) \\ &= m_{G_n}(G_n) m_{G/G_n}(W_j). \end{aligned}$$

Finally, $F \subseteq \bigcup_{j=1}^{\infty} U_j$ because

$$\bigcup W_j = \{\dot{x} : \Phi(\dot{x}) > 0\} = \pi(F). \quad \square$$

Remark. In the particular case that the subgroups X_j are all compact, the proof of Theorem 5.4.1 is a genuine triviality, as we noticed in Example 5.2.3(a).

As immediate corollaries of Theorem 5.4.1, we have the following theorems of Littlewood-Paley type to round off the chapter.

5.4.2. Theorem. *Let X and $(X_j)_{-\infty}^{\infty}$ be as in Theorem* 5.4.1. *For each integer j denote by Δ_j the set*

$$\Delta_j = X_{j+1} \backslash X_j.$$

Then if X is nondiscrete, the decomposition $(\Delta_j)_{j \in \mathbb{Z}}$ of X has the LP *property.*

If X is discrete, the decomposition $(\{0\}, (\Delta_j)_{j \in \mathbb{Z}})$ has the LP *property.*

Proof. We confine our attention to the nondiscrete case, leaving the remaining case to the reader.

Let $\mathscr{F}$ and $(\mathscr{F}_j)_{-\infty}^{\infty}$ be as in Theorem 5.4.1, and let $\mathsf{E}_j g$ denote the corresponding conditional expectation of g. By Theorems 5.4.1 and 5.3.8, there is a number B_p such that

$$\|\mathsf{Q}g\|_p \leqslant B_p \|g\|_p \tag{1}$$

for all Borel measurable functions g on G belonging to say $L^1 \cap L^\infty(G)$. Suppose that f is a Lebesgue measurable function in $L^1 \cap L^\infty(G)$. Then there is a Borel measurable function g on G which agrees a.e. with f.

Denote by $P_j f$ the function whose Fourier transform is $\xi_{X_j} \hat{f}$. Clearly $P_j f = P_j g$ and so we confine our attention to Borel measurable functions g in $L^1 \cap L^\infty(G)$. If we prove that

$$\mathsf{E}_j g = P_j g \tag{2}$$

it will follow that $S_{\Delta_j} g = \mathsf{S}_j g$ where

$$(S_{\Delta_j} g)^\wedge = \xi_{\Delta_j} \hat{g},$$

and (1) then translates into the inequality

$$\|(\sum |S_{\Delta_j}g|^2)^{1/2}\|_p \leqslant B_p\|g\|_p.$$

This inequality being established for all Borel functions g in $L^1 \cap L^\infty(G)$, it will follow from standard density arguments as in 1.2.6(ii) that $(\Delta_j)_{j\in\mathbb{Z}}$ has the LP property. It thus remains to prove (2).

By the definition of $\mathsf{E}_j g$, and of $\mathscr{F}_j$, $\mathsf{E}_j g$ is that essentially unique Borel measurable function, in $L^1 \cap L^\infty$, such that

$$\mathsf{E}_j g(x + y) = \mathsf{E}_j g(x) \tag{3}$$

for all y in G_j, and

$$\int_A \mathsf{E}_j g\, dm = \int_A g\, dm \tag{4}$$

for all sets A in $\mathscr{F}_j$ which have finite measure. From (3) it follows that $(\mathsf{E}_j g)^\wedge$ vanishes off X_j (cf. the proof of 5.1.1(ii)). Yet if $\chi \in X_j$, the function χ is constant on each coset of G_j since G_j is the annihilator of X_j; since $\bar{\chi}$ is also continuous on G, it is the uniform limit of a sequence $(\phi_k)_1^\infty$ of simple Borel functions, each of which is periodic modulo G_j. From (4) it follows that

$$\int_G \phi_k \mathsf{E}_j g\, dm = \int_G \phi_k g\, dm. \tag{5}$$

If we proceed to the limit in (5), we deduce that

$$\begin{aligned}(\mathsf{E}_j g)^\wedge(\chi) = \int_G \mathsf{E}_j g(x)\, \overline{\chi(x)}\, dm(x) &= \int_G g(x)\, \overline{\chi(x)}\, dm(x) \\ &= \hat{g}(\chi).\end{aligned}$$

The uniqueness theorem for Fourier transforms now assures us that $\mathsf{E}_j g = P_j g$. □

5.4.3. Theorem. *Let $(k_j)_1^\infty$ be a sequence of positive integers, each at least 2. Let $(\Delta_j)_{j=-1}^{-\infty}$ be the family of subsets of $\mathbb{Z}$ given by the formulas*

$$\Delta_{-1} = \mathbb{Z}\backslash k_1\mathbb{Z},$$
$$\Delta_{-j} = k_1 \cdots k_{j-1}\mathbb{Z}\backslash k_1 \cdots k_j\mathbb{Z}. \qquad (j \geqslant 2)$$

Then the decomposition $(\{0\}, (\Delta_j)_{-\infty}^{-1})$ is a Litlewood-Paley decomposition of $\mathbb{Z}$.

Proof. This is the particular case of Theorem 5.4.2 in which

$$X_j = \begin{cases} \mathbb{Z} & \text{when } j \geqslant 0 \\ k_1 \cdots k_{-j}\mathbb{Z} & \text{when } j < 0. \end{cases}$$ □

Chapter 6. The Theorems of M. Riesz and Stečkin for $\mathbb{R}$, $\mathbb{T}$ and $\mathbb{Z}$

6.1. Introduction

6.1.1. One of our primary aims in this chapter is to establish the M. Riesz multiplier theorem for $\mathbb{R}$, $\mathbb{T}$ and $\mathbb{Z}$. The formal statement of the result is as follows.

Theorem (M. Riesz). *Let G be any one of the groups $\mathbb{R}$, $\mathbb{T}$ and $\mathbb{Z}$, p an index in the range $(1, \infty)$, and Δ an interval (arc) of X. Then $\xi_\Delta \in M_p(X)$. Moreover, the norms of the multipliers ξ_Δ are uniformly bounded.*

In other words, there is a number $A_p > 0$ such that

$$\|S_\Delta f\|_p \leqslant A_p \|f\|_p$$

for all intervals (arcs) Δ in X and all f in $L^2 \cap L^p(G)$.

Another, equally important, aim is to prove a vector-valued version of the Riesz theorem.

Theorem. *Let G be any one of $\mathbb{R}$, $\mathbb{T}$ and $\mathbb{Z}$, and $(\Delta_j)_{j\in J}$ a countable family of intervals in X. Then the family $(\Delta_j)_{j\in J}$ has the* R *property.*

From the one-dimensional versions of these theorems we shall deduce corresponding results for finite products of $\mathbb{R}$, $\mathbb{T}$ and $\mathbb{Z}$.

A key step in proving both these theorems is the observation that the operators S_Δ can be neatly expressed in terms of the *conjugate function operator* T_c. The operator T_c is manufactured, as in 1.2.2, from the bounded function c on X, defined in the various cases as follows.

(i) When $G = \mathbb{R} = X$,

$$c(y) = -i \operatorname{sgn} y.$$

(ii) When $G = \mathbb{T}$ and $X = \mathbb{Z}$,

$$c(n) = -i \operatorname{sgn} n.$$

(iii) When $G = \mathbb{Z}$ and $X = \mathbb{T}$,

$$c(e^{iy}) = -i \operatorname{sgn} y$$

provided $-\pi \leqslant y < \pi$.

An idea of the way S_Δ is expressible in terms of T_c can be gained by looking at the simplest case, namely where $G = X = \mathbb{R}$, and $\Delta = (a, b)$. It is easy to verify that

$$ic(y) = \operatorname{sgn} y$$

and therefore

$$\begin{aligned} S_{(a,b)}f &= \frac{1}{2}\{\chi_a(iT_c(\chi_{-a}f)) - \chi_b(iT_c(\chi_{-b}f))\} \\ &= \frac{i}{2}\chi_a T_c(\chi_{-a}f) - \frac{i}{2}\chi_b T_c(\chi_{-b}f). \end{aligned}$$

Similar, though slightly more complicated, relations can be worked out in the cases $G = \mathbb{T}$ and $G = \mathbb{Z}$. Therefore, it should come as no surprise that the central result of the entire chapter is the following.

Theorem (Conjugate function theorem). *If* $1 < p < \infty$, *and* G *is* $\mathbb{R}$, $\mathbb{T}$ *or* $\mathbb{Z}$, *the function* c *is a multiplier of* L^p.

Notice that, unlike the Riesz theorems, the conjugate function theorem is a statement about the continuity of a *single* operator.

Once the conjugate function theorem, and with it the M. Riesz theorem, is established, we shall be able to deduce a stronger result, due to Stečkin.

Theorem (Stečkin). *Let* G *be any one of the groups* $\mathbb{R}$, $\mathbb{T}$ *and* $\mathbb{Z}$, *and let* ϕ *be a function of bounded variation on* X. *Then* $\phi \in M_p(X)$ *whenever* $1 < p < \infty$. *Moreover, there is for each* p *a number* C_p *independent of* ϕ *such that*

$$\|\phi\|_{p,p} \leqslant C_p \max(|\phi(0)|, \operatorname{Var} \phi).$$

Stečkin's theorem will itself be eventually superseded by the strong form of the Marcinkiewicz theorem. See Chapter 8.

In summary then, the main aims of the chapter are to prove the theorems of M. Riesz, the conjugate function theorem, and the theorem of Stečkin, in each of the three cases $\mathbb{R}$, $\mathbb{T}$ and $\mathbb{Z}$. This programme is carried out in Sections 6.2, 6.3 and 6.4 respectively. Vector versions of the Riesz theorems are established in Section 6.5; multi-dimensional versions of the various Riesz theorems are then derived in Section 6.6.

6.1.2. Everything we have said in 6.1.1 is expressed in the modern idiom; more importantly, the proofs of the main theorems will use purely real-variable methods, based on the Calderón-Zygmund theory of Chapter 2, and will sidestep the question of representing (in some sense) T_c as a singular convolution operator. It was not always so. In fact, this question of representation of T_c at one time amounted to an important part of the theory and the M. Riesz theorems took the form of statements about the so-called *Hilbert transform* H. Furthermore, many of the results were often expressed in terms of the continuity of the *bilinear* forms derived from certain singular kernels rather than in terms of the continuity of the corresponding linear mappings.

Accordingly we have decided to supplement the development in Sections 6.2–6.4 by proving the existence of the Hilbert transform; showing that the operator T_c is the same as H on the groups $\mathbb{R}$ and $\mathbb{T}$ (though not on $\mathbb{Z}$); and putting some of the results of Sections 6.2–6.4 into the old-fashioned form of statements about the continuity of certain bilinear forms. All this is presented in Section 6.7.

6.1.3. It should be pointed out that the term "conjugate function operator" arises from the fact that a number of proofs of the Riesz theorems depend in some measure on complex variable methods or the properties of harmonic functions. We shall have nothing further to say on this score.

6.1.4. Finally, we point out in Section 6.8 that in the cases $G = \mathbb{R}$ and $G = \mathbb{T}$, the Hilbert transform can be characterised to within a multiplicative constant as the only multiplier operator which commutes with dilations. In case $G = \mathbb{Z}$, the analogous statement is false, and a different though related characterisation obtains. These characterisations of H give an indication of an algebraic character why the Hilbert transform might be expected to play an important role in multiplier theory.

6.2. The M. Riesz, Conjugate Function, and Stečkin Theorems for $\mathbb{R}$

6.2.1. We identify $\mathbb{R}$ with its own character group in the usual way; i.e. we identify the element y of $\mathbb{R}$ with the character

$$\chi_y \colon x \to \exp(ixy)$$

which it generates.

In accordance with the conventions adopted in 1.2.2, if Δ is a subinterval of $\mathbb{R}$, $S_\Delta f$ is defined, for f in L^2, by the formula

$$(S_\Delta f)^\wedge = \xi_\Delta \hat{f}, \tag{1}$$

where ξ_Δ denotes the characteristic function of Δ relative to $\mathbb{R}$. The equality (1) is to be taken in the sense of L^2; that is, $S_\Delta f$ is the inverse transform, in the L^2 sense, of the function $\xi_\Delta \hat{f}$.

Here now are the statements and proofs of the M. Riesz and conjugate function theorems. As was foreshadowed in 6.1.1, the Riesz theorem will be seen to be a corollary of the conjugate function theorem.

6.2.2. Theorem (M. Riesz). *To each p in $(1, \infty)$ corresponds a number A_p such that*

$$\|S_\Delta f\|_p \leqslant A_p \|f\|_p \tag{2}$$

for every interval Δ in $\mathbb{R}$ and every f in $L^2 \cap L^p$.

6.2.3. Theorem (Conjugate function theorem). *Let c be the function on $\mathbb{R}$ defined by the formula*

$$c(y) = -i \operatorname{sgn} y.$$

Then to each p in $(1, \infty)$ there corresponds a number B_p such that

$$\|T_c f\|_p \leqslant B_p \|f\|_p \tag{3}$$

for all f in $L^2 \cap L^p$.

6.2.4. *Remark.* The assumption in 6.2.2 that $f \in L^2 \cap L^p$ avoids any complication over the a priori definition of $S_\Delta f$. It follows from (2), asserted for f in $L^2 \cap L^p$, that $S_\Delta f$ can be defined by L^p continuity for every f in L^p, and that (2) then holds for every f in L^p; see Appendix A. Furthermore, it also follows from (2) that $S_{\Delta_j} f \to S_\Delta f$ in L^p for every f in L^p and every increasing sequence (Δ_j) of intervals with union Δ. In particular, $S_{[-N,N]} f \to f$ in L^p as $N \to \infty$.

Proofs of 6.2.2 *and* 6.2.3. In proving (2), we may plainly suppose that Δ is of the form (a, b), where a and b are in $\mathbb{R} \cup \{\infty\} \cup \{-\infty\}$ and $a \leqslant b$. Then

$$S_\Delta = S_{(-\infty,b)} - S_{(-\infty,a)}$$

and it suffices to prove (2) in case $\Delta = (-\infty, a)$ for some a in $\mathbb{R}$ (the cases where $a = -\infty$ or $a = \infty$ being trivial). Furthermore,

$$S_{(-\infty,a)} f = \chi_a \cdot S_{(-\infty,0)}(\chi_{-a} f).$$

So, since multiplication by the character χ_a is an isometry of L^p onto itself, it suffices to prove (2) with $S_{(-\infty,0)}$ in place of S_Δ.

However, it is evident that

$$S_{(-\infty,0)} = \frac{1}{2}(\mathbf{1} - iT_c)$$

where $\mathbf{1}$ denotes the identity endomorphism of L^2. Thus 6.2.2 will be proved once we establish 6.2.3.

The proof of (3) will be accomplished by applying Corollary 2.4.5 to a suitable sequence (k_v) of integrable kernels for which $\hat{k}_v(y) \to -i \operatorname{sgn} y$. The following facts make it clear how the k_v should be selected. Facts (b) and (c) are moreover vital ingredients of the proof.

(a) If $f \in C_c(\mathbb{R})$, then by the inversion theorem

$$\begin{aligned}(2\pi)^{-1} \int_{(-\infty,\infty)} \hat{f}(y)(-i \operatorname{sgn} y)\, dy &= \lim_{N\to\infty} (2\pi)^{-1} \int_{[-N,N]} \hat{f}(y)(-i \operatorname{sgn} y)\, dy \\ &= \lim_{N\to\infty} \int_{(-\infty,\infty)} f(x)\left[\frac{1}{\pi x} - \frac{\cos Nx}{\pi x}\right] dx\end{aligned}$$

$$= \lim_{N\to\infty} \int_{(0,\infty)} \left[f(x) - f(-x)\right]\left[\frac{1}{\pi x} - \frac{\cos Nx}{\pi x}\right] dx$$

$$= \int_{(0,\infty)} [f(x) - f(-x)] \frac{1}{\pi x} dx \tag{4}$$

by the Riemann-Lebesgue lemma, since the function $x \to [f(x) - f(-x)]/x$ is integrable. Finally then,

$$(2\pi)^{-1} \int_{(-\infty,\infty)} \hat{f}(y)(-i \operatorname{sgn} y)\, dy = \lim_{\varepsilon\to 0} \int_{|x|\geqslant\varepsilon} f(x) \frac{1}{\pi x} dx. \tag{5}$$

The formula (5) shows that the function $1/\pi x$ gives rise, via a Cauchy principal value, to a distribution whose Fourier transform is $-i \operatorname{sgn} y$. Even if no reference to the generalised Fourier transformation is made, the two relations suggest strongly that the kernel $k(x) = 1/\pi x$ is close to the heart of the study of the conjugate function operator T_c. The next two facts help to bear out this claim.

(b)
$$\int_{0<\alpha\leqslant|x|\leqslant\beta} \frac{e^{-ixy}}{x} dx = -2i \operatorname{sgn} y \int_{(\alpha,\beta)} \frac{\sin t}{t} dt. \tag{6}$$

(c)
$$\lim_{\substack{\beta\to\infty \\ \alpha\to 0+}} \int_{(\alpha,\beta)} \frac{\sin t}{t} dt = \frac{\pi}{2}. \tag{7}$$

It follows from (6) and (7) that if we take k_ν to be the function

$$k_\nu(x) = \begin{cases} \dfrac{1}{\pi x} & \text{if } \dfrac{1}{\nu} \leqslant |x| < \nu \\ 0 & \text{elsewhere,} \end{cases} \tag{8}$$

then $\hat{k}_\nu(y) \to -i \operatorname{sgn} y$ as $\nu \to \infty$; moreover (6) implies that

$$\sup_\nu \|\hat{k}_\nu\|_\infty < \infty$$

since the integrals $\int_{(\alpha,\beta)}(\sin t)\, dt/t$ are uniformly bounded with respect to α and β ([9], 10.1.2).

Recall that the group $\mathbb{R}$ has the natural covering family $(U_\alpha) = ((-2^{-\alpha}, 2^{-\alpha}))_{\alpha\in\mathbb{Z}}$ with $\theta(\alpha) = \alpha - 1$: see 2.1.3(i). In order to be able to use Corollary 2.4.5 and so establish the theorem, we have only to show that the numbers $J(k_\nu)$, computed relative to the family (U_α), are bounded in ν, that is, that

$$\sup_\nu \sup_\alpha \int_{\mathbb{R}\setminus U_{\theta(\alpha)}} |k_\nu(x - y) - k_\nu(x)|\, dx < \infty.$$

Since

$$\sup_\alpha \int_{\mathbb{R}\setminus U_{\theta(\alpha)}} |k_\nu(x - y) - k_\nu(x)|\, dx \leqslant \sup_{y>0} \int_{|x|\geqslant 2y} |k_\nu(x - y) - k_\nu(x)|\, dx,$$

it will suffice to show that

$$\sup_{v} \sup_{y>0} \int_{|x|\geqslant 2y} |k_v(x-y) - k_v(x)|\, dx < \infty. \tag{9}$$

As a further reduction, notice that the mappings $\delta_\lambda: h \to \lambda^{-1}h(x/\lambda)$ are $\mathscr{L}^1$ isometries for all $\lambda > 0$ and that $k_v = (\delta_{1/v}h - \delta_v h)$, where

$$h(x) = \begin{cases} \dfrac{1}{\pi x} & \text{if } |x| \geqslant 1 \\ 0 & \text{elsewhere.} \end{cases}$$

Therefore

$$\begin{aligned} &\sup_{v} \sup_{y>0} \int_{|x|\geqslant 2y} |k_v(x-y) - k_v(x)|\, dx \\ &\quad \leqslant \sup_{v} \sup_{y>0} \int_{|x|\geqslant 2y} |\delta_{1/v}h(x-y) - \delta_{1/v}h(x)|\, dx \\ &\quad\quad + \sup_{v} \sup_{y>0} \int_{|x|\geqslant 2y} |\delta_v h(x-y) - \delta_v h(x)|\, dx \\ &\quad \leqslant 2 \sup_{y>0} \int_{|x|\geqslant 2y} |h(x-y) - h(x)|\, dx. \end{aligned}$$

It is plain then that the condition (9) will be fulfilled, and the numbers $J(k_v)$ uniformly bounded, if

$$\sup_{y>0} \int_{|x|\geqslant 2y} |h(x-y) - h(x)|\, dx < \infty. \tag{10}$$

The verification of (10) proceeds along elementary lines, as follows. (The reader may find it helpful to refer to a diagram.)

(i) If $1 \geqslant y > 0$, then

$$\pi \int_{x\geqslant 2y} |h(x-y) - h(x)|\, dx \leqslant 2\int_{[1,2]} \frac{1}{x}\, dx + \int_{[2,\infty)} \left(\frac{1}{x-y} - \frac{1}{x}\right) dx$$

and

$$\pi \int_{(-\infty,-2y)} |h(x-y) - h(x)|\, dx \leqslant \int_{[1,2]} \frac{1}{x}\, dx + \int_{[1,\infty)} \left(\frac{1}{x} - \frac{1}{x+y}\right) dx.$$

(ii) If $y > 1$, then $2y > 1 + y$, and so

$$\pi \int_{(2y,\infty)} |h(x-y) - h(x)|\, dx = \int_{(2y,\infty)} \left(\frac{1}{x-y} - \frac{1}{x}\right) dx = \ln 2,$$

while

$$\pi \int_{(-\infty,-2y)} |h(x-y) - h(x)|\, dx = \int_{(2y,\infty)} \left(\frac{1}{x} - \frac{1}{x+y}\right) dx = \ln\frac{3}{2}.$$

The hypotheses of Corollary 2.4.5 have now been verified, and so the proofs of the theorems are complete. □

Stečkin's theorem is a fairly easy consequence of the M. Riesz theorem.

6.2.5. Theorem (Stečkin). *If* $1 < p < \infty$, *and* ϕ *is a function of bounded variation on* $\mathbb{R}$, *then* $\phi \in M_p(\mathbb{R})$. *Moreover, there is a number* C_p *independent of* ϕ *such that*

$$\|\phi\|_{p,p} \leqslant C_p \max(|\phi(0)|, \operatorname{Var} \phi). \tag{11}$$

Proof. By 1.2.2(iii), the theorem will be established if we show that

$$\left| \int_{\mathbb{R}} \phi(y) \hat{f}(y)\, \hat{g}(y)\, dy \right| \leqslant C_p \max(|\phi(0)|, \operatorname{Var} \phi) \|f\|_p \|g\|_{p'} \tag{12}$$

for every pair of integrable functions f and g which have compactly supported Fourier transforms. Let f and g be a fixed but arbitrary pair of such functions throughout the remainder of the proof and suppose that $\hat{f}$ and $\hat{g}$ are both supported in the set $[-\lambda, \lambda]$. Define, for $y \geqslant -\lambda$,

$$\begin{aligned} \Psi(y) &= \int_{[-\lambda,y]} \hat{f}(t)\, \hat{g}(t)\, dt \\ &= \int \xi_{[-\lambda,y]}(t) \hat{f}(t)\, \hat{g}(t)\, dt. \end{aligned} \tag{13}$$

By the M. Riesz theorem, 1.2.2(iii), and this last expression (13) for $\Psi(y)$, we have that

$$|\Psi(y)| \leqslant A_p \|f\|_p \|g\|_{p'}, \tag{14}$$

A_p being the constant in the statement of Theorem 6.2.2. Hence

$$\begin{aligned} \left| \int \phi(y) \hat{f}(y)\, \hat{g}(y)\, dy \right| &= \left| \int_{[-\lambda,\lambda]} \phi(y)\, \Psi'(y)\, dy \right| \\ &= \left| \phi(\lambda)\Psi(\lambda) - \int_{[-\lambda,\lambda]} \Psi(y)\, d\phi(y) \right| \\ &\leqslant \left| \phi(\lambda)\, \Psi(\lambda) \right| + \left| \int_{[-\lambda,\lambda]} \Psi(y)\, d\phi(y) \right| \\ &\leqslant (|\phi(0)| + \operatorname{Var} \phi)\, |\Psi(\lambda)| + \operatorname{Var} \phi \,.\, \sup_{[-\lambda,\lambda]} |\Psi(y)| \\ &\leqslant A_p(|\phi(0)| + 2 \operatorname{Var} \phi) \|f\|_p \|g\|_{p'} \\ &\leqslant 3A_p \max(|\phi(0)|, \operatorname{Var} \phi) \|f\|_p \|g\|_{p'}, \end{aligned}$$

the integral with respect to ϕ being interpreted in the Riemann-Stieltjes sense. This establishes (12) with $C_p = 3A_p$; so the proof is complete. □

6.3. The M. Riesz, Conjugate Function, and Stečkin Theorems for $\mathbb{T}$

6.3.1. We remind the reader that the group $\mathbb{Z}$ can be identified with the character group of $\mathbb{T}$ by associating with each integer n the character

$$\chi_n \colon e^{it} \to e^{int}$$

which it generates.

Recall that if Δ is an interval in $\mathbb{Z}$, the sectional partial sum $S_\Delta f$ of the Fourier series of the function f in $L^2(\mathbb{T})$ is defined by the rule

$$(S_\Delta f)^\wedge = \xi_\Delta \hat{f} \tag{1}$$

where ξ_Δ denotes the characteristic function of Δ relative to $\mathbb{Z}$; otherwise expressed,

$$S_\Delta f = \sum_{n \in \Delta} \hat{f}(n)\chi_n, \tag{2}$$

the series converging in L^2 whenever $f \in L^2(\mathbb{T})$.

The Riesz theorem and the conjugate function theorem are now as follows.

6.3.2. Theorem (M. Riesz). *To each p in $(1, \infty)$ corresponds a number A_p such that*

$$\|S_\Delta f\|_p \leqslant A_p \|f\|_p \tag{3}$$

for every subinterval Δ of $\mathbb{Z}$ and every f in $L^2 \cap L^p(\mathbb{T})$.

6.3.3. Theorem (Conjugate function theorem). *Let c be the function on $\mathbb{Z}$ defined by the formula*

$$c(n) = -i \operatorname{sgn} n.$$

Then to each p in $(1, \infty)$ corresponds a number B_p such that

$$\|T_c f\|_p \leqslant B_p \|f\|_p \tag{4}$$

for all f in $L^2 \cap L^p(\mathbb{T})$.

6.3.4. *Remark.* A comment analogous to that in 6.2.4 applies in the present instance.

Proofs of 6.3.2 *and* 6.3.3. Notice that if a and b belong to $\mathbb{Z} \cup \{-\infty\} \cup \{\infty\}$, and $a \leqslant b$, then

$$S_{(a,b)}f = S_{(-\infty,b)}f - S_{(-\infty,a]}f;$$

that if a is an integer

$$S_{(-\infty,a)}f = \chi_a S_{(-\infty,0)}(\chi_{-a}f),$$

and

$$S_{(-\infty,a]}f = \chi_a S_{(-\infty,0]}(\chi_{-a}f);$$

that

$$S_{(-\infty,0)}f = \frac{1}{2}(\mathbf{1} - iT_c)f - \frac{1}{2}\hat{f}(0)$$

and

$$S_{(-\infty,0]}f = \frac{1}{2}(\mathbf{1} - iT_c)f + \frac{1}{2}\hat{f}(0).$$

It is plain therefore that Theorem 6.3.2 is a corollary of Theorem 6.3.3. We proceed to prove 6.3.3.

We prove (4) by using Corollary 2.4.5; note that $\mathbb{T}$ has the natural covering family (U_α) given in 2.1.3(ii). It is necessary to select a sequence (k_ν) of integrable kernels satisfying the hypotheses of Corollary 2.4.5 and for which $\hat{k}_\nu(n) \to -i \operatorname{sgn} n$ as $\nu \to \infty$. Now since, when $f \in C^\infty(\mathbb{T})$,

$$\begin{aligned} -i \sum \hat{f}(n) \operatorname{sgn} n &= \lim_{N\to\infty} -i \sum_{-N}^{N} \hat{f}(n) \operatorname{sgn} n \\ &= \lim_{N\to\infty} (2\pi)^{-1} \int_{[-\pi,\pi]} f(e^{it}) \frac{[\cos \frac{1}{2}t - \cos(N + \frac{1}{2})t]}{\sin \frac{1}{2}t} dt \\ &= (2\pi)^{-1} \int_{[0,\pi]} [f(e^{it}) - f(e^{-it})] \cot \tfrac{1}{2}t \, dt \\ &= \lim_{\varepsilon\to 0} (2\pi)^{-1} \int_{\pi \geqslant |t| \geqslant \varepsilon} f(e^{it}) \cot \tfrac{1}{2}t \, dt \end{aligned} \tag{5}$$

([9], 12.8.1; see also the corresponding argument in the proof of 6.2.3), it is clear that the kernel $k(e^{it}) = \cot(t/2)$ is right at the heart of the matter. Motivated by (5), we now make an observation which points the way to the choice of the k_ν.

For all $\varepsilon > 0$, $\varepsilon < \pi$,

$$\begin{aligned} (2\pi)^{-1} \int_{\pi \geqslant |t| \geqslant \varepsilon} \cot \tfrac{1}{2}t \, e^{-int} \, dt &= \frac{-i}{2\pi} \int_{[\varepsilon,\pi]} \frac{\cos \frac{1}{2}t}{\sin \frac{1}{2}t} \sin nt \, dt \\ &= \frac{-i}{2\pi} \int_{[\varepsilon,\pi]} \left\{ \frac{\sin(n + \frac{1}{2})t}{\sin \frac{1}{2}t} + \frac{\sin(n - \frac{1}{2})t}{\sin \frac{1}{2}t} \right\} dt. \end{aligned} \tag{6}$$

Since, for $n > 0$,

$$(2\pi)^{-1} \int_{[\varepsilon,\pi]} \frac{\sin (n + \frac{1}{2})t}{\sin \frac{1}{2} t} \, dt \to \frac{1}{2} \int_{[-\pi,\pi]} \mathsf{D}_n(e^{it}) \, dt$$

as $\varepsilon \to 0+$, where D_n denotes the n-th Dirichlet kernel, it follows that for all n,

$$\lim_{\varepsilon \to 0+} (2\pi)^{-1} \int_{\pi \geqslant |t| \geqslant \varepsilon} \cot \tfrac{1}{2} t \, e^{-int} \, dt = -i \operatorname{sgn} n. \tag{7}$$

Moreover, it is well known ([9], 10.1.2) that the numbers

$$\int_{[\varepsilon,\pi]} \frac{\sin (n + \frac{1}{2})t}{\sin \frac{1}{2} t} \, dt$$

are uniformly bounded with respect to ε and n.

It is therefore clear that a natural choice of k_ν is the function

$$k_\nu(e^{it}) = \begin{cases} \cot \frac{1}{2} t & \text{for } \frac{1}{\nu} \leqslant |t| \leqslant \pi \\ 0 & \text{for } |t| < \frac{1}{\nu}. \end{cases}$$

Then (see (7))

$$\hat{k}_\nu(n) \to -i \operatorname{sgn} n$$

as $\nu \to \infty$, and

$$\sup_\nu \|\hat{k}_\nu\|_\infty < \infty.$$

Conditions (20) and (21) of Corollary 2.4.5 having been established for the sequence (k_ν), it remains to prove the boundedness of the numbers $J(k_\nu)$ for the given covering family (U_α).

Write p for the 2π-periodic function such that

$$p(t) = \begin{cases} \frac{2}{t} & \text{for } -\pi \leqslant t < \pi, \ t \neq 0 \\ 0 & \text{for } t = 0, \end{cases}$$

and denote by p_ν the 2π-periodic function such that

$$p_\nu(t) = \begin{cases} p(t) & \text{for } \pi \geqslant |t| \geqslant \frac{1}{\nu} \\ 0 & \text{for } |t| < \frac{1}{\nu}. \end{cases}$$

Then if $\pi/2 \geqslant s > 0$,

$$\begin{aligned}\int_{\pi\geqslant|t|\geqslant 2s} |k_\nu(t-s)-k_\nu(t)|\,dt &\leqslant \int_{\pi\geqslant|t|\geqslant 2s} |k_\nu(t-s)-p_\nu(t-s)|\,dt \\ &\quad + \int_{\pi\geqslant|t|\geqslant 2s} |p_\nu(t-s)-p_\nu(t)|\,dt \\ &\quad + \int_{\pi\geqslant|t|\geqslant 2s} |p_\nu(t)-k_\nu(t)|\,dt \\ &\leqslant 2\int_{[-\pi,\pi)} |p(t)-k(t)|\,dt + \int_{\pi\geqslant|t|\geqslant 2s} |p_\nu(t-s)-p_\nu(t)|\,dt.\end{aligned}$$

But the function $p - k$ is integrable over $[-\pi, \pi)$ and so the numbers $J(k_\nu)$ will be bounded if

$$\sup_\nu \sup_{\pi/2\geqslant s>0} \int_{\pi\geqslant|t|\geqslant 2s} |p_\nu(t)-p_\nu(t-s)|\,dt < \infty. \tag{8}$$

The proof of (8) is somewhat similar to that of the corresponding assertion in Section 6.2. We leave it to the reader to fill in the details. □

Stečkin's theorem now follows as a corollary of the M. Riesz theorem.

6.3.5. Theorem (Stečkin). *If ϕ is a function of bounded variation on $\mathbb{Z}$, then $\phi \in M_p$ whenever $1 < p < \infty$. Moreover there is to each such p a number C_p independent of ϕ such that*

$$\|\phi\|_{p,p} \leqslant C_p \max(|\phi(0)|, \operatorname{Var}\phi). \tag{9}$$

Proof. This is a virtual copy of the proof for $\mathbb{R}$, the main difference being that summation by parts is used in place of integration by parts. We leave the details to the reader. □

6.4. The M. Riesz, Conjugate Function, and Stečkin Theorems for $\mathbb{Z}$

6.4.1. We identify the character group of $\mathbb{Z}$ with the group $\mathbb{T}$ or with the interval $[-\pi, \pi)$ in the customary manner, viz. we associate with each element t of $[-\pi, \pi)$ or correspondingly each element e^{it} of $\mathbb{T}$ the character

$$n \to e^{int}.$$

By an *arc in* $\mathbb{T}$ we mean a subset Δ of $\mathbb{T}$ of the form

$$\Delta = \{e^{it} : t \in \delta\}$$

where δ is an interval in $\mathbb{R}$; and by $S_\Delta f$, for f in $\ell^2(\mathbb{Z})$, we mean the ℓ^2 function whose Fourier transform is $\xi_\Delta \hat{f}$.

The conjugate function operator T_c on $\mathbb{T}$ is defined in terms of the function c on $\mathbb{T}$ such that

$$c(e^{it}) = -i \operatorname{sgn} t \tag{1}$$

when $-\pi \leqslant t < \pi$. Notice the simplifying feature, compared with the cases $G = \mathbb{R}$ and $G = \mathbb{T}$, that since $X = \mathbb{T}$ is of finite measure, c is integrable, and its Fourier transform can be computed directly. In fact

$$\hat{c}(n) = \begin{cases} 0 & \text{if } n \text{ is even} \\ -\dfrac{1}{\pi n} & \text{if } n \text{ is odd.} \end{cases} \tag{2}$$

6.4.2. Theorem (M. Riesz). *To each p in $(1, \infty)$ corresponds a number A_p such that*

$$\|S_\Delta f\|_p \leqslant A_p \|f\|_p \tag{3}$$

for every arc Δ in $\mathbb{T}$ and every f in $\ell^2 \cap \ell^p(\mathbb{Z})$.

6.4.3. Theorem (Conjugate function theorem). *Let c be the function defined in* (1). *Then to each p in $(1, \infty)$ corresponds a number B_p such that*

$$\|T_c f\|_p \leqslant B_p \|f\|_p \tag{4}$$

for all f in $\ell^2 \cap \ell^p(\mathbb{Z})$.

Proofs of 6.4.2 *and* 6.4.3. As in the two earlier cases, it is enough to prove (4). To see this, we make the following simple observations.

(i) Theorem 6.4.2 will be established once (3) is proved for intervals Δ of length π and those of length less than π.

(ii) If Δ is an interval of length at most π, which, without loss of generality, we may suppose to be of the form

$$\Delta = \{e^{it} : \alpha < t < \beta\}$$

where $\beta - \alpha \leqslant \pi$, then

$$S_\Delta = S_{\Delta_1} S_{\Delta_2}$$

where the intervals

$$\Delta_1 = \{e^{it} : \alpha < t < \alpha + \pi\}$$
$$\Delta_2 = \{e^{it} : \beta - \pi < t < \beta\}$$

are both of length π.

(iii) If Δ is an interval of length π, then for a suitable real number α

$$S_\Delta f = \chi_\alpha S_{\Delta_+}(\chi_{-\alpha} f).$$

Here

$$\Delta_+ = \{e^{it}: 0 \leqslant t < \pi\}.$$

(iv) $S_{\Delta_+} = (\mathbf{1} - iT_c)/2$, $\mathbf{1}$ denoting the identity mapping.

We proceed now to the proof of (4).

The group $\mathbb{Z}$ has the following natural covering family $(U_\alpha)_{\alpha\in\mathbb{Z}}$:

$$U_\alpha = \begin{cases} (-2^{-\alpha}, 2^{-\alpha}) & \text{if} \quad \alpha \leqslant 0 \\ \{0\} & \text{if} \quad \alpha > 0 \end{cases}$$

with associated function $\theta(\alpha) = \alpha - 1$ and constant $A = 3$. See 2.1.3(iv). It would be natural to apply Corollary 2.4.5 to the family (U_α) and the sequence $(k_\nu)_1^\infty$ of kernels defined by the formula

$$k_\nu(n) = \xi_{[-2\nu+1,2\nu-1]}(n)\hat{c}(-n).$$

In this case,

$$\begin{aligned} k_\nu(e^{it}) &= -\frac{1}{\pi}\sum_1^\nu \frac{1}{2m-1}\{e^{i(2m-1)t} - e^{-i(2m-1)t}\} \\ &= -\frac{2i}{\pi}\sum_1^\nu \frac{1}{2m-1}\sin(2m-1)t. \end{aligned} \tag{5}$$

Since the partial sums of the series $\sum_1^\infty(1/n)\sin nt$ are uniformly bounded and converge pointwise ([9], 7.2.2), the same is true of the partial sums of the series $\sum_1^\infty(1/2n)\sin 2nt$, and of the series $\sum_1^\infty(1/(2m-1))\sin(2m-1)t$. The Fourier coefficients of the sum function

$$-\frac{2i}{\pi}\sum_1^\infty \frac{1}{2m-1}\sin(2m-1)t$$

are, by Lebesgue's dominated convergence theorem, the same as those of c, and so

$$\hat{k}_\nu(e^{it}) \to c(e^{it}) \text{ a.e..}$$

One of the conditions of Corollary 2.4.5 is accordingly satisfied. On the other hand, $\hat{c}(n) = 0$ for all even integers n, and the translate of k_ν by amount 1 is "disjoint" from k_ν. If α is any negative integer, $m = 1$ belongs to U_α. Therefore

$$\begin{aligned}\sup_{m \in U_\alpha} \sum_{n \notin U_{\alpha-1}} &|k_\nu(n - m) - k_\nu(n)| \\ &\geqslant \sum_{n \notin U_{\alpha-1}} |k_\nu(n - 1) - k_\nu(n)| \\ &> \sum_{n \notin U_{\alpha-1}} |k_\nu(n)|\end{aligned}$$

if $2\nu - 1 > 2^{-\alpha+1}$. It is apparent then that the other condition in Corollary 2.4.5, viz.

$$\sup_\nu J(k_\nu) = \sup_\nu \sup_\alpha \sup_{m \in U_\alpha} \sum_{n \notin U_{\alpha-1}} |k_\nu(n - m) - k_\nu(n)| < \infty$$

does not hold. So we must think again.

Notice that $\hat{c}(n)$ can be written in the form

$$\hat{c}(n) = \{(-1)^n - 1\}/2\pi n \tag{6}$$

provided $n \neq 0$. In fact $\hat{c}(n)$ comes out in the form (6) when one does the computation of $\hat{c}(n)$ in the natural way. In view of the difficulty encountered a moment ago because $\hat{c}(n) = 0$ for even n, it is suggested we look now at the kernel h on $\mathbb{Z}$ given by the formula

$$h(n) = \begin{cases} 0 & \text{if } n = 0 \\ \dfrac{1}{\pi n} & \text{if } n \neq 0 \end{cases} \tag{7}$$

and the corresponding sequence $(h_\nu)_1^\infty$ defined by the rule

$$h_\nu = \xi_{[-\nu,\nu]} h.$$

Since

$$\hat{h}_\nu(e^{it}) = -\frac{2i}{\pi} \sum_1^\nu \frac{1}{n} \sin nt,$$

it is apparent from [9], 7.2.2 that

$$\sup_\nu \|\hat{h}_\nu\|_\infty < \infty \tag{8}$$

and that the sequence $(\hat{h}_\nu)_1^\infty$ converges pointwise on $\mathbb{T}$ to a function ω, say. The crucial point is that if we prove that $\omega \in M_p(\mathbb{T})$, we shall be finished. This is the case because the function

$$\omega'(e^{it}) = \omega(e^{i(t-\pi)}) \tag{9}$$

will then be in $M_p(\mathbb{T})$ since $M_p(\mathbb{T})$ is translation-invariant; and the Fourier coefficient of ω' at n is $(-1)^n\hat{\omega}(n)$. By (7) and (6), it will follow that $c \in M_p(\mathbb{T})$.

We proceed to prove that $\omega \in M_p(\mathbb{T})$ by applying Corollary 2.4.5. In virtue of (8), the condition (20) of Corollary 2.4.5 will hold, and ω will belong to $M_p(\mathbb{T})$ if

$$\sup_{\nu} \sup_{m \neq 0} \sum_{|n| \geqslant 2|m|} |h_\nu(n-m) - h_\nu(n)| < \infty. \tag{10}$$

The verification of (10) is elementary but tedious. We leave it to the reader as an exercise, with the suggestion that he consider the sum

$$\sum_{|n| \geqslant 2m} |h_\nu(n-m) - h_\nu(n)|$$

in each of the cases $0 < \nu \leqslant m$, $m \leqslant \nu \leqslant 2m$ and $2m < \nu$. □

Remark. The kernel h is the *Hilbert kernel* on $\mathbb{Z}$. See Section 6.7. One of the intriguing aspects of the proof just given is that although it is easy to compute the Fourier coefficients of c and so produce the kernel k on $\mathbb{Z}$ which defines the conjugate function operator, the Calderón-Zygmund theory is quite ineffective when applied to k itself (or rather, to the k_ν). Instead, it was necessary to work through the Hilbert kernel h. No such problem arose in the cases $G = \mathbb{R}$ and $G = \mathbb{T}$. It seems fair to say therefore that, granted that the Calderón-Zygmund techniques are the right ones to use, the Hilbert transform occupies a more fundamental position in the Riesz theory than the conjugate function operator.

6.4.4. Theorem (Stečkin). *If* $1 < p < \infty$ *and* ϕ *is of bounded variation on* $\mathbb{T}$, *then* $\phi \in M_p(\mathbb{T})$. *Moreover there is a number* C_p *independent of* ϕ *such that*

$$\|\phi\|_{p,p} \leqslant C_p \max(|\phi(0)|, \operatorname{Var} \phi).$$

Proof. This is a straightforward adaptation of the proof of Theorem 6.2.5. □

6.5. The Vector Version of the M. Riesz Theorem for $\mathbb{R}$, $\mathbb{T}$ and $\mathbb{Z}$

6.5.1. For use in Chapter 7, we shall need the vector version of the M. Riesz theorem. We state the theorem simultaneously for the underlying groups $G = \mathbb{R}$, $\mathbb{T}$ and $\mathbb{Z}$, assuming (Δ_i) to be any countable family of subintervals of $X = \mathbb{R}$, $\mathbb{Z}$ or $\mathbb{T}$, as the case may be.

6.5.2. Theorem. *To each* p *in* $(1, \infty)$ *corresponds a number* D_p *such that*

$$\left\{\int_G \left(\sum_i |S_{\Delta_i} f_i|^2\right)^{p/2} dx\right\}^{1/p} \leqslant D_p \left\{\int_G \left(\sum_i |f_i|^2\right)^{p/2} dx\right\}^{1/p} \tag{1}$$

for every countable family (f_i) of elements of $L^2 \cap L^p(G)$ and every countable family (Δ_i) of intervals of the dual group X.

Proofs for $G = \mathbb{R}$ and $G = \mathbb{T}$. These are exactly the same, and we deal with just the case of $\mathbb{R}$.

By the reduction argument used at the outset of the proofs of Theorems 6.2.2 and 6.2.3, it suffices to deal with the case in which every S_{Δ_i} is replaced by the conjugate function operator T_c. Thus we are reduced to proving that

$$\int \Big(\sum_i |T_c f_i|^2 \Big)^{p/2} dx \leqslant D_p^p \int \Big(\sum_i |f_i|^2 \Big)^{p/2} dx \tag{2}$$

with D_p a number depending only on p.

To do this, we appeal to Corollary 3.6.4 applied to the sequence of kernels (k_v) given by equation 6.2(8) above.

Proof for $G = \mathbb{Z}$. Once again, it is enough to establish (2). The new feature is that we do not now have the function c expressed directly as a limit of a sequence $(\hat{k}_v)$ where the k_v are such that Corollary 3.6.4 can be applied. However, in the notation of the proof of Theorems 6.4.2 and 6.4.3, if $n \neq 0$,

$$\begin{aligned} \hat{c}(n) &= \{(-1)^n - 1\}/2\pi n \\ &= \frac{1}{2}\{\hat{\omega}(n) - \hat{\omega}'(n)\}. \end{aligned}$$

The same relation holds when $n = 0$ since each side is then 0. Hence

$$\begin{aligned} \Big\{ \int \Big(\sum_i |T_c f_i|^2 \Big)^{p/2} dx \Big\}^{1/p} &= \frac{1}{2} \Big\{ \int \Big(\sum_i |T_\omega f_i - T_{\omega'} f_i|^2 \Big)^{p/2} dx \Big\}^{1/p} \\ &\leqslant \frac{1}{2} \Big\{ \int \Big(\sum_i |T_\omega f_i|^2 \Big)^{p/2} dx \Big\}^{1/p} + \frac{1}{2} \Big\{ \int \Big(\sum_i |T_{\omega'} f_i|^2 \Big)^{p/2} dx \Big\}^{1/p} \end{aligned} \tag{4}$$

by Minkowski's inequality. On the other hand, by 6.4(9),

$$T_{\omega'} f = \chi_{-\pi} T_\omega (\chi_\pi f)$$

where χ_π is the character $n \to e^{in\pi}$. So it follows from (4) that

$$\Big\{ \int \Big(\sum_i |T_c f_i|^2 \Big)^{p/2} dx \Big\}^{1/p} \leqslant \frac{1}{2} \Big\{ \Big(\sum_i |T_\omega f_i|^2 \Big)^{p/2} \Big\}^{1/p} + \frac{1}{2} \Big\{ \int \Big(\sum_i |T_\omega (\chi_\pi f_i)|^2 \Big)^{p/2} \Big\}^{1/p}.$$

Therefore the proof will be complete if we prove (2) with T_ω in place of T_c. Now recall that (see the proof of 6.4.3)

$$\omega = \lim \hat{h}_v \text{ a.e.}$$

where (6.4(8))

$$\sup_{\nu} \|\hat{h}_\nu\|_\infty < \infty$$

and (6.4(10))

$$\sup_{\nu} J(h_\nu) < \infty.$$

So we can deduce (2) with T_ω in place of T_c by applying Corollary 3.6.4 just as in the two preceding cases. □

6.6. The M. Riesz Theorem for $\mathbb{R}^k \times \mathbb{T}^m \times \mathbb{Z}^n$

6.6.1. Let k, m and n be nonnegative integers, G the group $\mathbb{R}^k \times \mathbb{T}^m \times \mathbb{Z}^n$ and X its dual, $X = \mathbb{R}^k \times \mathbb{Z}^m \times \mathbb{T}^n$. By a *rectangle in* X we mean a set Δ of the form

$$\Delta = (\Delta_1^1 \times \cdots \times \Delta_k^1) \times (\Delta_1^2 \times \cdots \times \Delta_m^2) \times (\Delta_1^3 \times \cdots \times \Delta_n^3)$$

where the sets Δ_i^1, Δ_j^2 and Δ_k^3 are intervals in $\mathbb{R}$, $\mathbb{Z}$ and $\mathbb{T}$ respectively.

By applying Theorem 1.3.5 in conjunction with Theorem 6.5.2, we obtain the multidimensional form of the vector version of the M. Riesz theorem.

6.6.2. Theorem. *Let $(\Delta_j)_{j\in J}$ be a countable family of rectangles in $\mathbb{R}^k \times \mathbb{Z}^m \times \mathbb{T}^n$. Then (Δ_j) has the* R *property.*

An immediate corollary is the multidimensional version of the (scalar) M. Riesz theorem.

6.6.3. Theorem (M. Riesz). *Let Δ be a rectangle in $\mathbb{R}^k \times \mathbb{Z}^m \times \mathbb{T}^n$. Then $\xi_\Delta \in M_p$ whenever $1 < p < \infty$. Moreover, there is a constant A_p such that*

$$\|\xi_\Delta\|_{p,p} \leqslant A_p$$

for all rectangles Δ.

6.7. The Hilbert Transform

6.7.1. As we pointed out in 6.1.2, it is a relatively recent development in outlook to prove the M. Riesz theorems by using general ideas from the theory of Fourier multipliers together with the Calderón-Zygmund theory of Chapter 2. It is our aim now to introduce the Hilbert transform, to prove the existence (in some sense) of the appropriate principal value integrals, and to point out that some of the basic inequalities for the Hilbert transform can be extracted, with a little extra work, from the proofs of the M. Riesz theorems already given. This is to reverse the historical development; but either way, the Hilbert transform is so fundamental that its first properties should be written down explicitly at this point.

6.7.2. The Hilbert kernels. (i) When $G = \mathbb{R}$, the *Hilbert kernel* h is the function

$$h(x) = \frac{1}{\pi x}. \tag{1}$$

Notice that h is not integrable, but that

$$\int_{|x| \geqslant \varepsilon} |h(x)|^q \, dx < \infty$$

for every $\varepsilon > 0$ and every q in $(1, \infty)$. In particular, h is locally integrable on $\mathbb{R} \backslash \{0\}$.

(ii) When $G = \mathbb{T}$, the *Hilbert kernel* is defined to be the function

$$k(e^{it}) = \cot \tfrac{1}{2} t. \tag{2}$$

In this case h is once more not integrable, but is bounded on the set

$$\{e^{it} \colon \varepsilon \leqslant |t| \leqslant \pi\}$$

for every $\varepsilon > 0$, $\varepsilon < \pi$.

(iii) Finally, when $G = \mathbb{Z}$,

$$h(n) = \begin{cases} 0 & \text{when} \quad n = 0 \\ \dfrac{1}{\pi n} & \text{when} \quad n \neq 0 \end{cases} \tag{3}$$

is the *Hilbert kernel.* Although h is not integrable, it is q-th power integrable over the *whole* of $G = \mathbb{Z}$ whenever $q > 1$. This feature leads to a considerable simplification in the definition of the Hilbert transform on $\mathbb{Z}$.

6.7.3. The definition of the Hilbert transform. It is plain from 6.7.2 that only in the case $G = \mathbb{Z}$ is the convolution $h * f$ of the Hilbert kernel with a general L^p function f defined when $1 < p < \infty$. It is therefore necessary to give a definition of the Hilbert transform on $\mathbb{R}$ and $\mathbb{T}$ in terms of principal value integrals. We set down the definition first and then give a theorem which establishes the existence of the limits in question, together with the continuity of the Hilbert transform.

Definition. Suppose that $1 < p < \infty$.

(i) Define the family $(h_\varepsilon)_{0 < \varepsilon < 1}$ of kernels on $\mathbb{R}$ by the rule

$$h_\varepsilon(x) = \begin{cases} \dfrac{1}{\pi x} & \text{if} \quad \varepsilon \leqslant |x| \leqslant 1/\varepsilon \\ 0 & \text{otherwise.} \end{cases}$$

The *Hilbert transform* Hf of the function f in $L^p(\mathbb{R})$ is defined to be the limit in the L^p norm

$$\lim_{\varepsilon \to 0} h_\varepsilon * f. \tag{4}$$

(ii) Let $(h_\varepsilon)_{0<\varepsilon<\pi}$ be the family of kernels on $\mathbb{T}$ defined by the rule

$$h_\varepsilon(e^{it}) = \begin{cases} \cot \frac{1}{2} t & \text{if} \quad \varepsilon \leqslant |t| \leqslant \pi \\ 0 & \text{elsewhere.} \end{cases}$$

The *Hilbert transform* Hf of a function f in $L^p(\mathbb{T})$ is the limit in L^p norm

$$\lim_{\varepsilon \to 0} h_\varepsilon * f. \tag{5}$$

(iii) If $f \in \ell^p(\mathbb{Z})$, the *Hilbert transform* Hf is the function

$$Hf = h * f. \tag{6}$$

6.7.4. Theorem. *Suppose that* $1 < p < \infty$. *Then*

(i) *if* $f \in L^p(\mathbb{R})$, *the limit* (4) *exists in* L^p *and the function* H *is a continuous mapping of* L^p *into itself. In fact* H *is the unique continuous extension of* T_c, *the conjugate function operator, to all of* $L^p(\mathbb{R})$.

(ii) *If* $f \in L^p(\mathbb{T})$, *the limit* (5) *exists in* L^p, *and the Hilbert transform is a continuous mapping of* L^p *into itself. Moreover,* H *is the unique continuous extension of the conjugate function operator* T_c *to all of* $L^p(\mathbb{T})$.

(iii) *If* $f \in \ell^p(\mathbb{Z})$, *the element* Hf *defined by* (6) *actually belongs to* $\ell^p(\mathbb{Z})$ *and the Hilbert transform is* ℓ^p-*continuous. Furthermore,* H *is the unique continuous extension to all of* $\ell^p(\mathbb{Z})$ *of the multiplier operator* T_ω, *where*

$$\omega(e^{it}) = -i(\pi - t)/\pi \quad \text{if} \quad 0 < t < 2\pi. \tag{7}$$

Proof. It will be simplest, and will probably help in understanding the proofs in the other two cases, if we begin with (iii).

(iii) *The case* $G = \mathbb{Z}$. If $f \in \ell^p(\mathbb{Z})$, and

$$f_N = \xi_{[-N,N]} f,$$

then $f_N \to f$ in ℓ^p as $N \to \infty$, and so

$$h * f_N \to h * f$$

pointwise at least. Moreover $\|f_N\|_p \leqslant \|f\|_p$. So if we prove the existence of a constant C such that

$$\|h * f_N\|_p \leqslant C \|f_N\|_p \tag{8}$$

for all N, the fact that $h * f \in \ell^p$ and

$$\|h * f\|_p \leqslant C\|f\|_p$$

will follow from (8) by using Fatou's lemma. In other words, it is enough to prove the ℓ^p continuity of the mapping $f \to h * f$ on the set of finitely supported functions f in order to establish the first part of the assertion in (iii). If f is such a function and we set

$$h_\nu = \xi_{[-\nu,\nu]}h$$

for each positive integer ν, it is clear that $h_\nu * f \to f$ in $\ell^p(\mathbb{Z})$. But we established in the course of the proofs of Theorems 6.4.2 and 6.4.3 that (see 6.4(7), (8) and (10))

$$h_\nu(e^{it}) = -\sum_1^\nu \frac{2i}{\pi n} \sin nt, \tag{9}$$

$$\sup_\nu \|\hat{h}_\nu\|_\infty < \infty,$$

$$\sup_\nu J(h_\nu) < \infty,$$

and that consequently (Corollary 2.4.5)

$$C = \sup_\nu \|L_{h_\nu}\|_{p,p} < \infty.$$

So

$$\|h_\nu * f\|_p \leqslant C\|f\|_p \tag{10}$$

for all ν and all finitely supported functions f. If we take the limit on ν in (10), we deduce that

$$\|h * f\|_p \leqslant C\|f\|_p.$$

This establishes the mapping properties of H. It remains to identify H with T_ω.

This is quite straightforward. For by 7.2.2 of [9], the sum $\hat{h}_\nu(e^{it})$ in (9) converges pointwise and boundedly to $\phi(e^{it})$, say. By Lebesgue's dominated convergence theorem,

$$\hat{\phi}(n) = \begin{cases} -\dfrac{1}{\pi n} & \text{if } \ n \neq 0 \\ 0 & \text{if } \ n = 0. \end{cases}$$

But the function ω given in (7) has these same Fourier coefficients. So $\phi = \omega$.

If $f \in \ell^2 \cap \ell^p(\mathbb{Z})$,

$$h_\nu * f \to h * f = Hf \tag{11}$$

pointwise on $\mathbb{Z}$. At the same time, since $\hat{h}_\nu \to \omega$ pointwise a.e. and boundedly, and $f \in \ell^2(\mathbb{Z})$,

$$\hat{h}_\nu \hat{f} \to \omega \hat{f}$$

in $L^2(\mathbb{T})$. By Plancherel's theorem,

$$h_\nu * f \to T_\omega f \quad \text{in} \quad \ell^2(\mathbb{Z}).$$

Referring back to (11), we see that $T_\omega f = Hf$ on $\ell^2 \cap \ell^p(\mathbb{Z})$, which completes the proof of (iii).

(i) *The case* $G = \mathbb{R}$. We begin by establishing the existence of the limit (4) when $g \in C_c^\infty(\mathbb{R})$ and the existence of a constant A_p such that

$$\|Hg\|_p \leqslant A_p \|g\|_p \tag{12}$$

for all g in $C_c^\infty(\mathbb{R})$.

If $0 < \varepsilon < 1$, we can write $h_\varepsilon = h_\varepsilon^{(1)} + h_\varepsilon^{(2)}$ where

$$h_\varepsilon^{(1)} = \xi_{[-1,1]} h_\varepsilon.$$

Since the family $(h_\varepsilon^{(2)})_{0<\varepsilon<1}$ has as limit in L^p, as $\varepsilon \to 0$, the function

$$h^{(2)}(x) = \begin{cases} \dfrac{1}{\pi x} & \text{if} \quad |x| \geqslant 1 \\ 0 & \text{elsewhere,} \end{cases}$$

and since $g \in L^1$, it is clear that the limit

$$\lim_{\varepsilon \to 0} h_\varepsilon^{(2)} * g$$

exists in L^p. As to $h_\varepsilon^{(1)} * g$, notice that since $h_\varepsilon^{(1)}$ is an odd function,

$$\begin{aligned} h_\varepsilon^{(1)} * g(x) &= \int [g(x-y) - g(x)] h_\varepsilon^{(1)}(y)\, dy \\ &= \frac{1}{\pi} \int_{\varepsilon \leqslant |y| \leqslant 1} [g(x-y) - g(x)]/y\, dy. \end{aligned} \tag{13}$$

By the mean value theorem however,

$$|[g(x-y) - g(x)]/y| \leqslant \|g'\|_\infty$$

for all x. So it follows from (13) that the family $(h_\varepsilon^{(1)} * g)_{\varepsilon>0}$ converges uniformly on $\mathbb{R}$. Yet if g is supported in the interval $[-\lambda, \lambda]$, the functions $h_\varepsilon^{(1)} * g$ are all supported in the set $[-\lambda - 1, \lambda + 1]$. Therefore uniform convergence implies convergence in L^p of the family $(h_\varepsilon^{(1)} * g)_{0<\varepsilon<1}$ as $\varepsilon \to 0$. In summary then, since both of the families $(h_\varepsilon^{(1)} * g)$ and $(h_\varepsilon^{(2)} * g)$ converge in L^p as $\varepsilon \to 0$, so too does the family $(h_\varepsilon * g)$. In other words, the Hilbert transform Hg of g exists as an element of $L^p(\mathbb{R})$. (Actually, as we have just shown, Hg is also a continuous function when $g \in C_c^\infty(\mathbb{R})$.)

We come next to the proof of (12). All the extra information necessary to the proof of (12) is already contained in the proofs of Theorems 6.2.2 and 6.2.3. For it was shown there that

$$\sup_\varepsilon \|\hat{h}_\varepsilon\|_\infty < \infty, \tag{14}$$

$$\sup_\varepsilon \|J(h_\varepsilon)\| < \infty \tag{15}$$

and that

$$\hat{h}_\varepsilon(y) \to -i \operatorname{sgn} y \tag{16}$$

as $\varepsilon \to 0$, h_ε being the same as the function $k_{1/\varepsilon}$ defined in 6.2(8). By (14), (15) and Corollary 2.4.5,

$$A_p = \sup_\varepsilon \|L_{h_\varepsilon}\|_{p,p} < \infty, \tag{17}$$

whence

$$\|h_\varepsilon * g\|_p \leqslant A_p \|g\|_p \tag{18}$$

for all $\varepsilon > 0$ and all g in C_c^∞. Now (12) follows from (18) if we pass to the limit as $\varepsilon \to 0$.

Observe that it follows from (14), (16) and Plancherel's theorem that

$$h_\varepsilon * g \to T_c g$$

in L^2 when $g \in C_c^\infty$. Therefore

$$Hg = T_c g$$

on the dense subspace C_c^∞ of L^p; and if H is well defined by (4) and is a continuous linear mapping of L^p into itself, it must be the unique continuous extension of the conjugate function operator T_c.

To finish the proof of (i), suppose that f is an arbitrary element of L^p, and that $\tau > 0$. Choose g in C_c^∞ so that

$$\|g - f\|_p < \tau.$$

Then if $0 < \varepsilon < \varepsilon' < 1$, we deduce from (17) that

$$\begin{aligned}\|h_\varepsilon * f - h_{\varepsilon'} * f\|_p &\leqslant \|h_\varepsilon * g - h_{\varepsilon'} * g\|_p + \|h_\varepsilon * (f - g)\|_p + \|h_{\varepsilon'} * (f - g)\|_p \\ &\leqslant \|h_\varepsilon * g - h_{\varepsilon'} * g\|_p + 2A_p\tau. \end{aligned}\tag{19}$$

Since $(h_\varepsilon * g)_{\varepsilon>0}$ converges in L^p as $\varepsilon \to 0$, there is a number $\delta > 0$ such that

$$\|h_\varepsilon * g - h_{\varepsilon'} * g\|_p < \tau$$

for $0 < \varepsilon < \varepsilon' < \delta$. So we conclude from (19) that when $0 < \varepsilon < \varepsilon' < \delta$,

$$\|h_\varepsilon * f - h_{\varepsilon'} * f\|_p \leqslant (2A_p + 1)\tau;$$

therefore the family $(h_\varepsilon * f)_{\varepsilon>0}$ is Cauchy, so converges, as $\varepsilon \to 0$. Since, by definition,

$$Hf = \lim h_\varepsilon * f$$

and (see (17))

$$\|h_\varepsilon * f\|_p \leqslant A_p\|f\|_p,$$

we conclude that

$$\|Hf\|_p \leqslant A_p\|f\|_p$$

and so complete the proof.

(ii) *The case $G = \mathbb{T}$.* This is quite similar to, though much simpler than, that just given for $\mathbb{R}$. We leave the details to the reader. □

6.7.5. Remarks. (i) It is possible to prove that on each of the groups $\mathbb{R}$ and $\mathbb{T}$, the limit of the family $(h_\varepsilon * f)_{\varepsilon>0}$ exists pointwise a.e. as $\varepsilon \to 0$ when $f \in L^p$ (even if $p = 1$). The argument for $\mathbb{R}$ is given in [38], Chapter 2, and that for $\mathbb{T}$ is similar.

(ii) It is possible to strengthen Theorem 6.7.4(i) by proving that for f in $L^p(\mathbb{R})$

$$\lim_{\substack{\varepsilon\to 0\\ v\to\infty}} h_{\varepsilon,v} * f = Hf,$$

where, for $0 < \varepsilon < v$,

$$h_{\varepsilon,v}(x) = \begin{cases} \dfrac{1}{\pi x} & \text{if} \quad \varepsilon \leqslant |x| \leqslant v \\ 0 & \text{otherwise.} \end{cases}$$

The proof is very little different from the one already given for 6.7.4(i) itself, except that it is notationally much more cumbersome.

6.7.6. Bilinear versions of the main theorems. It will be clear to the reader who has analysed the proofs of the M. Riesz theorem and Theorem 6.4.7 on the existence and continuity of the Hilbert transform, that the key component of each proof is the statement that the convolution operators L_{h_ε}, $0 < \varepsilon < 1$, have uniformly bounded (p, p) norms. This property of the h_ε can be put into bilinear form. We do this just for the case of $\mathbb{R}$. It is not unreasonable to call the statement a version of the M. Riesz theorem.

Theorem (M. Riesz: bilinear version). *Suppose that* $1 < p < \infty$ *and that* $(h_\varepsilon)_{0<\varepsilon<1}$ *is the family of kernels on* $\mathbb{R}$ *defined in* 6.7.3(i). *Then there is a constant* A_p *such that*

$$\sup_{0<\varepsilon<1} \left| \iint_{\varepsilon \leqslant |x-y| \leqslant 1/\varepsilon} \frac{f(x)g(y)}{x-y}\, dxdy \right| \leqslant A_p \|f\|_p \|g\|_{p'} \tag{20}$$

for every f *in* $L^p(\mathbb{R})$ *and every* g *in* $L^{p'}(\mathbb{R})$.

Proof. By Fubini's theorem

$$\begin{aligned} \iint_{\varepsilon \leqslant |x-y| \leqslant 1/\varepsilon} \frac{f(x)g(y)}{x-y}\, dxdy &= \iint h_\varepsilon(x-y) f(x) g(y)\, dxdy \\ &= -\int h_\varepsilon * f(y) g(y)\, dy \end{aligned} \tag{21}$$

because h_ε is odd. So (20) follows from (21) by applying Hölder's inequality and taking note of (17). □

In a similar way, the Stečkin theorem can be put into a bilinear form. For instance, on $G = \mathbb{R}$, a bilinear version of the theorem is as follows.

Theorem (Stečkin: bilinear form). *Let* ϕ *be an integrable function of bounded variation on* $\mathbb{R}$. *If* $1 < p < \infty$, $f \in L^1 \cap L^p(\mathbb{R})$, *and* $g \in L^1 \cap L^{p'}(\mathbb{R})$, *then*

$$\left| \iint \hat{\phi}(x-y) f(x) g(y)\, dxdy \right| \leqslant C_p' \max(|\phi(0)|, \operatorname{Var} \phi) \|f\|_p \|g\|_{p'} \tag{22}$$

where C_p' *is independent of* ϕ, f *and* g.

Proof. Let $(\mathrm{V}_N)_1^\infty$ be the sequence of de la Vallée-Poussin kernels. Then $f_N = \mathrm{V}_N * f \to f$ in both L^1 and L^p, $g_N = \mathrm{V}_N * g \to g$ in both L^1 and $L^{p'}$, while both f_N and g_N have Fourier transforms in $C_c(\mathbb{R})$. Since $\phi \in L^1$, $\hat{\phi}$ is bounded. Therefore we can apply Fubini's theorem to prove that

$$\iint \hat{\phi}(x-y) f_N(x) g_N(y)\, dxdy \to \iint \hat{\phi}(x-y) f(x) g(y)\, dxdy$$

as $N \to \infty$. Since moreover $\|f_N\|_p \to \|f\|_p$ and $\|g_N\|_{p'} \to \|g\|_{p'}$, we have shown that it will suffice to prove (22) for functions f and g whose Fourier transforms have compact supports. In this case, however,

$$\iint \hat{\phi}(x-y) f(x) g(y)\, dxdy = \int \hat{\phi}_r * f(y) g(y)\, dy \tag{23}$$

by Fubini's theorem, where $\hat{\phi}_r(t) = \hat{\phi}(-t)$. An application of Parseval's formula and the inversion theorem to (23) now shows that

$$\left|\iint \hat{\phi}(x - y)f(x)g(y)\, dxdy\right| = \left|\int \phi(s)\hat{f}(s)\hat{g}(-s)\, ds\right|. \tag{24}$$

Theorem 6.2.5 tells us that $\phi\hat{f}$ is the Fourier transform of a function h in $L^2 \cap L^p(\mathbb{R})$ such that

$$\|h\|_p \leqslant C_p \max(|\phi(0)|, \operatorname{Var}\phi)\|f\|_p.$$

Going on from (24), using the Parseval formula and Hölder's inequality, we conclude that

$$\begin{aligned}\left|\iint \hat{\phi}(x - y)f(x)g(y)\, dxdy\right| &= 2\pi\left|\int h(x)g(x)\, dx\right| \\ &\leqslant 2\pi C_p \max(|\phi(0)|, \operatorname{Var}\phi)\|f\|_p\|g\|_{p'}.\end{aligned}$$

This completes the proof. □

Remark. It is a straightforward matter to formulate and prove similar bilinear versions of the M. Riesz theorem and of the Stečkin theorem on each of the groups $\mathbb{T}$ and $\mathbb{Z}$.

6.8. A Characterisation of the Hilbert Transform

The following characterisation of the multiplier operators T_ϕ is well known and easy to establish. (See [3].)

6.8.1. Theorem. *The continuous linear operator T on $L^2(G)$ is of the form T_ϕ for some function ϕ in $L^\infty(X)$ if and only if it commutes with translations; i.e. if and only if*

$$\tau_a(Tf) = T(\tau_a f)$$

for all f in $L^2(G)$ and all a in G, where

$$\tau_a f(x) = f(x - a).$$

It follows from Theorem 6.7.4 that the Hilbert transform is a multiplier operator. Our aim in this section is to distinguish the Hilbert transform among all multiplier operators. The characterisation will be expressed in terms of the *dilation operators.*

6.8.2. Definition. (i) If $a \in \mathbb{R}\backslash\{0\}$, and f is a function on $\mathbb{R}$, the *dilation* $D_a f$

of f by amount a is the function

$$D_a f = f \circ \sigma_a$$

where σ_a is the continuous homomorphism of $\mathbb{R}$ such that

$$\sigma_a(x) = ax.$$

(ii) If $a \in \mathbb{Z}\backslash\{0\}$, and f is a function on $\mathbb{T}$, the *dilation $D_a f$* is defined to be the function

$$D_a f = f \circ \sigma_a$$

where

$$\sigma_a(x) = x^a$$

for all x in $\mathbb{T}$.

(iii) On $\mathbb{Z}$, the *dilation $D_a f$ of a function f by amount a*, where $a \in \mathbb{Z}\backslash\{0\}$, is the function

$$D_a f = f \circ \sigma_a$$

where

$$\sigma_a(x) = ax.$$

The characterisation of the Hilbert transform on each of $\mathbb{R}$ and $\mathbb{T}$ is straightforward, and we present it immediately below. The corresponding result on the group $\mathbb{Z}$ is less straightforward, and we give it separate treatment.

6.8.3. Theorem. (i) *The operator H on $L^2(\mathbb{R})$ is, up to a numerical factor, the only multiplier operator T_ϕ on $L^2(\mathbb{R})$ with the property that*

$$T_\phi D_a = \operatorname{sgn} a \, D_a T_\phi \tag{1}$$

for all a in $\mathbb{R}\backslash\{0\}$.

(ii) *The operator H on $L^2(\mathbb{T})$ is, apart from a numerical factor, the only multiplier operator T_ϕ on $L^2(\mathbb{T})$ having the property that*

$$T_\phi D_a = \operatorname{sgn} a \, D_a T_\phi$$

for all a in $\mathbb{Z}\backslash\{0\}$.

Proof. If $f \in L^2(\mathbb{R})$ and $a \in \mathbb{R}\backslash\{0\}$,

$$(D_a f)^\wedge(y) = |a|^{-1}\hat{f}(y/a).$$

So the relation (1) is equivalent to saying that

$$\phi(y)|a|^{-1}\hat{f}(y/a) = \operatorname{sgn} a\,|a|^{-1}\phi(y/a)\hat{f}(y/a)$$

for all f in $L^2(\mathbb{R})$ and all a in $\mathbb{R}\backslash\{0\}$; i.e. to the requirement that

$$\phi(y) = \operatorname{sgn} a\,\phi(y/a) \tag{2}$$

for almost every y and each a in $\mathbb{R}\backslash\{0\}$. Define

$$\Phi(t) = \int_0^t \phi(y)\,dy \tag{3}$$

for $t > 0$. Then (2) shows that for $a > 0$,

$$\begin{aligned}\Phi(t) &= \operatorname{sgn} a \int_0^t \phi(y/a)\,dy = \int_0^t \phi(y/a)\,dy \\ &= a\int_0^{t/a} \phi(s)\,ds = a\Phi(t/a)\end{aligned}$$

for every $t > 0$. Taking $t = a > 0$, we infer that

$$\Phi(t) = t\Phi(1). \tag{4}$$

So by differentiation of Φ we conclude, from Lebesgue's theorem, (3) and (4) that

$$\phi(y) = \Phi(1)$$

for almost all $y > 0$. But then (2) implies, if we take $a = -1$, that

$$\phi(y) = \Phi(1)\operatorname{sgn} y$$

for almost all y. Hence ϕ is equal almost everywhere to a constant multiple of c, where

$$c(y) = -i\operatorname{sgn} y.$$

So T_ϕ is a constant multiple of T_c which, as we saw in Theorem 6.7.4, is the same as H.

Conversely, straightforward calculations show that $H = T_c$ satisfies (1).

(ii) If $f \in L^2(\mathbb{T})$ and $a \in \mathbb{Z}\backslash\{0\}$, then

$$(D_a f)^\wedge(n) = \hat{f}(n/a)\xi_{a\mathbb{Z}}(n)$$

for all n in $\mathbb{Z}$, and so the condition (1) is equivalent to the requirement on ϕ that

$$\phi(n)\hat{f}(n/a)\xi_{a\mathbb{Z}}(n) = \operatorname{sgn} a\,\phi(n/a)\hat{f}(n/a)\xi_{a\mathbb{Z}}(n) \tag{5}$$

for all a in $\mathbb{Z}\backslash\{0\}$, and all n in $\mathbb{Z}$. Taking $n = 0$ and $a = -1$ in (5), we see that $\phi(0) = -\phi(0)$ and hence $\phi(0) = 0$; taking then $n = a$ to be an element of $\mathbb{Z}\backslash\{0\}$, we conclude that $\phi(n) = \phi(1)$ sgn n. Thus T_ϕ is a constant multiple of T_c. Conversely, (1) is easily seen to be satisfied when $H = T_c = T_\phi$. □

We pass on now to the remaining case of $G = \mathbb{Z}$ and show for a start that there is no nontrivial multiplier operator on $L^2(\mathbb{Z})$ which satisfies (1). In fact we prove a little more, as follows.

6.8.4. Lemma. *Let T_ϕ be a multiplier operator on $\ell^2(\mathbb{Z})$ such that*

$$T_\phi D_a = \delta(a) D_a T_\phi \tag{6}$$

for all a in $\mathbb{Z}\backslash\{0\}$, where δ is some complex-valued function on $\mathbb{Z}\backslash\{0\}$. Then T_ϕ is a scalar operator.

Proof. Let e be the function on $\mathbb{Z}$ such that

$$\mathsf{e}(n) = \begin{cases} 1 & \text{if} \quad n = 0 \\ 0 & \text{elsewhere.} \end{cases}$$

Let $k = T_\phi \mathsf{e}$. Then if f is a finitely supported function on $\mathbb{Z}$,

$$T_\phi f = k * f. \tag{7}$$

For

$$\begin{aligned} (T_\phi f)^\wedge &= \hat{\phi}\hat{f} = (\hat{\phi}.1)\hat{f} = (\hat{\phi}\hat{\mathsf{e}})\hat{f} \\ &= (k * f)^\wedge. \end{aligned}$$

If T_ϕ satisfies (6), then (7) shows that

$$k * D_a f(n) = \delta(a) k * f(an) \tag{8}$$

for every finitely supported f on $\mathbb{Z}$, all n in $\mathbb{Z}$, and all a in $\mathbb{Z}\backslash\{0\}$. If $\delta(a) = 0$ for some a, we could take $f = \mathsf{e}$ in (8) and conclude that $k = 0$ and so $T_\phi = 0$. If we neglect this case, we can assume δ is nonvanishing. By setting $n = 0$ in (8), we see that

$$\sum_m k(m) f(-am) = \delta(a) \sum k(m) f(-m) \tag{9}$$

for all finitely supported f. Suppose that $a \neq \pm 1$, and that f is an arbitrary finitely supported function vanishing on $a\mathbb{Z}$. Then (9) shows that

$$\delta(a) \sum_{m \notin a\mathbb{Z}} k(m) f(-m) = 0.$$

Since $\delta(a) \neq 0$, we conclude that

$$\sum_{m \notin a\mathbb{Z}} k(m) f(-m) = 0$$

for all f as specified; hence $k = 0$ off $a\mathbb{Z}$. This is the case for every a different from ± 1. So $k = 0$ except possibly at 0, and T_ϕ is a constant multiple of the identity mapping. □

Here now is a characterisation of the Hilbert transform on $\mathbb{Z}$.

6.8.5. Theorem. *The Hilbert transform on $\mathbb{Z}$ is, to within a multiplicative constant, the only multiplier operator T_ϕ on $\ell^2(\mathbb{Z})$ with the property that for every a in $\mathbb{Z}\backslash\{0\}$,*

$$T_\phi D_a f = a D_a T_\phi f \tag{10}$$

for all functions f in $\ell^2(\mathbb{Z})$ which are supported in $a\mathbb{Z}$.

Proof. The Hilbert transform H is a multiplier operator, and

$$Hf = h * f$$

where h is the Hilbert kernel. In other words,

$$Hf(n) = \sum_{m \neq 0} f(n - m) \frac{1}{m}. \tag{11}$$

If $a \in \mathbb{Z}\backslash\{0\}$, $f \in \ell^2(\mathbb{Z})$, and f is supported in $a\mathbb{Z}$, (11) shows that

$$\begin{aligned} HD_a f(n) &= \sum_{m \neq 0} \frac{1}{m} D_a f(n - m) \\ &= \sum_{m \neq 0} \frac{1}{m} f(an - am) \\ &= a \sum_{m' \in a\mathbb{Z}\backslash\{0\}} \frac{1}{m'} f(an - m') \\ &= a \sum_{m' \neq 0} \frac{1}{m'} f(an - m') \\ &= ah * f(an) = a D_a Hf(n), \end{aligned}$$

the second from last step valid because $f(an - m') \neq 0$ only when $m' \in a\mathbb{Z}$ if f is supported in $a\mathbb{Z}$. So (10) is verified for $T_\phi = H$.

Conversely, if T_ϕ is a multiplier operator on $\ell^2(\mathbb{Z})$, and

$$\mathsf{e}(n) = \begin{cases} 1 & \text{if } n = 0 \\ 0 & \text{otherwise,} \end{cases}$$

then

$$T_\phi f = k * f \tag{12}$$

at least for all finitely supported f (actually for all f in $\ell^2(\mathbb{Z})$) where $k = T_\phi \mathrm{e}$. For

$$\begin{aligned}(T_\phi f)^\wedge &= \phi \hat{f} = (\phi.1)\hat{f} \\ &= \hat{k}\hat{f}\end{aligned}$$

for such f. If T_ϕ has the property (10), then in particular,

$$T_\phi D_a \mathrm{e} = a D_a T_\phi \mathrm{e} \tag{13}$$

for all a in $\mathbb{Z}\backslash\{0\}$; that is, by (12), and the fact that $D_a \mathrm{e} = \mathrm{e}$,

$$k = a D_a k$$

for all a in $\mathbb{Z}\backslash\{0\}$. So

$$k(n) = a k(an)$$

for all integers n and all nonzero integers a. Hence $k(0) = 0$ and $k(n) = k(1)/n$ if $n \neq 0$. The operator T_ϕ is therefore a constant multiple of the Hilbert transform. □

6.8.6. Corollary. *The Hilbert transform on $\mathbb{Z}$ is the only multiplier operator T_ϕ on $\ell^2(\mathbb{Z})$ such that*

$$T_\phi D_a \mathrm{e} = a D_a T_\phi \mathrm{e}$$

for all a in $\mathbb{Z}\backslash\{0\}$.

Proof. This follows from an inspection of the proof of Theorem 6.8.5. □

Chapter 7. The Littlewood-Paley Theorem for $\mathbb{R}$, $\mathbb{T}$ and $\mathbb{Z}$: Dyadic Intervals

7.1. Introduction

7.1.1. In this chapter we establish the Littlewood-Paley theorem for $\mathbb{R}$, $\mathbb{T}$ and $\mathbb{Z}$ in the case of dyadic intervals and the corresponding dyadic partial sum operators. We present two approaches, one of which is, formally speaking, vectorial in nature, the other being partly scalar, partly vectorial. We also discuss the case of finite products of $\mathbb{R}$, $\mathbb{T}$ and $\mathbb{Z}$.

7.1.2. The dyadic decompositions. (a) The *dyadic decomposition* $(\Delta_j)_{j\in\mathbb{Z}}$ of $\mathbb{R}$ is defined by the formulas

$$\Delta_j = \begin{cases} [2^{j-1}, 2^j) & \text{if } j > 0 \\ (-1, 1) & \text{if } j = 0 \\ (-2^{|j|}, -2^{|j|-1}] & \text{if } j < 0. \end{cases} \tag{1}$$

The corresponding partial sum operators $S_j = S_{\Delta_j}$ are defined as follows. For f in $L^2(\mathbb{R})$, $S_j f$ is the class of the continuous function

$$x \to (2\pi)^{-1} \int_{\Delta_j} \hat{f}(y) \exp(ixy)\, dy. \tag{2}$$

Notice that when $f \in L^2$, the formula (2) makes sense since the function $\hat{f}\xi_{\Delta_j}$ is integrable.

(b) When $G = \mathbb{T}$, the *dyadic decomposition* $(\Delta_j)_{j\in\mathbb{Z}}$ of the dual group $\mathbb{Z}$ is given by the formulas

$$\Delta_j = \begin{cases} \{n \in \mathbb{Z} : 2^{j-1} \leqslant n < 2^j\} & \text{if } j > 0 \\ \{0\} & \text{if } j = 0 \\ \{n \in \mathbb{Z} : -2^{|j|} < n \leqslant -2^{|j|-1}\} & \text{if } j < 0. \end{cases} \tag{3}$$

For f in $L^2(\mathbb{T})$, $S_j f = S_{\Delta_j} f$ is the class of the partial sum function

$$\sum_{n\in\Delta_j} \hat{f}(n)\chi_n, \tag{4}$$

χ_n denoting as usual the function $e^{it} \to e^{int}$.

(c) In the case of $G = \mathbb{Z}$, we first of all decompose the interval $[-\pi, \pi)$ "dyadically" into the family $(I_j)_{j\in\mathbb{Z}}$ of sets as follows

$$I_j = \begin{cases} \left[-\pi, -\frac{\pi}{2}\right] \cup \left[\frac{\pi}{2}, \pi\right) & \text{when } j = 0 \\ \left[\frac{\pi}{2^{j+1}}, \frac{\pi}{2^j}\right) & \text{when } j > 0 \\ \left(-\frac{\pi}{2^{|j|}}, -\frac{\pi}{2^{|j|+1}}\right] & \text{when } j < 0. \end{cases} \tag{5}$$

Now denote by Δ_j the arc or interval in $\mathbb{T}$

$$\Delta_j = \{e^{it} : t \in I_j\}. \tag{6}$$

For f in $\ell^2(\mathbb{Z})$, $S_j f = S_{\Delta_j} f$ is the function in $\ell^2(\mathbb{Z})$ for which

$$(S_j f)(n) = \frac{1}{2\pi} \int_{I_j} \hat{f}(e^{it}) e^{int}\, dt. \tag{7}$$

We aim to prove that in all three cases, the dyadic decomposition $(\Delta_j)_{j\in\mathbb{Z}}$ has the LP property 1.2.4.

7.1.3. Preamble to the proofs. To prove the Littlewood-Paley theorem, it would suffice to prove that the decomposition in question has the WM property, viz. that if ϕ is a bounded function on X which is constant on each Δ_j then ϕ is an L^p multiplier for every p in $(1, \infty)$ and $\|\phi\|_{p,p}$ is majorised by a constant multiple of $\|\phi\|_\infty$. This is the same as showing that the operator T_ϕ on $L^2(G)$ determined by ϕ as in 1.2.2 is of type (p, p) on $L^2 \cap L^p(G)$ and has corresponding norm at most a constant multiple of $\|\phi\|_\infty$. Since it is evident that T_ϕ is of type (2, 2) and $\|T_\phi\|_{2,2} \leqslant \|\phi\|_\infty$, this would follow from the Marcinkiewicz interpolation theorem if T_ϕ could be shown to be of weak type (1, 1) and to have corresponding norm majorised in the desired fashion. This is the approach adopted with such simplicity and success in Chapter 4.

The rather surprising thing is that, although it can be shown that ϕ is indeed an L^p multiplier for p in $(1, \infty)$ when (Δ_j) is any one of the dyadic decompositions in 7.1.2, and that $\|\phi\|_{p,p}$ is majorised by a constant multiple of $\|\phi\|_\infty$, T_ϕ is generally NOT of weak type (1, 1): this feature, pointed out to us by John Fournier, will be discussed in Section 7.5. If in fact one tries to prove that T_ϕ is of weak type (1, 1) by using the methods of Chapter 2, one rapidly reaches the conclusion that the kernel of T_ϕ is too singular for the conditions (20) and (21) of Corollary 2.4.5 to be established for any reasonable sequence (k_ν). Fournier's example is therefore reassuring.

To sum up, there is no alternative but to adopt a different line of attack from that involving weak (1, 1) estimates on the T_ϕ.

We present two proofs (for each group) both of which deal with a modification of ϕ, and hence of T_ϕ. The LP result is in both instances finally captured by applying the R property of the family (Δ_j).

7.2. The Littlewood-Paley Theorem: First Approach

7.2.1. Theorem. *Let X be any one of the groups $\mathbb{R}$, $\mathbb{Z}$ and $\mathbb{T}$ and (Δ_j) the dyadic decomposition of X defined in* 7.1.2. *Then (Δ_j) has the* LP *property.*

Proof. We give a detailed proof for the case $X = \mathbb{Z}$ and indicate the modifications necessary to make the proof work in the other two settings.

Proof for $G = \mathbb{T}$, $X = \mathbb{Z}$. It suffices to prove the existence of a constant C_p such that

$$\left\| \left(\sum_{j=1}^{\infty} |S_j f|^2 \right)^{1/2} \right\|_p \leqslant C_p \|f\|_p \tag{1}$$

for all trigonometric polynomials f. (The index p is of course a fixed number in the range $(1, \infty)$.) For if (1) holds and we replace f by f_r: $e^{it} \to f(e^{-it})$ we deduce that

$$\left\| \left(|\hat{f}(0)|^2 + \sum_{-\infty}^{-1} |S_j f|^2 \right)^{1/2} \right\|_p \leqslant (C_p + 1)\|f\|_p$$

and hence

$$\left\| \left(\sum_{j\in\mathbb{Z}} |S_j f|^2 \right)^{1/2} \right\|_p \leqslant (2C_p + 1)\|f\|_p$$

Suppose now that, in order to attempt to prove (1), we start with a bounded function ϕ, constant on each Δ_j, and 0 off $\Delta_1 \cup \Delta_2 \cup \ldots$. We have noted in 7.1.3 the futility of trying to prove a weak (1, 1) inequality for the operator T_ϕ. Note that the kernel of T_ϕ is formally

$$\sum_{j=1}^{\infty} \phi(2^{j-1}) K_j$$

where

$$K_j = \sum_{n\in\Delta_j} \chi_n.$$

These kernels K_j are badly behaved and we shall seek to replace them by sufficiently well behaved new functions M_j defined as follows: $M_1 = K_1$; and for $j > 1$, M_j is such that $\hat{M}_j$ is affine on each of the intervals $[2^j, 2^j + 2^{j-2}]$ and $[2^{j-1} - 2^{j-2}, 2^{j-1}]$ and elsewhere satisfies the relations

$$\hat{M}_j(n) = \begin{cases} 1 & \text{if} \quad 2^{j-1} \leqslant n \leqslant 2^j \\ 0 & \text{if} \quad n \leqslant 2^{j-2} \quad \text{or} \quad n \geqslant 2^j + 2^{j-2}. \end{cases} \tag{2}$$

The idea behind the replacement is that, whereas K_j behaves like $\chi_{3.2^{j-2}}$ times the Dirichlet kernel of order 2^{j-2}, M_j is, for $j > 1$, the same character times the de la Vallée-Poussin kernel

$$\mathsf{V}_{2^{j-2}} = 2\mathsf{F}_{2^{j-1}-1} - \mathsf{F}_{2^{j-2}-1},$$

where F_k denotes the Fejér kernel of order k. The sequence

$$\left(\sum_{j=1}^{\nu} \phi(2^{j-1})M_j\right)_{\nu=1}^{\infty}$$

may therefore be expected to be less singular than the sequence

$$\left(\sum_{j=1}^{\nu} \phi(2^{j-1})K_j\right)_{\nu=1}^{\infty}.$$

This expectation proves to be justified, as is made more precise by statement (M) below.

If $f \in L^2(\mathbb{T})$, the series

$$\sum_{j=1}^{\infty} \phi(2^{j-1})M_j * f$$

converges in L^2, as follows from the Parseval formula and the facts that

$$\lim_{\nu \to \infty} \sum_{j=1}^{\nu} \phi(2^{j-1})\hat{M}_j(n) = \psi(n) \tag{3}$$

exists for each n in $\mathbb{Z}$, and

$$\begin{aligned} \sup_{n \in \mathbb{Z}} \left| \sum_{j=1}^{\nu} \phi(2^{j-1})\hat{M}_j(n) \right| &\leqslant \|\phi\|_\infty \sup_{n \in \mathbb{Z}} \sum_{j=1}^{\infty} \hat{M}_j(n) \\ &\leqslant 3\|\phi\|_\infty \end{aligned} \tag{4}$$

since (2) shows that for any n in $\mathbb{Z}$, $\hat{M}_j(n)$ is nonzero for at most three indices j. The operator T_ψ therefore has the explicit description on $L^2(\mathbb{T})$ in the formula

$$T_\psi f = \sum_1^\infty \phi(2^{j-1})M_j * f,$$

which is analogous to the formula

$$T_\phi f = \sum_1^\infty \phi(2^{j-1})K_j * f;$$

the convergence of the series defining $T_\psi f$ follows from (3) and (4). Recall that we are assuming that ϕ vanishes off $\bigcup_{j>0} \Delta_j$ and is constant on each Δ_j. We now claim that

(M) *T_ψ is of weak type* (1, 1) *on* $L^1 \cap L^2$; *and*

$$\|T_\psi f\|_p \leqslant C'_p \|\phi\|_\infty \|f\|_p \tag{5}$$

for p in $(1, \infty)$ *and f in* $L^2 \cap L^p$, *where C'_p is independent of ϕ and f.*

Once (M) is established, we may proceed as follows. If $t \in [0, 1]$, the sequence $(r_j(t))_{j=0}^\infty$ of Rademacher functions takes only the values ± 1 and so (5) shows that for p in $(1, \infty)$ and f a trigonometric polynomial

$$\int_{\mathbb{T}} \left| \sum_{j=0}^{\infty} r_j(t) M_j * f(x) \right|^p dx \leqslant C_p'^p \|f\|_p^p. \tag{6}$$

The integrand in (6) is plainly measurable on $[0, 1] \times \mathbb{T}$ and so the Fubini-Tonelli theorem permits us to integrate (6) with respect to t and then to rewrite the result in the form

$$\int_{\mathbb{T}} \left\{ \int_{[0,1]} \left| \sum_{j=0}^{\infty} r_j(t) M_j * f(x) \right|^p dt \right\} dx \leqslant C_p'^p \|f\|_p^p. \tag{7}$$

Since the Rademacher functions form a $\Lambda(q)$ set for every q ([9], 14.2.1), (7) implies that

$$\int_{\mathbb{T}} \left\{ \int_{[0,1]} \left| \sum_{j=0}^{\infty} r_j(t) M_j * f(x) \right|^2 dt \right\}^{1/2} dx \leqslant C_p''^p \|f\|_p^p;$$

that is, by orthonormality of the r_j, that

$$\int_{\mathbb{T}} \left(\sum_{j=0}^{\infty} \left| M_j * f(x) \right|^2 \right)^{p/2} dx \leqslant C_p''^p \|f\|_p^p. \tag{8}$$

At this point we use Theorem 6.5.2 (the R property) to deduce from (8) that

$$\int_{\mathbb{T}} \left(\sum_{j=0}^{\infty} \left| S_j(M_j * f) \right|^2 \right)^{p/2} dx \leqslant C_p^p \|f\|_p^p.$$

But $S_j(M_j * f) = S_j f$, as follows from (2). Thus we obtain the inequality

$$\left\{ \int_{\mathbb{T}} \left(\sum_{j=1}^{\infty} \left| S_j f(x) \right|^2 \right)^{p/2} dx \right\}^{1/p} \leqslant C_p \|f\|_p$$

valid for all trigonometric polynomials f. This is (1).

We have now to go back and prove statement (M).

Proof of (M). Write, for $\nu \geqslant 1$,

$$k_\nu = \sum_{j=1}^{\nu} \phi(2^{j-1}) M_j. \tag{9}$$

The group $\mathbb{T}$ has the covering family $(U_\alpha)_{\alpha\in\mathbb{Z}}$ described in 2.1.3(ii). Reference to (3), (4) and Corollary 2.4.5 shows that (M) will follow at once if we prove that the numbers $J(k_\nu)$, computed relative to $(U_\alpha)_{\alpha\in\mathbb{Z}}$, are majorised by $C\|\phi\|_\infty$, where C is independent of ϕ and ν. To achieve this, it suffices to show that there exists a number M, independent of ϕ, ν and τ, such that

$$I = \int_{\pi \geqslant |t| \geqslant 2\tau} \left| \sum_{j=1}^{\nu} \phi(2^{j-1})\{M_j(e^{i(t-s)}) - M_j(e^{it})\} \right| dt \leqslant M\|\phi\|_\infty \tag{10}$$

for every ν, every τ in $(0, \pi/2]$ and every s such that $|s| \leqslant \tau$. To this end we note that

$$\begin{aligned} I &\leqslant \sum_{j=1}^{\nu} \|\phi\|_\infty \int_{\pi \geqslant |t| \geqslant 2\tau} |M_j(e^{i(t-s)}) - M_j(e^{it})|\, dt \\ &= \|\phi\|_\infty \sum_{j=1}^{\nu} I_j, \end{aligned} \tag{11}$$

say. We proceed to majorise each I_j.

As a start,

$$\begin{aligned} I_j &\leqslant 2 \int_{\pi \geqslant |t| \geqslant \tau} |M_j(e^{it})|\, dt \\ &= 2 \int_{\pi \geqslant |t| \geqslant \tau} |\mathsf{V}_{2^{j-2}}(e^{it})|\, dt, \end{aligned}$$

where the de la Vallée-Poussin kernel $\mathsf{V}_{2^{j-2}}$ is such that

$$\mathsf{V}_{2^{j-2}}(e^{it}) = \frac{2}{2^{j-1}} \frac{\sin^2 2^{j-1}t/2}{\sin^2 t/2} - \frac{1}{2^{j-2}} \frac{\sin^2 2^{j-2}t/2}{\sin^2 t/2},$$

and hence

$$\begin{aligned} \int_{\pi \geqslant |t| \geqslant \tau} |\mathsf{V}_{2^{j-2}}(e^{it})|\, dt &\leqslant M' 2^{-j} \int_{|t| \geqslant \tau} t^{-2}\, dt \\ &\leqslant M'' 2^{-j} \tau^{-1}, \end{aligned}$$

M' and M'' being absolute constants. Thus

$$I_j \leqslant 2M'' 2^{-j} \tau^{-1}. \tag{12}$$

The estimate (12) leads to a poor majorisation for I when τ is small. To cope with the problem of small τ, we make use of Bernstein's inequality (Theorem D.2.1) and deduce that ($M^{(3)}$, $M^{(4)}$ and $M^{(5)}$ denoting absolute constants)

$$\begin{aligned} I_j &= \int_{\pi \geqslant |t| \geqslant 2\tau} |M_j(e^{i(t-s)}) - M_j(e^{it})| \, dt \\ &\leqslant \int_{|t| \leqslant \pi} |\cdots| \, dt \\ &\leqslant M^{(3)} 2^j \tau \|M_j\|_1 \\ &\leqslant M^{(4)} 2^j \tau \end{aligned} \tag{13}$$

for $|s| \leqslant \tau$, since the spectrum of M_j lies in the interval $[-(2^j + 2^{j-2}), 2^j + 2^{j-2}]$.

From (11), (12) and (13) we conclude that

$$\begin{aligned} I &\leqslant M^{(5)} \|\phi\|_\infty \sum_{j=1}^{\infty} \min(2^j \tau, 2^{-j}\tau^{-1}) \\ &\leqslant M \|\phi\|_\infty, \end{aligned} \tag{14}$$

since

$$\sum_{j=1}^{\infty} \min(2^j \tau, 2^{-j}\tau^{-1}) < \sum_{-\infty}^{m_0} 2^j \tau + \sum_{m_0+1}^{\infty} 2^{-j}\tau^{-1} \leqslant 4 \tag{15}$$

where $m_0 = [-(\ln \tau / \ln 2)]$. This proves (10) and so completes the proof for $G = \mathbb{T}$.

Proof for $G = \mathbb{R}$. A proof for $\mathbb{R}$ can be given by making the obvious changes in the proof for $\mathbb{T}$. The main supplementary observations to be made are the following. (See also the proof for $G = \mathbb{Z}$ below, which effectively amplifies these points.)

(a) In showing that it suffices to prove (1) for suitably restricted functions f, use has to be made of the fact that

$$\|S_0 f\|_p \leqslant A_p \|f\|_p,$$

which is a particular instance of the M. Riesz theorem 6.2.2.

(b) If $\lambda > 0$ and V_λ is the de la Vallée-Poussin kernel of order λ on $\mathbb{R}$, then

$$\mathsf{V}_\lambda(x) = 2\mathsf{F}_{2\lambda} - \mathsf{F}_\lambda$$

where F_λ is the Fejér kernel of order λ; and

$$\mathsf{F}_\lambda(x) = \frac{1}{2\pi} \frac{\sin^2 \lambda x/2}{\lambda (x/2)^2}.$$

Therefore the estimates (12) and (13) are as before.

(c) The appropriate form of Bernstein's inequality to use in proving (13) is Theorem D.1.1.

Proof for $G = \mathbb{Z}$, $X = \mathbb{T}$. Once more, making the obvious changes in the proof for $G = \mathbb{T}$ results in a proof for $G = \mathbb{Z}$. Here is an outline of the main steps, highlighting the new features.

It suffices to prove that for some constant C_p,

$$\left\| \left(\sum_{j=1}^{\infty} |S_j f|^2 \right)^{1/2} \right\|_p \leqslant C_p \|f\|_p \tag{16}$$

for all f in $\ell^2 \cap \ell^p(\mathbb{Z})$; for

$$\|S_0 f\|_p \leqslant A_p \|f\|_p \tag{17}$$

by the M. Riesz theorem (6.3.2); and since

$$S_{-j} f = [S_j(f_r)]_r$$

where $f_r(n) = f(-n)$, (16) implies that

$$\left\| \left(\sum_{-\infty}^{-1} |S_j f|^2 \right)^{1/2} \right\|_p \leqslant C_p \|f\|_p. \tag{18}$$

The LP property then follows by collecting (16), (17) and (18) together.

For each integer $j \geqslant 1$, write M_j for the function on $\mathbb{Z}$ such that $\hat{M}_j(e^{it}) = 1$ when $\pi/2^{j+1} \leqslant t \leqslant \pi/2^j$, 0 when $t \in [-\pi, \pi) \backslash (\pi/2^{j+1} - \pi/2^{j+2}, \pi/2^j + \pi/2^{j+2})$ and is affine in t on each of the intervals $[\pi/2^{j+2}, \pi/2^{j+1}]$ and $[\pi/2^j, \pi/2^j + \pi/2^{j+2}]$. Let ϕ be a bounded function on $\mathbb{T}$ and write

$$k_\nu(n) = \sum_{j=1}^{\nu} \phi(e^{i\pi/2^{j+1}}) M_j(n).$$

Just as in the case of $\mathbb{T}$, the inequality (16) will follow from Corollary 2.4.5 and the R property of the family $(\Delta_j)_1^\infty$ of arcs in $\mathbb{T}$ if it is shown that the numbers $J(k_\nu)$, computed relative to the covering family $(U_\alpha)_{\alpha \in \mathbb{Z}}$ of $\mathbb{Z}$ given in 2.1.3(iv), are bounded by $C\|\phi\|_\infty$ where C is independent of ϕ and ν, that is, if

$$\sum_{|n| \geqslant 2^{\alpha+1}} \left| \sum_{j=1}^{\nu} \phi(e^{i\pi/2^{j+1}})(M_j(n-m) - M_j(n)) \right| \leqslant C\|\phi\|_\infty \tag{19}$$

for $|m| < 2^\alpha$, and all integers $\alpha \geqslant 0$.

For $0 < \lambda < \pi/2$, let V_λ be the de la Vallée-Poussin kernel of order λ on $\mathbb{Z}$, so that

$$\mathsf{V}_\lambda = 2\mathsf{F}_{2\lambda} - \mathsf{F}_\lambda$$

where

$$\mathsf{F}_\lambda(n) = \begin{cases} \lambda/2\pi & \text{if } n = 0 \\ \dfrac{2\sin^2 n\lambda/2}{\pi n^2 \lambda} & \text{if } n \neq 0 \end{cases} \tag{20}$$

is the Fejér kernel of order λ. The function M_j is a certain character times $V_{\pi/2^{j+2}}$ and so if $\alpha \geqslant 0$ and $|m| \leqslant 2^j$,

$$\sum_{|n| \geqslant 2^{\alpha+1}} |M_j(n-m) - M_j(n)| \leqslant 2 \sum_{|n| \geqslant 2^{\alpha}} |V_{\pi/2^{j+1}}(n)| \leqslant C' 2^{j-\alpha} \tag{21}$$

by (20). On the other hand, Bernstein's inequality (Theorem D.3.1) leads us to the estimate

$$\sum_{|n| \geqslant 2^{\alpha+1}} |M_j(n-m) - M_j(n)| \leqslant C'' 2^{\alpha-j} \tag{22}$$

for $|m| < 2^{\alpha}$ since $\hat{M}_j$ has spectrum in the set

$$\left\{e^{it}: |t| \leqslant \frac{\pi}{2^j} + \frac{\pi}{2^{j+2}}\right\}.$$

Finally, combining (21) and (22) leads to the conclusion that if $\alpha \geqslant 0$ and $|m| < 2^{\alpha}$, then

$$\begin{aligned}\sum_{|n| \geqslant 2^{\alpha+1}} &\left| \sum_{j=1}^{\nu} \phi(e^{i\pi/2^{j+1}})(M_j(n-m) - M_j(n)) \right| \\ &\leqslant \|\phi\|_{\infty} \max(C', C'') \left(\sum_{j=1}^{\alpha+1} 2^{j-\alpha} + \sum_{j=\alpha+2}^{\infty} 2^{\alpha-j} \right) \\ &\leqslant C\|\phi\|_{\infty}.\end{aligned}$$

This is the estimate (19) which we sought. □

7.2.2. Suppose that instead of decomposing the group $\mathbb{R}$ as we did in 7.1.2(a), we form the decomposition comprising the intervals $(\Delta'_j)_{-\infty}^{\infty}$ and $(\Delta''_j)_{-\infty}^{\infty}$, where

$$\Delta'_j = (2^j, 2^{j+1}]$$

and

$$\Delta''_j = -\Delta'_j.$$

It should cause the reader no difficulty to go through the proof of Theorem 7.2.1 and show that the ideas used there can be employed with very little change to prove that this new decomposition of $\mathbb{R}$ which collapses dyadically down towards 0 from both sides also has the LP property. One would need to establish the appropriate bound on $J(k'_\nu)$ where

$$k'_\nu = \sum_{-\nu}^{\nu} \phi(2^j) M_j(x)$$

and $\hat{M}_j$ is the translate of $\hat{V}_{2^{j-1}}$ by amount 3.2^{j-1}, V_λ denoting as always the de la Vallée-Poussin kernel of order λ. The estimates analogous to (12) and (13) hold, with exactly the same reasoning; and then it follows that

$$\int_{|t| \geqslant 2\tau} \left| \sum_{-v}^{v} \phi(2^j)(M_j(t-s) - M_j(t)) \right| dt \leqslant C\|\phi\|_\infty$$

for all $\tau > 0$, all $|s| \leqslant \tau$ and all v since, as noted in (15),

$$\sum_{-\infty}^{\infty} \min(2^j\tau, 2^{-j}\tau^{-1}) \leqslant 4.$$

7.3. The Littlewood-Paley Theorem: Second Approach

In this section, we present a slightly different proof of the Littlewood-Paley theorem. The main contrast with the proofs given in Section 7.2 is that we argue directly on a vectorial convolution operator and so avoid the argument involving Rademacher functions. We give the proof for $\mathbb{R}$ only. The proofs for the other groups can be left to the reader since it is only a matter of going through the proof for $\mathbb{R}$ and making routine changes.

Proof for $\mathbb{R}$. Let M_j $(j \geqslant 1)$ be the function on $\mathbb{R}$ whose Fourier transform is 1 on $[2^{j-1}, 2^j]$, 0 outside $(2^{j-2}, 2^j + 2^{j-2})$ and affine on each of the intervals $[2^{j-2}, 2^{j-1}]$ and $[2^j, 2^j + 2^{j-2}]$. It will suffice to prove the existence of a constant C_p such that

$$\left\| \left(\sum_1^N |M_j * f|^2 \right)^{1/2} \right\|_p \leqslant C_p \|f\|_p \tag{1}$$

for all positive integers N and all f in $L^2 \cap L^p$. For if (1) holds, the R property of the family $(\Delta_j)_1^\infty$ of intervals and the fact that $S_j(M_j * f) = S_j f$ permit us to deduce that

$$\left\| \left(\sum_1^\infty |S_j f|^2 \right)^{1/2} \right\|_p \leqslant C_p' \|f\|_p$$

whence the LP property is derived as in the proof of Theorem 7.2.1.

Now

$$M_j = \chi_{2^{j-1}+2^{j-2}} V_{2^{j-2}} \tag{2}$$

where, for $\lambda > 0$,

$$V_\lambda(x) = \frac{1}{2\pi}\left\{ \frac{2 \sin^2 \lambda x}{\lambda x^2} - \frac{4 \sin^2 \lambda x/2}{\lambda x^2} \right\}. \tag{3}$$

Since each point y of $\mathbb{R}$ belongs to at most three of the intervals $[2^{j-2}, 2^j + 2^{j-2}]$,

$$\sum_{j=1}^{\infty} |\hat{M}_j(y)|^2 \leqslant 3 \tag{4}$$

for all y in $\mathbb{R}$.

To establish (1), we apply Theorem 3.6.2, taking $\mathscr{H}_1 = \mathbb{C}$, $\mathscr{H}_2 = \mathbb{C}^N$, and K the kernel such that $K(x)$ carries $a \in \mathbb{C}$ to the vector

$$(M_j(x)a)_{1 \leqslant j \leqslant N}$$

of $\mathbb{C}^N$; cf. 3.3.2(ii). The inequality (1) will follow if we prove that the numbers $J(K)$ (computed for the natural covering family on $\mathbb{R}$; see 2.1.3(i)) and $\|\hat{K}\|_\infty$ are both majorised by something independent of N.

Since $\hat{K}(y)$ is the operator carrying the complex number a to the vector

$$\left(\int M_j(x) \exp(-ixy)\, dx.a\right)_{1 \leqslant j \leqslant N}$$

of $\mathbb{C}^N$, it appears that

$$|\hat{K}(y)| = \left(\sum_{j=1}^{N} |\hat{M}_j(y)|^2\right)^{1/2}$$

and so (4) guarantees that

$$\|\hat{K}\|_\infty \leqslant 3^{1/2},$$

which is visibly independent of N.

To majorise $J(K)$ in the desired fashion, we will show that

$$\int_{|x| \geqslant 2\tau} \left(\sum_1^N |M_j(x-y) - M_j(x)|^2\right)^{1/2} dx \leqslant D \tag{5}$$

for $\tau > 0$, and $|y| \leqslant \tau$, where D is independent of N and τ. Since $\|\cdot\|_{\ell^2} \leqslant \|\cdot\|_{\ell^1}$, and therefore

$$\begin{aligned}\int_{|x| \geqslant 2\tau} \left(\sum_1^N |M_j(x-y) - M_j(x)|^2\right)^{1/2} dx &\leqslant \int_{|x| \geqslant 2\tau} \sum_1^N |M_j(x-y) - M_j(x)|\, dx, \\ &= \sum_1^N \int_{|x| \geqslant 2\tau} |M_j(x-y) - M_j(x)|\, dx, \quad (6)\end{aligned}$$

it will be enough to majorise suitably the integrals

$$\int_{|x| \geqslant 2\tau} |M_j(x-y) - M_j(x)|\, dx.$$

Now we have already remarked in the proof of Theorem 7.2.1 that the following analogues of 7.2(12) and 7.2(13) hold for $\tau > 0$, $|y| \leqslant \tau$ and all j:

$$\int_{|x| \geqslant 2\tau} |M_j(x - y) - M_j(x)|\, dx \leqslant A 2^{-j} \tau^{-1}$$

and

$$\int_{|x| \geqslant 2\tau} |M_j(x - y) - M_j(x)|\, dx \leqslant B 2^{j} \tau,$$

A and B denoting absolute constants. Consequently, 7.2(15) shows that

$$\begin{aligned} \int_{|x| \geqslant 2\tau} \sum_1^N |M_j(x - y) - M_j(x)|\, dx &\leqslant C \sum_1^N \min(2^j \tau, 2^{-j} \tau^{-1}) \\ &\leqslant C \sum_1^\infty \min(2^j \tau, 2^{-j} \tau^{-1}) \\ &\leqslant 4C = D \end{aligned}$$

where $C = \max(A, B)$. In view of (6), (5) is now proved, and with it, the theorem. □

7.4. The Littlewood-Paley Theorem for Finite Products of $\mathbb{R}$, $\mathbb{T}$ and $\mathbb{Z}$: Dyadic Intervals

Suppose that G is as in Section 5.6, i.e., that

$$G = \prod_{i=1}^m G_i,$$

where each G_i is either $\mathbb{R}$, $\mathbb{T}$ or $\mathbb{Z}$, so that

$$X = \prod_{i=1}^m X_i,$$

where X_i is $\mathbb{R}$, $\mathbb{Z}$ or $\mathbb{T}$. Consider the family of dyadic intervals $(\Delta_{j_1,\ldots,j_m})_{j_1,\ldots,j_m \in \mathbb{Z}}$ in X defined by the formula

$$\Delta_{j_1,\ldots,j_m} = \prod_{i=1}^m \Delta_{i,j_i},$$

where Δ_{i,j_i} is the j_i-th dyadic interval in X_i, as defined in 7.1.2.

Theorem 7.2.1 and repeated application of Theorem 1.3.4 lead to the conclusion that the decomposition $(\Delta_{j_1,\ldots,j_m})$ of X has the LP property. If we take account of 1.2.6(i)–(iii), the full result may be stated formally as follows.

Theorem. *If $1 < p < \infty$, there are positive numbers A_p and B_p such that*

$$A_p\|f\|_p \leqslant \left\|\left(\sum_{j_1,\ldots,j_m\in\mathbb{Z}} |S_{j_1,\ldots,j_m}f|^2\right)^{1/2}\right\|_p \leqslant B_p\|f\|_p$$

for all f in $L^p(G) = L^p(\prod_1^m G_i)$ where

$$S_{j_1,\ldots,j_m} = S_{\Delta_{j_1,\ldots,j_m}} = T_{\xi_{\Delta_{j_1,\ldots,j_m}}}.$$

7.5. Fournier's Example

7.5.1. As was remarked in 7.1.3, if ϕ is a bounded function constant on each of the members Δ_j of the dyadic decomposition of $X (= \mathbb{R}, \mathbb{Z}$ or $\mathbb{T})$, the multiplier operator T_ϕ is generally not of weak type (1, 1). We give a counterexample due to John Fournier.

7.5.2. Let ϕ be the function on $\mathbb{Z}$ which is equal to 1 on $\Delta_2 \cup \Delta_4 \cup \cdots$ and 0 elsewhere. Let

$$Uf = T_\phi f - \chi_1 T_\phi(\chi_1^{-1}f).$$

If f is a trigonometric polynomial,

$$Uf = \sum_{j=1}^{\infty} \{\hat{f}(2^{2j-1})\chi_{2^{2j-1}} - \hat{f}(2^{2j})\chi_{2^{2j}}\}, \tag{1}$$

which is a trigonometric polynomial with spectrum contained in the set

$$E = \{2^i : i = 1, 2, \ldots\}$$

which by [9], 15.2.4, is a Sidon set.

Suppose that T_ϕ is of weak type (1, 1) on the set of trigonometric polynomials. Then so too is U, and there is therefore a constant A such that

$$\mu_{Uf}(y) = m(\{x : |Uf(x)| > y\}) \leqslant \frac{A}{y}\|f\|_1$$

for all $y > 0$ and all trigonometric polynomials f. It follows that if $0 < p < 1$, and $\|f\|_1 = 1$,

$$\|Uf\|_p^p = p\int_0^\infty y^{p-1}\mu_{Uf}(y)\,dy \leqslant 1 + p\int_1^\infty y^{p-1}\frac{A}{y}\,dy$$

$$= 1 + A\frac{p}{1-p} = B^p, \text{ say.}$$

Hence

$$\|Uf\|_p \leqslant B\|f\|_1 \tag{2}$$

for all trigonometric polynomials f. Since E is a Sidon set, it is also a $\Lambda(2)$ set ([9], 15.3.1) and so there is a constant C such that

$$\|Uf\|_2 \leqslant C\|Uf\|_p. \tag{3}$$

By (2) and (3),

$$\|Uf\|_2 \leqslant BC\|f\|_1. \tag{4}$$

Apply (4) with $f = \mathsf{F}_N$, the Fejér kernel of order N, and use (1) and the Parseval formula to conclude that

$$\sum_{j=1}^{\infty} (|\hat{\mathsf{F}}_N(2^{2j-1})|^2 + |\hat{\mathsf{F}}_N(2^{2j})|^2) \leqslant B^2C^2 \tag{5}$$

since $\|\mathsf{F}_N\|_1 = 1$. Now (5) is plainly false since the sum on the left tends to infinity with N.

This contradiction shows that T_ϕ is not of weak type (1, 1) on the set of trigonometric polynomials.

It is an easy matter to conclude that T_ϕ is not of weak type (1, 1) on other natural domains. For instance, it is plain that T_ϕ is not of weak type (1, 1) on $L^1 \cap L^2(\mathbb{T})$; nor is it of weak type (1, 1) on the set of simple functions on $\mathbb{T}$. (This last fact follows from an easy approximation argument: see [9], Exercise 13.18.)

Chapter 8. Strong Forms of the Marcinkiewicz Multiplier Theorem and Littlewood-Paley Theorem for $\mathbb{R}$, $\mathbb{T}$ and $\mathbb{Z}$

8.1. Introduction

8.1.1. To date, we have proved the weak Marcinkiewicz theorem and the Littlewood-Paley theorem for (i) the disconnected groups of Chapters 4 and 5; and (ii) the groups $\mathbb{R}$, $\mathbb{T}$ and $\mathbb{Z}$ with the associated dyadic decompositions of their duals. Cf. Chapter 7. It is our intention now to strengthen the results in (ii).

The major new results (Theorems 8.2.1, 8.3.1 and 8.4.2) state that if a bounded function ϕ has uniformly bounded variations over the dyadic intervals, then it is a multiplier of L^p for all p in $(1, \infty)$. From this it will be deduced (Theorem 8.2.7, 8.3.2 and Theorem 8.4.4) that if $(\alpha_k)_0^\infty$ is a Hadamard sequence, and ϕ is a bounded function, constant on each of the intervals δ_j of the decomposition determined by $(\alpha_k)_0^\infty$, then ϕ is a multiplier of L^p for all p in the range $(1, \infty)$. In other words, the Hadamard decomposition $(\delta_j)_{j\in\mathbb{Z}}$ has the WM property. As a consequence, it has the LP property.

8.1.2. It is amusing to note the strategy involved here: whereas in Chapter 7 we were forced to prove first the Littlewood-Paley theorem and then deduce the equivalent weak Marcinkiewicz theorem, now we do just the opposite. However, the dyadic form of the Littlewood-Paley theorem is used in the proof that a Hadamard decomposition has the WM property; so there is no paradox after all!

8.1.3. The Littlewood-Paley theorems we establish for $\mathbb{R}$, $\mathbb{T}$ and $\mathbb{Z}$ all deal with decompositions by intervals whose lengths grow or shrink exponentially. In Section 8.5 we show that decompositions by intervals δ_j whose lengths grow or shrink like a fixed power of the (integral) index j do not have the LP property. We give also an example of a sequence $(n_k)_0^\infty$ quite different from a Hadamard sequence which determines a decomposition of $\mathbb{Z}$ having the LP property.

8.2. The Strong Marcinkiewicz Multiplier Theorem for $\mathbb{T}$

8.2.1. Theorem. *Let $(\Delta_j)_{j\in\mathbb{Z}}$ be the usual dyadic decomposition of $\mathbb{Z}$ (see 7.1.2) and suppose ϕ is a function on $\mathbb{Z}$ such that*

$$|\phi(n)| \leqslant B \tag{1}$$

and

$$\sup_j \sum_{n\in\Delta_j} |\Delta\phi(n)| \leqslant B \tag{2}$$

where $\Delta\phi(n) = \phi(n+1) - \phi(n)$. *Then for every* p *in* $(1, \infty)$, $\phi \in M_p(\mathbb{Z})$ *and*

$$\|\phi\|_{p,p} \leqslant C_p B \tag{3}$$

where C_p *is a number depending only on* p.

Proof. We have to show that, if ϕ satisfies (1) and (2), $p \in (1, \infty)$ and $f \in L^2 \cap L^p$, then

$$\|T_\phi f\|_p \leqslant C_p \|f\|_p; \tag{4}$$

actually, it suffices to prove this for trigonometric polynomials f. By the dyadic form of the Littlewood-Paley theorem for $\mathbb{T}$ (Theorem 7.2.1), it is equivalent to show that

$$\left\| \left(\sum_{j\in\mathbb{Z}} |S_j(T_\phi f)|^2 \right)^{1/2} \right\|_p \leqslant C_p' B \left\| \left(\sum_{j\in\mathbb{Z}} |S_j f|^2 \right)^{1/2} \right\|_p; \tag{5}$$

and in order to prove (5), it is sufficient to show that

$$\left\| \left(\sum_{j=1}^{\infty} |S_j(T_\phi f)|^2 \right)^{1/2} \right\|_p \leqslant D_p B \left\| \left(\sum_{j=1}^{\infty} |S_j f|^2 \right)^{1/2} \right\|_p. \tag{6}$$

For if we have (6) and we replace ϕ by $\tilde{\phi}: n \to \overline{\phi(-n)}$, and f by $\bar{f}$, we deduce the corresponding inequality in which the summations are over $(-\infty, -1]$; and moreover,

$$|S_0(T_\phi f)| = |\phi(0)\hat{f}(0)| \leqslant B|\hat{f}(0)| = B|S_0 f|.$$

To prove (6), notice that when $j \geqslant 1$,

$$\begin{aligned} S_j(T_\phi f) &= \sum_{2^{j-1}}^{2^j-1} \phi(n)\hat{f}(n)\chi_n \\ &= \sum_{2^{j-1}}^{2^j-1} \phi(n)[S_{(n-1,2^j)}f - S_{(n,2^j)}f]. \end{aligned} \tag{7}$$

A standard summation by parts puts (7) into the form

$$S_j(T_\phi f) = -\phi(2^j - 1)S_{(2^j-1,2^j)}f + \sum_{2^{j-1}}^{2^j-2} S_{(n,2^j)}f\Delta\phi(n) + \phi(2^{j-1})S_{(2^{j-1}-1,2^j)}f.$$

If we observe that $S_{(2^{j-1},2^j)}f = 0$, this can be re-expressed more neatly:

$$S_j(T_\phi f) = \sum_{n\in\Delta_j} S_{(n,2^j)} f \Delta\phi(n) + \phi(2^{j-1}) S_j f. \tag{8}$$

Finally, since $S_{(n,2^j)}f = S_{(n,2^j)}(S_j f)$ whenever $n \in \Delta_j$, (8) can be rewritten in the more complicated, but more useful, form

$$S_j(T_\phi f) = \sum_{n\in\Delta_j} S_{(n,2^j)}(S_j f)\Delta\phi(n) + \phi(2^{j-1}) S_j f.$$

Hence, by Schwarz' inequality, (1) and (2),

$$\begin{aligned} |S_j(T_\phi f)| &\leqslant \Big(\sum_{n\in\Delta_j} |\Delta\phi(n)|\Big)^{1/2} \cdot \Big(\sum_{n\in\Delta_j} |\Delta\phi(n)||S_{(n,2^j)}(S_j f)|^2\Big)^{1/2} + B|S_j f| \\ &\leqslant B^{1/2}\Big(\sum_{n\in\Delta_j} |\Delta\phi(n)||S_{(n,2^j)}(S_j f)|^2\Big)^{1/2} + B|S_j f|. \end{aligned}$$

Now apply Minkowski's inequality twice, first for ℓ^2, then for L^p, to conclude that

$$\begin{aligned} \left\|\Big(\sum_1^\infty |S_j(T_\phi f)|^2\Big)^{1/2}\right\|_p &\leqslant B^{1/2}\left\|\Big(\sum_{j=1}^\infty \sum_{n\in\Delta_j} |\Delta\phi(n)||S_{(n,2^j)}(S_j f)|^2\Big)^{1/2}\right\|_p \\ &\quad + B\|(\sum |S_j f|^2)^{1/2}\|_p \\ &= B^{1/2}(\mathrm{I}) + B\left\|\Big(\sum_1^\infty |S_j f|^2\Big)^{1/2}\right\|_p. \end{aligned} \tag{9}$$

To estimate (I), notice that for every n in $\bigcup_{j=1}^\infty \Delta_j = \{1, 2, \ldots\}$, there is a unique index j_n such that $n \in \Delta_{j_n}$; write $f_j = S_j f$,

$$g_n = |\Delta\phi(n)|^{1/2} f_{j_n}$$

and

$$\Delta_n^* = (n, 2^{j_n}).$$

Then

$$(\mathrm{I}) \leqslant \left\|\Big(\sum_{n=1}^\infty |S_{\Delta_n^*} g_n|^2\Big)^{1/2}\right\|_p. \tag{10}$$

The family of intervals $(\Delta_n^*)_1^\infty$ has the R property, by Theorem 6.5.2. So we deduce from (10) and (2) that

$$\begin{aligned}
(\mathrm{I}) &\leqslant A_p\left\|\left(\sum_{n=1}^{\infty}|g_n|^2\right)^{1/2}\right\|_p = A_p\left\|\left(\sum_{n=1}^{\infty}|\Delta\phi(n)||f_{j_n}|^2\right)^{1/2}\right\|_p \\
&= A_p\left\|\left(\sum_{j=1}^{\infty}\sum_{n\in\Delta_j}|\Delta\phi(n)||S_jf|^2\right)^{1/2}\right\|_p \\
&\leqslant A_pB^{1/2}\left\|\left(\sum_{j=1}^{\infty}|S_jf|^2\right)^{1/2}\right\|_p. \qquad (11)
\end{aligned}$$

All together then, (9) and (11) lead to (6) with $D_p = A_p + 1$. □

8.2.2. An equivalent form of the strong Marcinkiewicz theorem. The following lemma makes it clear that the condition (2) can be rephrased in terms of the means, over intervals symmetric about 0, of the function $n \to |n\Delta\phi(n)|$.

Lemma. *Let ϕ be a function on $\mathbb{Z}$ such that* (1) *and* (2) *hold. Then*

$$\sup_{R\geqslant 1}\frac{1}{R}\sum_{|n|\leqslant R}|n\Delta\phi(n)| \leqslant B' \qquad (12)$$

with $B' \leqslant 8B$.

Conversely, if

$$|\phi(n)| \leqslant B' \qquad (13)$$

and (12) *holds, then* (1) *and* (2) *are satisfied, with $B \leqslant 2B'$.*

Proof. If $2^{j-1} \leqslant R < 2^j$ for some positive integer j, then

$$\begin{aligned}
\sum_{|n|\leqslant R}|n\Delta\phi(n)| &\leqslant \sum_{|r|\leqslant j}\sum_{n\in\Delta_r}|n\Delta\phi(n)| \\
&\leqslant \sum_{|r|\leqslant j}2^{|r|}\sum_{n\in\Delta_r}|\Delta\phi(n)| \\
&\leqslant \sum_{|r|\leqslant j}2^{|r|}B \leqslant B2^{j+2} \leqslant 8BR.
\end{aligned}$$

Conversely, if (12) and (13) hold and $j \neq 0$,

$$\begin{aligned}
\sum_{n\in\Delta_j}|\Delta\phi(n)| &\leqslant 2^{-|j|+1}\sum_{n\in\Delta_j}|n\Delta\phi(n)| \\
&\leqslant 2^{-j+1}\sum_{|n|\leqslant 2^{|j|}}|n\Delta\phi(n)| \\
&\leqslant 2^{-|j|+1}. B'2^{|j|} = 2B';
\end{aligned}$$

and if $j = 0$,

$$\sum_{n\in\Delta_0}|\Delta\phi(n)| = |\Delta\phi(0)| \leqslant 2B'. \quad □$$

Theorem 8.2.1 can therefore be given the following equivalent form.

8.2.3. Theorem. *If ϕ is a function on $\mathbb{Z}$ such that*

$$|\phi(n)| \leqslant B'$$

and

$$\sup_{R \geqslant 1} R^{-1} \sum_{|n| \leqslant R} |n\Delta\phi(n)| \leqslant B',$$

then $\phi \in M_p(\mathbb{Z})$ for every p in $(1, \infty)$, and

$$\|\phi\|_{p,p} \leqslant C_p' B',$$

C_p' being a number depending only on p.

8.2.4. The Hörmander-Mihlin version. The natural analogue of the Hörmander-Mihlin theorem for $\mathbb{R}^n$ ([21], Theorem 2.5) is as follows. We show it to be a corollary of Theorem 8.2.1.

8.2.5. Theorem. *Let ϕ be a function on $\mathbb{Z}$ such that for every $R \geqslant 1$,*

$$\sum_{R/2 \leqslant |n| \leqslant R} |\phi(n)|^2 \leqslant B^2 R \tag{14}$$

and

$$\sum_{R/2 \leqslant |n| \leqslant R} |\Delta\phi(n)|^2 \leqslant B^2/R. \tag{15}$$

Then

$$|\phi(n)| \leqslant 3B \tag{16}$$

for all $n \neq 0$; $\phi \in M_p(\mathbb{Z})$ whenever $1 < p < \infty$; and

$$\|\phi\|_{p,p} \leqslant C_p'' \max(|\phi(0)|, B) \tag{17}$$

where C_p'' is a number depending only on p.

Proof. We begin by establishing (16).

Suppose that k and m are integers such that $2^{j-1} \leqslant k < m < 2^j$ for some positive integer j. By applying (15) with $R = 2^j$ along with the Cauchy-Schwarz inequality, we deduce that

$$\begin{aligned} |\phi(m) - \phi(k)| &\leqslant \sum_{2^{j-1} \leqslant |n| \leqslant 2^j} |\Delta\phi(n)| \\ &\leqslant \left(\sum_{2^{j-1} \leqslant |n| \leqslant 2^j} |\Delta\phi(n)|^2 \right)^{1/2} 2^{j/2} \leqslant B. \end{aligned} \tag{18}$$

It follows that if

$$M_j = \max_{2^{j-1} \leqslant m \leqslant 2^j} |\phi(m)|,$$

then

$$|\phi(n)| \geqslant |M_j - B| \geqslant 0$$

for all n in Δ_j, $j > 0$. A similar result holds if $j < 0$.

Now apply (14) for $R = 2^j$ to conclude that if $j > 0$,

$$(M_j - B)^2 2^{j-1} \leqslant \sum_{2^{j-1} \leqslant |n| \leqslant 2^j} |\phi(n)|^2 \leqslant B^2 2^j,$$

whence

$$|M_j - B| \leqslant 2B,$$

and

$$M_j \leqslant 3B.$$

A similar result obtains if $j < 0$. This establishes (16), and proves that

$$\|\phi\|_\infty \leqslant \max(3B, |\phi(0)|) \leqslant 3 \max(B, |\phi(0)|). \tag{19}$$

On the other hand, if we look back to (18) we see that it contains the fact that when $j \neq 0$,

$$\sum_{n \in \Delta_j} |\Delta\phi(n)| \leqslant B. \tag{20}$$

The final assertion (17) of the theorem follows from (19), (20) and Theorem 8.2.1. □

8.2.6. The strong Marcinkiewicz theorem for Hadamard decompositions. It is an easy matter to deduce from Theorem 8.2.3 the following more general version of the theorem, dealing with so-called Hadamard decompositions.

We say that a sequence $(\alpha_k)_0^\infty$ of positive integers is a *Hadamard sequence* if

$$\inf_{k \geqslant 0} \alpha_{k+1}/\alpha_k = \lambda > 1. \tag{21}$$

The associated *Hadamard decomposition* of $\mathbb{Z}$ is formed of the intervals $(\delta_j)_{j \in \mathbb{Z}}$ defined by the formulas

$$\delta_j = \begin{cases} [\alpha_{j-1}, \alpha_j) & \text{if } j > 0 \\ (-\alpha_0, \alpha_0) & \text{if } j = 0 \\ (-\alpha_{|j|}, -\alpha_{|j|-1}] & \text{if } j < 0. \end{cases}$$

8.2.7. Theorem (Marcinkiewicz). *Let $(\delta_j)_{j\in\mathbb{Z}}$ be a Hadamard decomposition of $\mathbb{Z}$ and let ϕ be a function on $\mathbb{Z}$ such that*

$$|\phi(n)| \leqslant A \tag{22}$$

and

$$\sup_j \sum_{n\in\delta_j} |\Delta\phi(n)| \leqslant A. \tag{23}$$

Then $\phi \in M_p$ for every p in the range $(1, \infty)$, and

$$\|\phi\|_{p,p} \leqslant C'_p A$$

where C'_p is a number depending only on p.

Proof. It will be enough to show that conditions (22) and (23) imply conditions (12) and (13). Only condition (12) needs verification.

If $1 \leqslant R < \alpha_0$, then

$$\frac{1}{R}\sum_{|m|\leqslant R} |m\Delta\phi(m)| \leqslant \sum_{\delta_0} |\Delta\phi(m)| \leqslant A.$$

If, on the other hand, $\alpha_0 \leqslant R$, then there is a unique index k for which

$$\alpha_k \leqslant R < \alpha_{k+1}. \tag{24}$$

Then

$$\begin{aligned} R^{-1}\sum_{|m|\leqslant R} |m\Delta\phi(m)| &\leqslant R^{-1}\Big\{\sum_{|m|<\alpha_k}\cdots + \sum_{\alpha_k\leqslant|m|<\alpha_{k+1}}\cdots\Big\} \\ &\leqslant R^{-1}\sum_{|m|<\alpha_k} |m\Delta\phi(m)| + \sum_{\delta_k\cup\delta_{-k}} |\Delta\phi(m)| \\ &\leqslant \frac{1}{\alpha_k}\sum_{|m|<\alpha_k} |m\Delta\phi(m)| + 2A \end{aligned}$$

by virtue of (23) and (24). Hence, by (23) and (21),

$$\begin{aligned} R^{-1}\sum_{|m|\leqslant R} |m\Delta\phi(m)| &\leqslant 2A + \frac{1}{\alpha_k}\Big\{\sum_{|j|\leqslant k-1}\sum_{m\in\delta_j} |m\Delta\phi(m)|\Big\} \\ &\leqslant 2A + \frac{1}{\alpha_k}\Big\{\alpha_1\sum_{m\in\delta_0} |\Delta\phi(m)| + \alpha_2\sum_{m\in\delta_1\cup\delta_{-1}} |\Delta\phi(m)| \\ &\quad + \cdots + \alpha_k\sum_{m\in\delta_{k-1}\cup\delta_{-(k-1)}} |\Delta\phi(m)|\Big\} \end{aligned}$$

$$\leqslant 2A + \frac{A}{\alpha_k}\{\alpha_1 + 2\alpha_2 + \cdots + 2\alpha_k\}$$

$$\leqslant 2A + \frac{2A}{\alpha_k}\{\lambda^{-k+1} + \cdots + 1\}\alpha_k$$

$$\leqslant 2A + 2A\sum_0^\infty \lambda^{-s}.$$

In both cases then, condition (12) holds. □

Remark. It is an easy matter to show that the conditions (12) and (13) imply (22) and (23), so that in fact the "Hadamard" and "dyadic" versions of the strong Marcinkiewicz theorem are equivalent.

8.2.8. Corollary (LP and WM properties). *The Hadamard decomposition* $(\delta_j)_{j\in\mathbb{Z}}$ *has the* WM *property and hence the* LP *property.*

Proof. If ϕ is a bounded function, constant on each member of the decomposition, then for each j,

$$\sum_{n\in\delta_j} |\Delta\phi(n)| \leqslant 2\|\phi\|_\infty.$$

So by Theorem 8.2.7, $\phi \in M_p(\mathbb{Z})$, and

$$\|\phi\|_{p,p} \leqslant 2C_p'\|\phi\|_\infty. \quad \square$$

8.3. The Strong Marcinkiewicz Multiplier Theorem for $\mathbb{R}$

Theorems 8.2.1 and 8.2.3 have natural analogues in the real line setting. As was the case in Section 8.2, the key step is proving the dyadic form of the theorem.

8.3.1. Theorem. *Let* $(\Delta_j)_{j\in\mathbb{Z}}$ *be the usual dyadic decomposition of* $\mathbb{R}$ *(see* 7.1.2). *If* ϕ *is a function on* $\mathbb{R}$ *such that*

$$|\phi(y)| \leqslant B \tag{1}$$

and

$$\sup_j \operatorname*{Var}_{\Delta_j} \phi \leqslant B, \tag{2}$$

then $\phi \in M_p$ *for all* p *in* $(1, \infty)$, *and*

$$\|\phi\|_{p,p} \leqslant C_p B$$

where C_p *is a number depending only on* p.

Proof. In order to simplify some of the technical problems, let us observe that it will suffice to prove the theorem in the case where ϕ vanishes off $[1, \infty)$. For if ϕ is a quite general function satisfying (1) and (2), then the functions

$$\psi = \xi_{[1,\infty)}\phi,$$
$$\psi' = (\xi_{(-\infty,-1]}\phi)_r,$$

and

$$\psi'' = (\xi_{(-1,1)}\phi)_2$$

all vanish off $[1, \infty)$ and satisfy the same conditions as ϕ. (The suffices "r" and "2" denote "reflection" and "translation by amount 2" respectively.) It is a trivial matter to deduce that ϕ itself is in M_p with

$$\|\phi\|_{p,p} \leqslant 3C_pB$$

if it is known that ψ, ψ' and ψ'' are multipliers with norms at most C_pB.

Assuming then that $\phi(y) = 0$ for $y < 1$, let us observe that it will now suffice to prove that

$$\left\|\left(\sum_1^\infty |S_j(T_\phi f)|^2\right)^{1/2}\right\|_p \leqslant C_p'B\left\|\left(\sum_1^\infty |S_jf|^2\right)^{1/2}\right\|_p \tag{3}$$

for all f in $L^2 \cap L^p$, C_p' denoting a number depending only on p. For since $\phi = 0$ off $[1, \infty)$, (3) implies that

$$\begin{aligned}\left\|\left(\sum_{j\in\mathbb{Z}} |S_j(T_\phi f)|^2\right)^{1/2}\right\|_p &= \left\|\left(\sum_1^\infty |S_j(T_\phi f)|^2\right)^{1/2}\right\|_p \\ &\leqslant C_p'B\left\|\left(\sum_{j\in\mathbb{Z}} |S_jf|^2\right)^{1/2}\right\|_p.\end{aligned} \tag{4}$$

By the Littlewood-Paley theorem (7.2.1), the left side of (4) is at least $A_p\|T_\phi f\|_p$, while the right side is at most $B_p\|f\|_p$.

In establishing (3), we shall follow the train of the corresponding argument in 8.2.1.

Recall that if Δ is any bounded interval in $\mathbb{R}$, and $f \in L^2(\mathbb{R})$, $S_\Delta f$ is the class of the everywhere-defined, continuous function

$$x \to (2\pi)^{-1}\int_\Delta \hat{f}(y)e^{ixy}\,dy; \tag{5}$$

we continue to denote this function by $S_\Delta f$.

If $[\alpha, \beta]$ is a closed and bounded interval in $\mathbb{R}$, and $x \in \mathbb{R}$, the function θ:

$$y \to S_{[\alpha,y]}f(x)$$

is a continuous function of bounded variation on $[\alpha, \beta]$. This follows from the observation that, by (5),

$$|\theta(y') - \theta(y)| = \left|(2\pi)^{-1} \int_{[y,y']} \hat{f}(s)e^{isx}\, ds\right| \leqslant (2\pi)^{-1} \int_{[y,y']} |\hat{f}(s)|\, ds$$

whenever $\alpha \leqslant y < y' \leqslant \beta$. Moreover θ is, by (5), the indefinite integral of the integrable function

$$y \to (2\pi)^{-1}\hat{f}(y)e^{ixy}$$

on $[\alpha, \beta]$.

If $j \geqslant 1$, we can therefore write

$$S_j(T_\phi f) = (2\pi)^{-1} \int_{[2^{j-1},2^j]} \phi(y)\hat{f}(y)e^{ixy}\, dy = \int_{[2^{j-1},2^j]} \phi(y)\, d(S_{[2^{j-1},y]}f(x)), \tag{6}$$

a Riemann-Stieltjes integral. If we integrate (6) by parts, and take account of the fact that

$$S_{[2^{j-1},y]}f = S_{[0,y]}(S_j f),$$

we deduce that

$$S_j(T_\phi f)(x) = \phi(2^j)S_j f(x) - \int_{[2^{j-1},2^j]} S_{[2^{j-1},y]}f(x)\, d\phi(y) = \phi(2^j)S_j f(x) - \int_{[2^{j-1},2^j]} S_{[0,y]}(S_j f)(x)\, d\phi(y). \tag{7}$$

Regard the last integral in (7) as the Riemann-Stieltjes integral of the continuous function

$$y \to S_{[0,y]}(S_j f)(x)$$

with respect to the function ϕ of bounded variation on $[2^{j-1}, 2^j]$. Define

$$y_m(j, k) = 2^{j-1} + \frac{(m-1)(2^j - 2^{j-1})}{k}$$

for $j, k = 1, 2, \ldots$ and $m = 1, \ldots, k + 1$. For a given j and k, the points $y_m(j, k)$ determine a partition of $[2^{j-1}, 2^j]$ of mesh $2^{j-1}/k$. Therefore, by (7),

$$S_j(T_\phi f) = \phi(2^j)S_j f(x) - \lim_{k\to\infty} \sum_{m=1}^{k+1} S_{[0,y_m(j,k)]}S_j f(x). c_m(j,k) \tag{8}$$

where

$$c_m(j,k) = \phi(y_m(j,k)) - \phi(y_{m-1}(j,k));$$

so, by (1) and (2),

$$\sum_{m=1}^{k+1} |c_m(j,k)| \leqslant \operatorname*{Var}_{\bar{\Delta}_j} \phi \leqslant 3B. \tag{9}$$

Applying Minkowski's inequality for ℓ^2 to (8), we see that

$$\left(\sum_{j=1}^{\infty} |S_j T_\phi f(x)|^2\right)^{1/2} \leqslant B\left(\sum_{j=1}^{\infty} |S_j f(x)|^2\right)^{1/2} + \left(\sum_{j=1}^{\infty} \left|\lim_{k\to\infty} \cdots\right|^2\right)^{1/2}, \tag{10}$$

by (1). By Fatou's lemma in ℓ^2, however,

$$\begin{aligned}
&\left(\sum_{j=1}^{\infty} \left|\lim_{k\to\infty} \sum_{m=1}^{k+1} S_{[0,y_m(j,k)]} S_j f(x).\, c_m(j,k)\right|^2\right)^{1/2} \\
&\leqslant \liminf_{k\to\infty} \left(\sum_{j=1}^{\infty} \left|\sum_{m=1}^{k+1} S_{[0,y_m(j,k)]} S_j f(x).\, c_m(j,k)\right|^2\right)^{1/2} \\
&\leqslant \liminf_{k\to\infty} \left(\sum_{j=1}^{\infty} \sum_{m=1}^{k+1} |S_{[0,y_m(j,k)]} S_j f(x)|^2 |c_m(j,k)|.\, \sum_{m=1}^{k+1} |c_m(j,k)|\right)^{1/2} \\
&\leqslant 3^{1/2} B^{1/2} \liminf_{k\to\infty} \left(\sum_{j=1}^{\infty} \sum_{m=1}^{k+1} |S_{[0,y_m(j,k)]} S_j f(x)|^2 |c_m(j,k)|\right)^{1/2},
\end{aligned} \tag{11}$$

the last but one step by use of the Cauchy-Schwarz inequality, and the last step by (9). Taking (10) and (11) into account, and applying the Minkowski inequality and Fatou's lemma in $L^p(\mathbb{R})$, we conclude that

$$\begin{aligned}
\left\|\left(\sum_{j=1}^{\infty} |S_j T_\phi f|^2\right)^{1/2}\right\|_p &\leqslant B\left\|\left(\sum_{j=1}^{\infty} |S_j f|^2\right)^{1/2}\right\|_p \\
&+ 3^{1/2} B^{1/2} \liminf_{k\to\infty} \left\|\left(\sum_{j=1}^{\infty} \sum_{m=1}^{k+1} |S_{[0,y_m(j,k)]} S_j f(x)|^2 |c_m(j,k)|\right)^{1/2}\right\|_p.
\end{aligned} \tag{12}$$

Now, for a fixed index k, the family $([0, y_m(j,k)])_{j,m}$ has the R property (Theorem 6.5.2) and thus

$$\begin{aligned}
&\left\|\left(\sum_{j=1}^{\infty} \sum_{m=1}^{k+1} |S_{[0,y_m(j,k)]} S_j f|^2 |c_m(j,k)|\right)^{1/2}\right\|_p \\
&\leqslant A_p \left\|\left(\sum_{j=1}^{\infty} \sum_{m=1}^{k+1} |S_j f|^2 |c_m(j,k)|\right)^{1/2}\right\|_p \\
&\leqslant A_p 3^{1/2} B^{1/2} \left\|\left(\sum_{j=1}^{\infty} |S_j f|^2\right)^{1/2}\right\|_p
\end{aligned}$$

by (9). Therefore (12) leads at once to (3) with $C_p' = 3A_p + 1$. □

8.3.2. Equivalent versions of the theorem; Hadamard decompositions.

(a) It is a straightforward exercise to show, following the ideas of the proof of Lemma 8.2.2, that a bounded function ϕ on $\mathbb{R}$ satisfies the condition

$$\sup_j \operatorname{Var}_{\Delta_j} \phi < \infty$$

if and only if

$$\sup_{R>0} R^{-1} \int_{|y| \leqslant R} |y| \, |d\phi| < \infty.$$

(b) Let $(\alpha_k)_0^\infty$ be a sequence of positive numbers. We say the sequence is a *Hadamard sequence* if

$$\frac{\alpha_{k+1}}{\alpha_k} \geqslant \lambda$$

for all k, where $\lambda > 1$. By a *Hadamard decomposition of* $\mathbb{R}$ we mean a decomposition $(\delta_j)_{j \in \mathbb{Z}}$ of $\mathbb{R}$ determined by a Hadamard sequence $(\alpha_k)_0^\infty$ according to the rules

$$\delta_j = \begin{cases} [\alpha_{j-1}, \alpha_j) & \text{if } j > 0 \\ (-\alpha_0, \alpha_0) & \text{if } j = 0 \\ (-\alpha_{|j|}, -\alpha_{|j|-1}] & \text{if } j < 0. \end{cases}$$

It is a routine matter to use (a) to establish, along the lines of Theorem 8.2.7, the version of the strong Marcinkiewicz multiplier theorem appropriate to a Hadamard decomposition of $\mathbb{R}$; and from there it is an immediate step to the WM and LP properties of a Hadamard decomposition, as in the case of $\mathbb{T}$: see Corollary 8.2.8.

8.4. The Strong Marcinkiewicz Multiplier Theorem for $\mathbb{Z}$

8.4.1. The theorems we have just proved for the groups $G = \mathbb{T}$ and $G = \mathbb{R}$ have their natural counterparts for the group $G = \mathbb{Z}$. The proofs follow broadly the same general principles as those already written out in full, and are in detail very similar to those for $\mathbb{R}$. To save repetition, we merely cite the main theorems, and leave to the interested reader the task of checking our claim that nothing new is involved in the proofs.

8.4.2. Theorem. *Let $(\Delta_j)_{j \in \mathbb{Z}}$ be the dyadic decomposition of $\mathbb{T}$ defined in* 7.1.2(c). *If ϕ is a function on $\mathbb{T}$ such that*

$$\sup_{t \in R} |\phi(e^{it})| \leqslant B$$

and

$$\sup_j \operatorname{Var}_{\Delta_j} \phi \leqslant B,$$

then $\phi \in M_p(\mathbb{T})$ *whenever* $1 < p < \infty$; *there is, moreover, a number* C_p *such that*

$$\|\phi\|_{p,p} \leqslant C_p B$$

for all such functions ϕ.

8.4.3. Hadamard decompositions of $\mathbb{T}$. Let $(\alpha_k)_0^\infty$ be a sequence of real numbers in the interval $(0, \pi]$ having the property that

$$\frac{\alpha_{k+1}}{\alpha_k} \leqslant \frac{1}{\lambda}$$

for all k, where $\lambda > 1$. By definition, a *Hadamard decomposition* $(\delta_j)_{j\in\mathbb{Z}}$ *of* $\mathbb{T}$ is one determined from such a sequence $(\alpha_k)_0^\infty$ according to the prescriptions

$$\delta_0 = \{e^{it}: \alpha_0 \leqslant |t| \leqslant \pi\}$$
$$\delta_j = \{e^{it}: \alpha_j \leqslant t < \alpha_{j-1}\}$$

if $j > 0$; and

$$\delta_j = \{e^{it}: \alpha_{-j} \leqslant -t < \alpha_{-j-1}\}$$

if $j < 0$.

Here now are the Marcinkiewicz and Littlewood-Paley theorems for Hadamard decompositions of $\mathbb{T}$.

8.4.4. Theorem (Marcinkiewicz). *Let* $(\delta_j)_{j\in\mathbb{Z}}$ *be a Hadamard decomposition of* $\mathbb{T}$ *and* ϕ *a function on* $\mathbb{T}$ *such that*

$$\sup_t |\phi(e^{it})| \leqslant B$$

and

$$\sup_j \operatorname{Var}_{\delta_j} \phi \leqslant B.$$

Then $\phi \in M_p(\mathbb{T})$ *whenever* $1 < p < \infty$; *and there is a number* C_p *depending only on* p *such that*

$$\|\phi\|_{p,p} \leqslant C_p B.$$

8.4.5. Theorem. *Every Hadamard decomposition of* $\mathbb{T}$ *has the* LP *and* WM *properties.*

8.4.6. *Remark.* Suppose that we form the decomposition of $\mathbb{R}$ comprising the intervals Δ'_j and Δ''_j defined in 8.2.2. It is a routine exercise to check that the methods used in the proof of Theorem 8.3.1 can be applied, with no more than minor modification, to prove that if ϕ is a function on $\mathbb{R}$ for which

$$\sup |\phi(y)| < \infty$$

and

$$\sup_j \max (\operatorname*{Var}_{\Delta'_j} \phi, \operatorname*{Var}_{\Delta''_j} \phi) < \infty,$$

then $\phi \in M_p(\mathbb{R})$ whenever $1 < p < \infty$. The aim of this comment is not so much to indicate an extension of Theorem 8.3.1, but rather to point out that one can *deduce* Theorems 8.4.4 and 8.4.5 from this extension. The arguments are not terribly difficult, though some work is needed. For an idea of the methods to be applied, the reader should consult [8] and [26].

8.5. Decompositions which are not Hadamard

8.5.1. All the versions of the Littlewood-Paley theorem which we have proved in this chapter and the preceding one deal with Hadamard decompositions of one sort or another. Suppose now that $(n_k)_0^\infty$ is an increasing sequence of positive integers, not necessarily Hadamard, and that the family of intervals $(\delta_j)_{j\in\mathbb{Z}}$ of $\mathbb{Z}$ is determined by the sequence $(n_k)_0^\infty$ in the customary way, viz.

$$\delta_j = \begin{cases} [n_{j-1}, n_j) & \text{if } j > 0 \\ (-n_0, n_0) & \text{if } j = 0 \\ (-n_{|j|}, -n_{|j|-1}] & \text{if } j < 0. \end{cases} \tag{1}$$

It is tempting to entertain the conjecture (even if only briefly!) that the sequence $(n_k)_0^\infty$ has to be not too different from a Hadamard sequence if $(\delta_j)_{j\in\mathbb{Z}}$ is to have the LP property. As it stands the conjecture is too vague. But here is a simple result which shows that a sequence $(n_k)_0^\infty$ can be "quite different from" a Hadamard sequence and still determine a decomposition with the LP and WM properties.

8.5.2. Theorem. *Let $(n'_k)_0^\infty$ and $(n''_k)_0^\infty$ be the sequences*

$$n'_k = 2^k$$
$$n''_k = 3^k,$$

and $(\delta'_j)_{j\in\mathbb{Z}}$ and $(\delta''_j)_{j\in\mathbb{Z}}$ the corresponding decompositions of $\mathbb{Z}$. Denote by $(\delta_j)_{j\in\mathbb{Z}}$ the decomposition of $\mathbb{Z}$ formed of all the sets of the form $\delta'_i \cap \delta''_k$ $(i, k \in \mathbb{Z})$. Then $(\delta_j)_{j\in\mathbb{Z}}$ has the LP *and* WM *properties.*

Proof. This follows from Theorems 1.4.1 and Corollary 8.2.8. □

Remark. The collection $(\delta_j)_{j\in\mathbb{Z}}$ contains the intervals of the form $[2^k, 3^m)$ in which $3^m > 2^k$ and $3^m < 2^{k+1}$. The ratio $3^m/2^k$ can be arbitrarily close to 1. Therefore the decomposition $(\delta_j)_{j\in\mathbb{Z}}$ is certainly not determined by a Hadamard sequence $(n_k)_0^\infty$.

The second author has proved recently that if (n'_k) and (n''_k) are Hadamard sequences, then the "sum" set $E = \{n'_j + n''_k : j, k \geqslant 1\}$, when written in increasing order, determines a decomposition of $\mathbb{Z}$ which has the LP property. Notice that this result is stronger than Theorem 9.2.3 *infra*. Cf. also the introductory remarks in Section 9.2. (Gaudry, G.I.: The Littlewood-Paley Theorem and Fourier multipliers for certain locally compact groups. Proc. Conf. INAM, Rome, 1976. To appear.)

This result and Theorem 8.5.2 suggest that it is likely to be very difficult, if not impossible, to specify necessary and sufficient conditions of a purely arithmetic character on a sequence $(n_k)_0^\infty$ in order that it determine a decomposition having the LP property.

If the decomposition $(\delta_j)_{j\in\mathbb{Z}}$ of Theorem 8.5.2 is determined by the sequence $(n_k)_0^\infty$ according to the rules in (1), then as we have just seen, the n_k grow very irregularly. To conclude these remarks here are two results, one for $G = \mathbb{T}$, the other for $G = \mathbb{Z}$, both of which deal with regularly, but rather slowly, growing/shrinking intervals. In both cases we conclude that the decompositions do not have the LP property.

8.5.3. Theorem. *Let $(\delta_j)_{j\in\mathbb{Z}}$ be a decomposition of $\mathbb{T}$ by intervals δ_j. Suppose that $J \supseteq \{1, 2, \ldots\}$ and that the length λ_j of δ_j satisfies the condition*

$$\lambda_j = \frac{a}{j^m} \tag{2}$$

for $j = 1, 2, \ldots$, where $a > 0$ and $m > 1$ are independent of j. Then the decomposition $(\delta_j)_{j\in J}$ does not have the LP property.

(The restriction $m > 1$ is necessary in order that $\sum_1^\infty \lambda_j$ be finite.)

Proof. Suppose the contrary. For each p in the range $(1, \infty)$, there is then a constant B_p such that

$$\left\| \left(\sum_{j\in J} |S_j f|^2 \right)^{1/2} \right\|_p \leqslant B_p \|f\|_p$$

for all f in $\ell^2 \cap \ell^p(\mathbb{Z})$. Taking $f = \mathrm{e}$, the function on $\mathbb{Z}$ which is 1 at 0 and 0 elsewhere, we deduce that

$$\left\| \left(\sum_{j\in J} |\hat{\xi}_{\delta_j}|^2 \right)^{1/2} \right\|_p \leqslant B_p;$$

since $J \supseteq \{1, 2, \ldots\}$, it follows that

$$\left\| \left(\sum_{j=1}^{\infty} |\hat{\xi}_{\delta_j}|^2 \right)^{1/2} \right\|_p \leqslant B_p. \tag{3}$$

But if δ_j is an interval in $\mathbb{T}$ of length λ_j, it is an easy matter to check that

$$|\hat{\xi}_{\delta_j}(n)| = \frac{1}{|n|} |\sin(\lambda_j n/2)|$$

if $n \neq 0$. By virtue of (3),

$$\sum_{n\neq 0} \frac{1}{|n|^p}\left(\sum_{j=1}^{\infty} |\sin(\lambda_j n/2)|^2\right)^{p/2} < \infty$$

for every p in $(1, \infty)$. That is, on account of (2),

$$\sum_{n\neq 0} \frac{1}{|n|^p}\left(\sum_{j=1}^{\infty} \sin^2 \frac{an}{2j^m}\right)^{p/2} < \infty. \tag{4}$$

However,

$$\begin{aligned}\sum_{j=1}^{\infty} \sin^2 \frac{an}{2j^m} &\geqslant \sum_{2j^m \geqslant a|n|} \frac{Aa^2n^2}{4j^{2m}} \\ &= \frac{Aa^2|n|^2}{4} \sum_{j \geqslant (a|n|/2)^{1/m}} \frac{1}{j^{2m}} \\ &\geqslant B|n|^2 \Big/ \left\{\left(\frac{a|n|}{2}\right)^{1/m}\right\}^{2m-1} = C|n|^{1/m}\end{aligned} \tag{5}$$

where A, B and C are positive constants. It follows from (5) that

$$\begin{aligned}\sum_{n\neq 0} \frac{1}{|n|^p}\left(\sum_{j=1}^{\infty} \sin^2 \frac{an}{2j^m}\right)^{p/2} &\geqslant C^{p/2} \sum_{n\neq 0} \frac{1}{|n|^p} |n|^{p/2m} \\ &= \infty\end{aligned}$$

if $p(1 - m/2) < 1$. This contradicts (4) and shows that the decomposition $(\delta_j)_{j\in J}$ does not have the LP property. □

8.5.4. Theorem. *Let $(n_k)_0^\infty$ be an increasing sequence of positive integers such that $n_k - n_{k-1} \to \infty$ as $k \to \infty$, and $(\delta_j)_{j\in\mathbb{Z}}$ the decomposition of $\mathbb{Z}$ defined by* (1). *If $(\delta_j)_{j\in\mathbb{Z}}$ has the* LP *property, there is to each index $p > 2$ a number C_p such that*

$$n_k^{1/p'} \leqslant C_p\left\{\sum_{j=1}^{k} (n_j - n_{j-1})^{2/p'}\right\}^{1/2} \tag{6}$$

for all integers $k \geqslant 1$.

Proof. If $(\delta_j)_{j\in\mathbb{Z}}$ has the LP property, there is a positive constant A_p such that

$$A_p\|f\|_p \leqslant \left\|\left(\sum_{j\in\mathbb{Z}} |S_j f|^2\right)^{1/2}\right\|_p \tag{7}$$

for all f in $L^2 \cap L^p(\mathbb{T})$. It follows from (7) and Minkowski's inequality for the index $p/2$ that

$$A_p\|f\|_p \leqslant \left(\sum_{j\in\mathbb{Z}} \|S_j f\|_p^2\right)^{1/2} \tag{8}$$

when $p > 2$. (See Lemma 9.1.1 for the full details of this step.)

Consider, for each positive integer k, the function $f = \mathsf{D}_{n_k}$, the Dirichlet kernel of order n_k. Then (8) and Minkowski's inequality in ℓ^2 imply that

$$A_p\|\mathsf{D}_{n_k}\|_p \leqslant \|\mathsf{D}_{n_0}\|_p + 2^{1/2}\left(\sum_{j=1}^{\infty} \|S_{[n_{j-1},n_j)}\mathsf{D}_{n_k}\|_p^2\right)^{1/2} \tag{9}$$

since

$$\|S_j\mathsf{D}_{n_k}\|_p = \|S_{-j}\mathsf{D}_{n_k}\|_p$$

for $1 \leqslant j \leqslant k$.

Observe now the following facts:

(a) if $1 < p < \infty$, then

$$\|\mathsf{D}_N\|_p \sim B_p N^{1/p'}$$

as $N \to \infty$ ([9], Exercise 7.5);

(b) if $1 \leqslant j \leqslant k$,

$$S_{[n_{j-1},n_j)}\mathsf{D}_{n_k} = \sum_{n_{j-1}}^{n_j-1} \chi_n; \tag{10}$$

(c) if $n_j - n_{j-1}$ is odd, the right side of (10) is a character times $\mathsf{D}_{(n_j-n_{j-1}-1)/2}$; and if $n_j - n_{j-1}$ is even, the right side of (10) is a character times $\{\mathsf{D}_{(n_j-n_{j-1})/2} - \chi_{(n_j-n_{j-1})/2}\}$;

(d) as a consequence of (c), (a) and the assumption that $n_j - n_{j-1} \to \infty$ as $j \to \infty$,

$$\left\|\sum_{n_{j-1}}^{n_j} \chi_n\right\|_p \sim B_p\left(\frac{n_j - n_{j-1}}{2}\right)^{1/p'}.$$

We conclude from facts (a)–(c) and (9) that there is a number C_p such that (6) holds, since $\|\mathsf{D}_{n_0}\|_p$ is a fixed number. □

8.5.5. Corollary. *Let $(n_k)_0^\infty$ be an increasing sequence of positive integers such that $n_j - n_{j-1} \sim aj^m$ as $j \to \infty$, where $a > 0$ and $m > 0$. Then the decomposition $(\delta_j)_{j\in\mathbb{Z}}$ determined by the sequence $(n_k)_0^\infty$ does not have the LP property.*

Proof. Since $n_j - n_{j-1} \sim aj^m$ as $j \to \infty$, there are positive constants A and B

such that

$$Aj^m \leqslant n_j - n_{j-1} \leqslant Bj^m \tag{11}$$

for all $j \geqslant 1$. Therefore, if $k > 1$,

$$\begin{aligned} n_k &= (n_k - n_{k-1}) + \cdots + (n_1 - n_0) + n_0 \\ &> A \sum_1^k j^m > A \int_{[0,k-1]} t^m\, dt \\ &= \frac{A(k-1)^{m+1}}{m+1}. \end{aligned} \tag{12}$$

If the decomposition $(\delta_j)_{j\in\mathbb{Z}}$ has the LP property, (6), (11) and (12) show that when $p > 2$,

$$\begin{aligned} A\frac{(k-1)^{m+1}}{m+1} &\leqslant C_p^{p'} B\left\{\int_{[0,k+1]} t^{2m/p'}\, dt\right\}^{p'/2} \\ &= C_p^{p'} B(k+1)^{m+p'/2} \cdot \left\{\frac{p'}{2m+p'}\right\}^{p'/2}. \end{aligned} \tag{13}$$

Clearly, (13) cannot hold as $k \to \infty$ since $p'/2 < 1$ when $p > 2$. This contradiction establishes the corollary. □

Chapter 9. Applications of the Littlewood-Paley Theorem

In this chapter, we show how the Littlewood-Paley theorem for $\mathbb{Z}$ can be used to construct (i) examples of sets which are $\Lambda(p)$ for every p; and (ii) an example of a multiplier of L^p which is in a certain sense "singular". The results are due to Meyer [30] and Figà-Talamanca and Gaudry [14] respectively.

We begin with a few simple observations which will be essential in the constructions.

9.1. Some General Results

Throughout this section it is supposed that G is a compact (Hausdorff) Abelian group and that $(\Delta_j)_{j\in J}$ is a decomposition of X having the LP property.

The basic idea running through this chapter is that it is possible to construct sets or functions on $\mathbb{Z}$ having desired L^p behaviour by constructing judiciously the pieces which sit in the dyadic or Hadamard blocks and then invoking the Littlewood-Paley theorem to guarantee that the individual pieces can be put together in the right way. The expressions of the type $\|(\sum|f_j|^2)^{1/2}\|_p$ which occur in the statement of the LP property are, however, difficult to handle, and it is easier to work with $(\sum\|f_j\|_p^2)^{1/2}$. Lemma 9.1.1 shows how the two expressions are related.

9.1.1. Lemma. *Suppose $p\in[1, 2]$ and that $(f_j)_{j\in J}$ is a family of measurable functions on G. Then*

(i) $$\left(\sum_{j\in J}\|f_j\|_p^2\right)^{1/2}\leqslant\left\|\left(\sum_{j\in J}|f_j|^2\right)^{1/2}\right\|_p; \tag{1}$$

and

(ii) $$\left\|\left(\sum_{j\in J}|f_j|^2\right)^{1/2}\right\|_{p'}\leqslant\left(\sum_{j\in J}\|f_j\|_{p'}^2\right)^{1/2}. \tag{2}$$

Proof. To establish (i), set $r = p/2$, so that $r\in[1/2, 1]$. By Minkowski's inequality in L^r ([18], p. 146),

$$\begin{aligned}\sum_{j\in J} \|f_j\|_p^2 &= \sum_{j\in J}\left\{\int |f_j^2|^r\right\}^{1/r}\\ &\leqslant \left\{\int\left(\sum_{j\in J} |f_j^2|\right)^r\right\}^{1/r}\\ &= \left\{\int\left(\sum_{j\in J} |f_j|^2\right)^{p/2}\right\}^{2/p}\\ &= \left\|\left(\sum_{j\in J} |f_j|^2\right)^{1/2}\right\|_p^2.\end{aligned}$$

Now take square roots, and (1) is proved.

In proving (ii), we set $q = p'/2$ so that $q \geqslant 1$. Suppose first that $q < \infty$. Then by Minkowski's inequality and the monotone convergence theorem,

$$\begin{aligned}\left\|\left(\sum_{j\in J} |f_j|^2\right)^{1/2}\right\|_{p'}^{p'} &= \int\left(\sum_{j\in J} |f_j|^2\right)^q\\ &\leqslant \left\{\sum_{j\in J}\left(\int |f_j|^{2q}\right)^{1/q}\right\}^q\\ &= \left\{\left(\sum_{j\in J} \|f_j\|_{p'}^2\right)^{1/2}\right\}^{2q}.\end{aligned} \tag{3}$$

The inequality (2) follows from (3) by extracting p'-roots.

When $p' = \infty$, (2) is a triviality. □

9.1.2. Corollary. *Suppose that $(\Delta_j)_{j\in J}$ is a decomposition of X having the* LP *property and that $p \in (1, 2]$. There exist positive numbers A_p and B_p such that*

(i)
$$\left(\sum_{j\in J} \|S_j f\|_p^2\right)^{1/2} \leqslant B_p \|f\|_p \tag{4}$$

for all f in $L^2 \cap L^p$; and

(ii)
$$A_p \|f\|_{p'} \leqslant \left(\sum_{j\in J} \|S_j f\|_{p'}^2\right)^{1/2} \tag{5}$$

for all f in $L^2 \cap L^{p'}$.

Proof. This follows at once from Lemma 9.1.1, the definition of the LP property, and 1.2.6(ii). □

9.1.3. Families uniformly of type $\Lambda(p)$. Suppose $p \in (2, \infty)$. Recall ([9], 15.5.4) that a subset F of X is of type $\Lambda(p)$ if and only if there is a number C_p such that

$$\|v\|_p \leqslant C_p \|v\|_2 \tag{6}$$

for every F-spectral trigonometric polynomial v on G. It is easy to check that this is equivalent to the condition

$$\sum_F |\hat{u}(\chi)|^2 \leqslant C_p^2 \|u\|_{p'}^2$$

for every trigonometric polynomial u on G.

Accordingly, a family $(F_j)_{j\in J}$ of subsets of X is said to be *uniformly of type* $\Lambda(p)$ if and only if there is a number C_p, independent of j, such that

$$\sum_{F_j} |\hat{u}(\chi)|^2 \leqslant C_p^2 \|u\|_{p'}^2 \tag{7}$$

for every j in J and every trigonometric polynomial u on G.

9.1.4. Theorem. *Suppose that $p \in (2, \infty)$ and that the family $(F_j)_{j\in J}$ of subsets of X is uniformly of type $\Lambda(p)$. Suppose further that $(\Delta_j)_{j\in J}$ is a decomposition of X having the* LP *property. The conclusion is that the set*

$$E = \bigcup_{j\in J} (F_j \cap \Delta_j)$$

is of type $\Lambda(p)$.

Proof. Apply (7) with $u = f_j = S_{\Delta_j} f$, f being any trigonometric polynomial on G; remember that $\hat{f}_j$ and $\hat{f}$ agree on Δ_j. It appears then that

$$\begin{aligned}\sum_{F_j\cap\Delta_j} |\hat{f}(\chi)|^2 &= \sum_{F_j\cap\Delta_j} |\hat{f}_j(\chi)|^2 \\ &\leqslant \sum_{F_j} |\hat{f}_j(\chi)|^2 \\ &\leqslant C_p^2 \|f_j\|_{p'}^2.\end{aligned}$$

If we sum over j and use Corollary 9.1.2(i), it follows that

$$\sum_E |\hat{f}(\chi)|^2 \leqslant C_p^2 \sum_{j\in J} \|f_j\|_{p'}^2 \leqslant C_p^2 B_{p'}^2 \|f\|_{p'}^2,$$

which shows that E is of type $\Lambda(p)$. □

9.2. Construction of $\Lambda(p)$ Sets in $\mathbb{Z}$

It is a standard fact ([9], 15.2.4 and 15.3.3) that a Hadamard set of positive integers is a Sidon set and hence a $\Lambda(p)$ set for every p in $(0, \infty)$. The first two results here are of the following type: if E and F are suitable Hadamard sets, then $E + F$ and $E - F$ are $\Lambda(p)$ sets for every p. The third result deals with the span Q_r over $\{-1, 1\}$ of r copies of the Hadamard set E. Under certain conditions it is shown that Q_r is also of type $\Lambda(p)$ for every p.

Apart from their intrinsic interest, these results are of significance because $E + E$ is *never* a Sidon set when E is an infinite set ([23], p. 61).

9.2.1. Theorem. *Suppose that* $(n_k)_{k=0}^{\infty}$ *is a sequence of positive integers satisfying the condition* $n_{k+1} \geqslant 2n_k$ *for every* $k \geqslant 0$. *Then*

$$E = \{n_j - n_i : i \in \mathbb{Z}, j \in \mathbb{Z}, 0 \leqslant i < j\}$$

is of type $\Lambda(p)$ *for every* p *in* $(0, \infty)$.

Proof. We may and will assume that $p \in (2, \infty)$. Write $F = \{n_k : k \geqslant 0\}$, and define E_j for $j \in \mathbb{Z}$ as follows:

$$E_j = \{n_j - n_i : i \in \mathbb{Z}, 0 \leqslant i < j\} \subseteq n_j - F$$

for $j > 0$, and

$$E_j = \varnothing$$

if $j \leqslant 0$. Now F is a Hadamard set, and so ([9], 15.3.3) to every p in $(2, \infty)$ corresponds a number C_p such that

$$\sum_{m \in F} |\hat{u}(m)|^2 \leqslant C_p^2 \|u\|_{p'}^2$$

and hence also

$$\sum_{E_j} |\hat{u}(m)|^2 \leqslant \sum_{n_j - F} |\hat{u}(m)|^2 \leqslant C_p^2 \|u\|_{p'}^2,$$

for every j in $\mathbb{Z}$ and every trigonometric polynomial u. Thus the family $(E_j)_{j \in \mathbb{Z}}$ is uniformly of type $\Lambda(p)$.

Also, if $j > 0$ and $0 \leqslant i < j$,

$$n_{j-1} \leqslant n_j - n_i < n_j,$$

so that $E_j \subseteq \Delta_j = [n_{j-1}, n_j)$ and therefore $E_j \cap \Delta_j = E_j$; the same is trivially true if $j \leqslant 0$. Since the decomposition of $\mathbb{Z}$ defined by the sequence $(n_k)_0^{\infty}$ is Hadamard, Theorem 9.1.4 shows that $E = \bigcup_{j \in J} E_j$ is of type $\Lambda(p)$. □

9.2.2. Corollary. *Let* $F = (n_j)_0^{\infty}$ *be as above. Then* $F - F$ *is a* $\Lambda(p)$ *set for every* p *in* $(0, \infty)$.

Proof. $F - F = E \cup (-E) \cup \{0\}$. Now use Exercise 15.10(1) of [9]. □

9.2.3. Theorem. *Suppose that* $(m_k)_{k=0}^{\infty}$ *and* $(n_k)_{k=0}^{\infty}$ *are sequences of positive integers such that* $m_{k+1} \geqslant \alpha m_k$ *and* $n_{k+1} \geqslant \alpha n_k$ *for every* $k \geqslant 0$ *and some* $\alpha > 1$. *Then the set*

$$F = \{m_i + n_j : i, j = 0, 1, 2, \ldots\}$$

is of type $\Lambda(p)$ *for every* p *in* $(0, \infty)$.

Proof. Again we may and will assume that $p \in (2, \infty)$.

Each of $M = (m_k)$ and $N = (n_k)$ can be split into a finite union of subsequences M_r and N_s for each of which the corresponding value of α is greater than 2. Then $F = M + N$ is the union of the sets $M_r + N_s$ and it will suffice ([9], Exercise 15.10(1)) to show that $M_r + N_s$ is of type $\Lambda(p)$. Thus we may assume that $\alpha > 2$.

By symmetry and the same exercise it will suffice to show that

$$F_0 = \{m_i + n_j : i, j = 0, 1, 2, \ldots; n_j \leqslant m_i\}$$

is of type $\Lambda(p)$. To do this, define E_i for i in $\mathbb{Z}$ as follows:

$$E_i = \{m_{i-1} + n_j : j = 0, 1, 2, \ldots; n_j \leqslant m_{i-1}\} \quad \text{if} \quad i > 0,$$

and

$$E_i = \varnothing \quad \text{if} \quad i \leqslant 0.$$

Define also

$$\Delta_i = \begin{cases} [m_{i-1}, m_i) & \text{if} \quad i > 0, \\ (-m_0, m_0) & \text{if} \quad i = 0, \\ -\Delta_{|i|} & \text{if} \quad i < 0; \end{cases}$$

these are the intervals of the Hadamard decomposition of $\mathbb{Z}$ defined by the sequence (m_k). If $i > 0$ and $n_j \leqslant m_{i-1}$, then

$$m_{i-1} < m_{i-1} + n_j \leqslant 2m_{i-1} < m_i$$

and so $E_i \subseteq \Delta_i$; the same is trivially true if $i \leqslant 0$. Thus $E_i \cap \Delta_i = E_i$ for every i, and so

$$F_0 = \bigcup_{i \in \mathbb{Z}} E_i = \bigcup_{i \in \mathbb{Z}} (E_i \cap \Delta_i).$$

In view of Theorem 9.1.4, it remains only to verify that $(E_i)_{i \in \mathbb{Z}}$ is uniformly of type $\Lambda(p)$. Since $E_i \subseteq m_{i-1} + N$ if $i > 0$ and $E_i = \varnothing$ otherwise, this follows (as in the proof of Theorem 9.2.1) from the fact that N is a Hadamard, hence a Sidon, set. □

9.2.4. Theorem. *Let $(n_k)_0^\infty$ be a sequence of positive integers such that $n_{k+1} \geqslant 3n_k$ for all $k \geqslant 0$. Let r be a positive integer. Then the set Q_r of sums*

$$\varepsilon_1 n_{k_1} + \cdots + \varepsilon_r n_{k_r}, \tag{1}$$

where $0 \leqslant k_1 < k_2 < \cdots < k_r$ and $\varepsilon_j \in \{-1, 1\}$ for each j in $\{1, \ldots, r\}$, is of type $\Lambda(p)$ for every p in $(0, \infty)$.

Proof. If we write P_r for the set of integers of the form (1), wherein $\varepsilon_r = 1$, then $Q_r = P_r \cup (-P_r)$ and it will therefore suffice to show that P_r is of type $\Lambda(p)$ for every p in $(2, \infty)$. This is done by induction on r.

If $r = 1$, P_r is a Hadamard set, hence a Sidon set, and the desired result follows from [9], 15.3.3.

Assume that P_r is of type $\Lambda(p)$, so that

$$\sum_{P_r} |\hat{f}(\chi)|^2 \leqslant C_{r,p}^2 \|f\|_{p'}^2,$$

and hence also

$$\sum_{Q_r} |\hat{f}(\chi)|^2 \leqslant 2C_{r,p}^2 \|f\|_{p'}^2, \tag{2}$$

for every trigonometric polynomial f. We write

$$P_{r+1} = \bigcup_{k>0} P_{r+1,k} \tag{3}$$

where $P_{r+1,k}$ is the set of integers of the form

$$\varepsilon_1 n_{k_1} + \cdots + \varepsilon_r n_{k_r} + n_k, \tag{4}$$

$0 \leqslant k_1 < \cdots < k_r < k$, and every ε_j belongs to $\{-1, 1\}$. We aim now to show that for large k, $P_{r+1,k}$ is a subset of the interval $[m_k, m_{k+1})$ where the sequence (m_k) to be defined below is Hadamard. Since $n_{j+1} \geqslant 3n_j$ for all j, we see immediately that

$$\varepsilon_1 n_1 + \cdots + \varepsilon_r n_r + n_k \geqslant n_k - \frac{n_k}{3} - \cdots - \frac{n_k}{3^r} = n_k\left(1 - \frac{1}{3} - \cdots - \frac{1}{3^r}\right) \tag{5}$$

and

$$\varepsilon_1 n_1 + \cdots + \varepsilon_r n_{k_r} + n_k \leqslant n_k\left(1 + \frac{1}{3} + \cdots + \frac{1}{3^r}\right). \tag{6}$$

Write $\lambda = 1 - 1/3 - \cdots - 1/3^r$, $\mu = 1 + 1/3 + \cdots + 1/3^r$ and define $m_k = [\lambda n_k]$. From (5) and (6),

$$m_k \leqslant \varepsilon_1 n_1 + \cdots + \varepsilon_r n_{k_r} + n_k \leqslant \mu n_k < m_{k+1} \tag{7}$$

if $[\lambda n_{k+1}] > \mu n_k$. This will certainly be the case if $\lambda n_{k+1} - 1 > \mu n_k$, i.e. if

$$\frac{n_{k+1}}{n_k} > \frac{1}{\lambda n_k} + \frac{\mu}{\lambda}.$$

But

$$\frac{\mu}{\lambda} = \frac{1 - (1/3)^{r+1}}{1 + (1/3)^{r+1}} \cdot \frac{4/3}{2/3} > \frac{2(1 - (1/9))}{1 + (1/9)} = \frac{16}{10}$$

and $n_{k+1}/n_k \geqslant 3$ by hypothesis. Therefore (7) will hold for all sufficiently large k; say for $k \geqslant h$.

Furthermore, we see from (5) and (7) that

$$\frac{m_{k+1}}{m_k} > \frac{\mu}{\lambda} > 1.$$

The sequence $(m_k)_0^\infty$ is therefore Hadamard, and the decomposition it determines has the LP property.

Finally then, by (3),

$$\begin{aligned} P_{r+1} &= \bigcup_{0<k<h} P_{r+1,k} + \bigcup_{k\geqslant h} P_{r+1,k} \cap [m_k, m_{k+1}) \\ &= F \cup \bigcup_{k\geqslant h} P_{r+1,k} \cap [m_k, m_{k+1}) \end{aligned}$$

say, where F is a finite set. In order to complete the proof of the theorem, it suffices, by Theorem 9.1.4 and Corollary 8.2.8, to show that the sets $P_{r+1,k}$ $(k = 1, 2, \ldots)$ are uniformly of type $\Lambda(p)$. But by (4) and (2)

$$\begin{aligned} \sum_{P_{r+1,k}} |\hat{f}(\chi)|^2 &\leqslant \sum_{n_k+Q_r} |\hat{f}(\chi)|^2 \\ &= \sum_{Q_r} |(\chi_{-n_k} f)^\wedge(\chi)|^2 \\ &\leqslant 2C_{r,p}^2 \|f\|_{p'}^2. \quad \square \end{aligned}$$

9.3. Singular Multipliers

9.3.1. The space $m_p(\mathbb{Z})$. We have on a number of occasions remarked on the well known fact that $M_1(X)$ is the space of Fourier-Stieltjes transforms, and that if $\mu \in M(G)$,

$$\|\hat{\mu}\|_{1,1} = \|\mu\|. \tag{1}$$

Another, simpler, fact is that $M_2(X) = L^\infty(X)$; and

$$\|\phi\|_{2,2} = \|\phi\|_\infty \tag{2}$$

if $\phi \in L^\infty(X)$.

Consider now the particular case $X = \mathbb{Z}$, and let $m_p(\mathbb{Z})$ denote the closure in $M_p(\mathbb{Z})$ of the space of *finitely supported* functions on $\mathbb{Z}$. It is clear from (1) that $m_1(\mathbb{Z})$ is just $A(\mathbb{Z})$, the space of Fourier transforms of $L^1(\mathbb{T})$ functions. For (1) shows that forming $m_1(\mathbb{Z})$ comes to the same thing as taking the closure in $M(\mathbb{T})$ of the space of trigonometric polynomials. Notice that

$$A(\mathbb{Z}) = m_1(\mathbb{Z}) \subseteq c_0(\mathbb{Z}) \cap M_1(\mathbb{Z}). \tag{3}$$

At the other end of the scale ($p = 2$), (2) shows that

$$m_2(\mathbb{Z}) = c_0(\mathbb{Z}). \tag{4}$$

In general, since $M_p(\mathbb{Z}) \subseteq \ell^\infty(\mathbb{Z})$, and

$$\|\phi\|_\infty \leqslant \|\phi\|_{p,p} \tag{5}$$

it is clear that

$$m_p(\mathbb{Z}) \subseteq c_0(\mathbb{Z}) \cap M_p(\mathbb{Z}). \tag{6}$$

Now in the case where $p = 1$, it is a standard fact ([40], Theorem (7.6)) that there exist *singular* measures on $\mathbb{T}$ whose Fourier-Stieltjes transforms belong to $c_0(\mathbb{Z})$. Therefore the inclusion in (3) is proper. This contrasts with the equality in (4).

Hörmander [21] posed the problem (actually in the setting of $\mathbb{R}^n$, though the problem is a general one) whether, in the case that $1 < p < 2$, $m_p(\mathbb{Z})$ is a *proper* subspace of $M_p \cap c_0(\mathbb{Z})$. By analogy with the case $p = 1$, we describe an element of $M_p \cap c_0(\mathbb{Z})\backslash m_p(\mathbb{Z})$ (if there are any!) as a *singular multiplier*.

Here now is a proof that singular multipliers do indeed exist. The proof constructs explicit examples of such multipliers.

9.3.2. A key lemma. The following result should be compared with Theorem 9.1.4, to which it is closely analogous.

Lemma. *Let p be in the range* $(1, 2)$. *Suppose ϕ is a function on $\mathbb{Z}$, zero for $n \leqslant 0$. Denote by ϕ_j the function $\xi_{\Delta_j}\phi$, where (Δ_j) is the dyadic decomposition of $\mathbb{Z}$. If*

$$\|T_{\phi_j} g\|_{p'} \leqslant A\|g\|_2 \tag{7}$$

for all trigonometric polynomials g and all indices $j \geqslant 1$, then $\phi \in M_p(\mathbb{Z})$.

Proof. We prove the equivalent assertion that $\phi \in M_{p'}(\mathbb{Z})$.

If f is an arbitrary trigonometric polynomial, apply (7) to each function $g_j = S_j f$, noticing that

$$T_{\phi_j}(S_j f) = T_\phi(S_j f) = S_j(T_\phi f).$$

It follows from (7) that

$$\begin{aligned} \|S_j(T_\phi f)\|_{p'}^2 &\leqslant A^2\|S_j f\|_2^2 \\ &\leqslant A^2 \sum_{\Delta_j} |\hat{f}(m)|^2. \end{aligned} \tag{8}$$

Add the inequalities (8) to conclude that

$$\Big(\sum_j \|S_j(T_\phi f)\|_{p'}^2\Big)^{1/2} \leqslant A\|f\|_2. \tag{9}$$

Taking note of Corollary 9.1.2(ii), the LP property of (Δ_j), and the fact that $\|f\|_2 \leqslant \|f\|_{p'}$ since $p' > 2$, we deduce from (9) that

$$\|T_\phi f\|_{p'} \leqslant A_p^{-1} (\sum_j \|S_j(T_\phi f)\|_{p'}^2)^{1/2} \leqslant A_p^{-1} A\|f\|_2 \leqslant A_p^{-1} A\|f\|_{p'}. \quad \square$$

9.3.3. The Rudin-Shapiro polynomials. The *Rudin-Shapiro polynomials* are the members of the sequences $(\rho_n)_0^\infty$ and $(\sigma_n)_0^\infty$ of trigonometric polynomials defined as follows.

$$\begin{cases} \rho_0 = \sigma_0 = 1; \\ \begin{cases} \rho_n(e^{it}) = \rho_{n-1}(e^{it}) + \exp(i2^{n-1}t)\sigma_{n-1}(e^{it}) \\ \sigma_n(e^{it}) = \rho_{n-1}(e^{it}) - \exp(i2^{n-1}t)\sigma_{n-1}(e^{it}). \end{cases} \quad (n \geqslant 1) \end{cases} \tag{10}$$

Since

$$\begin{aligned} |\rho_n|^2 + |\sigma_n|^2 &= 2(|\rho_{n-1}|^2 + |\sigma_{n-1}|^2) \\ &= \cdots \\ &= 2^{n+1}, \end{aligned}$$

it is plain that

$$\|\rho_n\|_\infty \leqslant 2^{(n+1)/2}. \tag{11}$$

Furthermore, the function $\hat{\rho}_n$ takes the values 0, $+1$ and -1 only, and is supported exactly on the set $\{0, \ldots, 2^n - 1\}$. Therefore, if $1 < r < \infty$,

$$\|\hat{\rho}_n\|_r = 2^{n/r}. \tag{12}$$

9.3.4. Construction of ϕ. Lemma 9.3.2 indicates that it is possible to construct a multiplier ϕ from the "pieces" ϕ_j if the operators T_{ϕ_j} are made to have uniform bounds on their $(L^2, L^{p'})$ norms. The following lemma shows how this can be arranged. The method, based on the Hausdorff-Young theorem, is crude, but is just good enough for the problem at hand.

Lemma. *Suppose ψ is a bounded function on $\mathbb{Z}$ and that $1 < p < 2$. Set $r = 2p/(2-p)$. Then*

$$\|T_\psi g\|_{p'} \leqslant \|\psi\|_r \|g\|_2 \tag{13}$$

for all trigonometric polynomials g.

Proof. In order to prove (13), it suffices to prove the dual inequality

$$\|T_\psi g\|_2 \leqslant \|\psi\|_r \|g\|_p.$$

The Hausdorff-Young inequality states that

$$\|\hat{g}\|_{p'} \leqslant \|g\|_{p}.$$

Hölder's inequality for the index $r/2$ shows therefore that

$$\|T_{\psi}g\|_2^2 = \sum |\psi(m)\hat{g}(m)|^2 \leqslant (\sum |\psi(m)|^r)^{2/r} (\sum |\hat{g}(m)|^{p'})^{2/p'} \leqslant \|\psi\|_r^2 \|g\|_p^2,$$

since $(r/2)' = p'/2$. This completes the proof. □

The construction of ϕ now proceeds as follows. Define

$$\phi(n) = \begin{cases} 0 & \text{if} \quad n \leqslant 0 \\ \hat{\rho}_{j-1}(n - 2^{j-1})/2^{(j-1)/r} & \text{if} \quad n \in \Delta_j, j \geqslant 1. \end{cases} \tag{14}$$

Then (12) and (13) imply that

$$\|T_{\phi_j}g\|_{p'} \leqslant \|g\|_2$$

for all trigonometric polynomials g, so that, by Lemma 9.3.2, $\phi \in M_p$. It is clear from (14) that $\phi \in c_0(\mathbb{Z})$. It remains to show that ϕ is not approximable in M_p by finitely supported functions. This last matter is tidied up in the proof of the theorem.

9.3.5. Theorem. *If* $1 < p < 2$, *there are singular multipliers of* $L^p(\mathbb{T})$. *In other words, the inclusion* $m_p(\mathbb{Z}) \subseteq M_p \cap c_0(\mathbb{Z})$ *is proper.*

Proof. We have only to prove that the function ϕ, constructed in 9.3.4, does not belong to $m_p(\mathbb{Z})$. Suppose then that ψ is a function on $\mathbb{Z}$ with finite support F. Choose j so large that F lies to the left of 2^{j-1}. Then

$$\begin{aligned} \|\phi - \psi\|_{p,p} &= \|\phi - \psi\|_{p',p'} \\ &= \sup \|T_{(\phi-\psi)}f\|_{p'}/\|f\|_{p'}, \end{aligned} \tag{15}$$

f being allowed to range over all nonzero trigonometric polynomials.

For each index $k \geqslant j$, define

$$f_k(e^{it}) = \exp(2^{k-1}it)\rho_{k-1}(e^{it}). \tag{16}$$

Then by (14), (16) and the fact that $\hat{\rho}_{k-1} = \pm 1$ on $[0, 2^{k-1})$

$$T_{(\phi-\psi)}f_k = T_{\phi_k}f_k = \left(\sum_{m \in \Delta_k} e^{imt}\right)\Big/2^{(k-1)/r}; \tag{17}$$

moreover,

$$\|f_k\|_{p'} \leqslant \|f_k\|_{\infty} \leqslant 2^{k/2} \tag{18}$$

by (11). It follows from (15), (17) and (18) that

$$\|\phi - \psi\|_{p',p'} \geqslant \liminf_{k\to\infty} \|T_{(\phi-\psi)}f_k\|_{p'}/\|f_k\|_{p'}$$

$$\geqslant \liminf_{k\to\infty} \|\sum_{m\in\Delta_k} e^{imt}\|_{p'}/2^{(k-1)/r}2^{k/2}. \tag{19}$$

However,

$$\begin{aligned}\sum_{m\in\Delta_k} e^{imt} &= \sum_{2^{k-1}}^{2^k-1} e^{imt} \\ &= \exp[(2^{k-1} + 2^{k-2})it] \sum_{-2^{k-2}}^{2^{k-2}} e^{imt} - \exp(2^k it) \\ &= \exp(3.2^{k-2}it)\mathsf{D}_{2^{k-2}}(e^{it}) - \exp(2^k it),\end{aligned} \tag{20}$$

where D_N stands for the Dirichlet kernel of order N. But it is a standard fact ([9], Exercise 7.5) that

$$\|\mathsf{D}_N\|_{p'} \sim A_{p'}N^{1/p}$$

as $N \to \infty$; so (19) and (20) show that

$$\begin{aligned}\|\phi - \psi\|_{p',p'} &\geqslant \lim_{k\to\infty} \|\mathsf{D}_{2^{k-2}}(e^{it})\|_{p'}/2^{k(1/r+1/2)-1/r} \\ &= A_{p'} \lim_{k\to\infty} 2^{(k-2)/p}/2^{k/p-1/r} \\ &= A_{p'}2^{-1/p-1/2}\end{aligned}$$

since (Lemma 9.3.4)

$$\frac{1}{r} = \frac{1}{p} - \frac{1}{2}.$$

In other words, every finitely supported function ψ on $\mathbb{Z}$ is distant at least $A_{p'}/2^{(1/2+1/p)}$ from ϕ. This completes the proof. □

Appendix A. Special Cases of the Marcinkiewicz Interpolation Theorem

In a number of places in Chapters 2, 3 and 5, we employ simple forms of the Marcinkiewicz interpolation theorem. The purpose of this appendix is to present for the reader's convenience statements and proofs of the theorems involved and to explain the concepts associated with them. More general versions of the Marcinkiewicz theorem can be found in [9], Section 13.8 and [33], Appendix B.

Let us agree that the measure spaces appearing below are always σ-finite.

A.1. The Concepts of Weak Type and Strong Type

A.1.1. Strong type. Let $(M, \mathscr{M}, \mu)$ and $(N, \mathscr{N}, \nu)$ be measure spaces and p an index in the range $[1, \infty]$. Suppose D is a subset of $L^p(\mu)$ and T is a mapping from D into the space of complex-valued measurable functions on N, or the space of nonnegative extended-real-valued measurable functions on N, or the space of equivalence classes of one of these. We say T is *of strong type* (p, p), or simply *of type* (p, p), *on* D if there is a constant B such that

$$\|Tf\|_p \leqslant B\|f\|_p \tag{1}$$

for all f in D. The smallest number B for which (1) holds is then termed the (p, p) *norm of* T *on* D and is denoted $\|T\|_{p,p}$, when D is understood.

Even if there may not exist a finite constant B for which (1) holds, it is customary to define

$$\|T\|_{p,p} = \sup\{\|Tf\|_p/\|f\|_p : f \in D, f \neq 0\}$$

(provided of course that $D \neq \{0\}$). So we may say that T is of type (p, p) on D if and only if $\|T\|_{p,p} < \infty$.

A.1.2. *Remarks.* (i) In practice, D is usually a linear subspace of $L^p(\mu)$, T takes its values in the space of complex measurable functions on N, and is linear.

(ii) Our concern in the text is mostly with operators T of the form T_ϕ introduced in 1.2.2 and 2.4.1. The operators T_ϕ are viewed as having the initial domain $L^2(G)$ and range in $L^2(G)$, G being an arbitrary LCA group. Our main interest is in knowing whether, when $1 \leqslant p \leqslant \infty$ and T_ϕ is restricted to $D = L^2 \cap L^p(G)$,

T_ϕ is of type (p, p) on D; in several instances, it is the *value* of $\|T_\phi\|_{p,p}$ which is more important.

(iii) The definition of strong type is clearly D-dependent, in general. However, in most practical instances, this poses no difficulty. To illustrate the point, suppose $\phi \in \mathscr{L}^\infty(X)$, X being the dual group of G, $T = T_\phi$ and T is of type (p, p) on D. Suppose that D is a linear subspace of $L^2 \cap L^p(G)$ and that for every f in $L^2 \cap L^p(G)$ there exists a sequence (f_n) extracted from D such that

$$\lim \|f_n - f\|_2 = 0 \text{ and } \lim \|f_n\|_p \leqslant \|f\|_p. \tag{2}$$

($D = L^1 \cap L^\infty(G)$ for example). Then if (1) holds for every f in D, it continues to hold for every f in $L^2 \cap L^p(G)$.

To see this, suppose $f \in L^2 \cap L^p(G)$ and that (f_n) is as above. Since $f_n \to f$ in $L^2(G)$ and every operator T_ϕ is continuous on $L^2(G)$ (cf. 1.2.2), it follows that $Tf_n \to Tf$ in $L^2(G)$. Hence there is a subsequence (Tf_{n_j}) which converges pointwise a.e. to Tf. Now

$$\|Tf\|_p \leqslant \liminf_{j\to\infty} \|Tf_{n_j}\|_p \tag{3}$$

by Fatou's lemma if $p < \infty$, and trivially otherwise. Since T is of type (p, p) on D,

$$\|Tf_{n_j}\|_p \leqslant B\|f_{n_j}\|_p \tag{4}$$

for all j; combining (2), (3) and (4), we deduce that T is of type (p, p) on $L^2 \cap L^p(G)$.

In the same way, T is continuously extendable into an operator of type (p, p) on $L^p(G)$.

A.1.3. Weak type. Let T and D be as in A.1.1, and denote by λ_{Tf} the distribution function of $|Tf|$; that is, define, for $t > 0$,

$$\lambda_{Tf}(t) = \nu(\{y \in N: |Tf(y)| > t\}).$$

If $p < \infty$, we say T is *of weak type* (p, p) *on* D if there is a nonnegative real number A such that

$$\lambda_{Tf}(t) \leqslant A^p t^{-p} \|f\|_p^p \tag{5}$$

for all f in D and all $t > 0$. If there exist such numbers A, there is a smallest, called the *weak* (p, p) *norm of* T *on* D. If no such number exists, the weak (p, p) norm of T on D is set equal to ∞.

The mapping T is said to be *of weak type* (∞, ∞) *on* D if and only if it is of type (∞, ∞) on D; its *weak* (∞, ∞) *norm* is declared to be the same as its (∞, ∞) norm.

It is very simple to see that a mapping T of type (p, p) on D is also of weak type there, but the converse is false, unless of course $p = \infty$.

A.1.4. *Remark.* The choice of D is again to some extent immaterial. For instance, suppose that $p \in [1, \infty)$, $T = T_\phi$, D is as in A.1.2(iii) and that (5) holds for all f in D and all $t > 0$. Then (5) holds for all f in $L^2 \cap L^p(G)$.

To see this, adopt the notation of A.1.2(iii). Then

$$\{y: |Tf(y)| > t\} \subseteq \bigcup_i \bigcap_{j \geqslant j} \{y: |Tf_{n_j}(y)| > t\}$$

and hence

$$\begin{aligned}\lambda_{Tf}(t) &\leqslant \lim_{i\to\infty} \nu(\bigcap_{j\geqslant i} \{y: |Tf_{n_j}(y)| > t\}) \\ &\leqslant \liminf_{j\to\infty} \nu(\{y: |Tf_{n_j}(y)| > t\}) \\ &\leqslant \liminf_{j\to\infty} A^p t^{-p} \|f_{n_j}\|_p^p \\ &\leqslant A^p t^{-p} \|f\|_p^p.\end{aligned}$$

At this point, it is possible to go one step further and extend T from $L^2 \cap L^p(G)$ into a mapping from $L^p(G)$ into the set of classes of measurable functions in such a way that (5) continues to hold for every f in $L^p(G)$. For, given f in $L^p(G)$, select any sequence (f_n) from $L^2 \cap L^p(G)$ converging in L^p to f. Write g_n for any function of the class Tf_n. Then (5) shows that the sequence (g_n) is Cauchy in measure and therefore converges in measure to some function g. It also follows from (5) that the class of g does not depend on the choice of the sequence (f_n) (provided $f_n \to f$ in $L^p(G)$, of course). So we may define Tf to be the class of g. Once again, there is a subsequence (g_{n_j}) converging a.e. to g and so the same argument as before leads to (5).

It follows from (5) that Tf, although it may not belong to $L^p(G)$, does belong locally to $L^q(G)$ for every $q < p$ ([9], Exercise 13.16).

A.2. The Interpolation Theorems

Let f be a measurable function and $t > 0$. Denote by f_t and f^t the following functions.

$$f_t(x) = \begin{cases} f(x) & \text{if } |f(x)| \leqslant t \\ 0 & \text{otherwise} \end{cases}$$

$$f^t(x) = \begin{cases} f(x) & \text{if } |f(x)| > t \\ 0 & \text{otherwise.} \end{cases}$$

With this notation fixed, we can state and prove the first theorem.

A.2.1. Theorem (Marcinkiewicz). *Suppose that $r \in (1, \infty)$, D is a linear subspace of $L^1 \cap L^r(M)$ and T is an operator mapping D into the set of equivalence* *classes of complex measurable functions or of nonnegative extended-real-valued measurable functions on N. Assume that D and T satisfy the following conditions.*

(i) *If $f \in D$ and $t > 0$, then f_t and f^t are in D.*

(ii) *T is subadditive in the sense that*

$$|T(f + g)| \leqslant |Tf| + |Tg|$$

for f and g in D.

(iii) *T is of weak type* (1, 1) *on D with weak* (1, 1) *norm at most A_1, so that*

$$\lambda_{Tf}(t) \leqslant A_1 t^{-1} \|f\|_1 \tag{1}$$

for f in D and $t > 0$.

(iv) *T is of weak type (r, r) on D with weak (r, r) norm at most A_r, so that*

$$\lambda_{Tf}(t) \leqslant A_r^r t^{-r} \|f\|_r^r \tag{2}$$

for f in D and $t > 0$.

Suppose that $1 < p < r$. Then T is of type (p, p) on D and

$$\|Tf\|_p \leqslant A_p \|f\|_p \tag{3}$$

for all f in D, where

$$A_p^p = \{2A_1(p - 1)^{-1} + (2A_r)^r (r - p)^{-1}\} p. \tag{4}$$

In other words, the (p, p) norm of T on D is at most A_p, where A_p is given by (4).

Proof. Suppose $f \in D$ and $t > 0$. Since $f = f_t + f^t$, the subadditivity of T and (i) show that

$$|Tf| \leqslant |Tf_t| + |Tf^t|,$$

whence it follows that

$$\lambda_{Tf}(t) \leqslant \nu(\{y: |Tf^t(y)| > t/2\}) + \nu(\{y: |Tf_t(y)| > t/2\}).$$

Applying (1) and (2) to f^t and f_t respectively, we deduce that

$$\begin{aligned} \lambda_{Tf}(t) &\leqslant 2A_1 t^{-1} \int_M |f^t| \, d\mu + (2A_r)^r t^{-r} \int_M |f_t|^r \, d\mu \\ &= 2A_1 \int_{\{x: |f(x)| > t\}} |f| \, d\mu + 2^r A_r^r t^{-r} \int_{\{x: |f(x)| \leqslant t\}} |f|^r \, d\mu. \end{aligned} \tag{5}$$

Now

$$\|Tf\|_p^p = p \int_0^\infty t^{p-1} \lambda_{Tf}(t) \, dt \tag{6}$$

and so we deduce from (5) that

$$\begin{aligned} p^{-1}\|Tf\|_p^p &\leqslant \int_0^\infty t^{p-1}\left\{2A_1 t^{-1}\int_{\{x:\,|f(x)|>t\}} |f|\,d\mu\right\}dt \\ &\quad + \int_0^\infty t^{p-1}\left\{2^r A_r^r t^{-r}\int_{\{x:\,|f(x)|\leqslant t\}} |f|^r\,d\mu\right\}dt \\ &= 2A_1\int_0^\infty t^{p-2}\left\{\int_M |f(x)|\phi(x,t)\,d\mu(x)\right\}dt \\ &\quad + 2^r A_r^r\int_0^\infty t^{p-1-r}\left\{\int_M |f(x)|^r\,\psi(x,t)\,d\mu(x)\right\}dt, \end{aligned} \tag{7}$$

where ϕ is the characteristic function of the set

$$E = \{(x,t)\colon |f(x)| > t\} \subseteq M \times (0,\infty)$$

and ψ is the characteristic function of the set

$$F = \{(x,t)\colon |f(x)| \leqslant t\} \subseteq (M \times (0,\infty))\backslash E.$$

If $(s_n)_{n\geqslant 1}$ is an enumeration of the positive rationals,

$$E = \bigcup_{n\geqslant 1} (\{x\colon |f(x)| > s_n\} \times (0, s_n)),$$

which shows that E is measurable in the pair of variables. The same is therefore true of F and so, by the Fubini theorem (recall that M is σ-finite) we may invert the order of the integrations in (7) to conclude that

$$\begin{aligned} p^{-1}\|Tf\|_p^p &\leqslant 2A_1\int_M\left\{\int_0^{|f(x)|} t^{p-2}\,dt\right\}|f(x)|\,d\mu(x) \\ &\quad + 2^r A_r^r\int_M\left\{\int_{|f(x)|}^\infty t^{p-1-r}\,dt\right\}|f(x)|^r\,d\mu(x) \\ &= 2A_1\int_M (p-1)^{-1}|f(x)|^{p-1}|f(x)|\,d\mu(x) \\ &\quad + 2^r A_r^r\int_M (r-p)^{-1}|f(x)|^{p-r}|f(x)|^r\,d\mu(x), \end{aligned}$$

which is equivalent to (3) and (4). □

The second case of the Marcinkiewicz theorem deals with operators simultaneously of weak types (r, r) and (∞, ∞), where $1 \leqslant r < \infty$.

A.2.2. Theorem (Marcinkiewicz). *Suppose that* $1 \leqslant r < \infty$, *and that* D *and* T *satisfy the assumptions in the statement of Theorem* A.2.1 *save for condition* (iii). *In place of* (iii), *assume that* (iii′) *T is of (weak) type* (∞, ∞) *on D, with (weak)* (∞, ∞) *norm at most* A_∞, *so that*

$$\|Tf\|_\infty \leqslant A_\infty \|f\|_\infty \tag{8}$$

for all f in D.

If $r < p < \infty$, *then T is of type* (p, p) *on D, and*

$$\|Tf\|_p \leqslant A_p \|f\|_p \tag{9}$$

for all f in D, where

$$A_p^p = \frac{p 2^p A_r^r A_\infty^{p-r}}{(p - r)}.$$

Proof. Suppose $f \in D$ and $t > 0$. We may assume that $A_\infty > 0$ and so write

$$f = f_{t/2A_\infty} + f^{t/2A_\infty}.$$

The condition (8) shows that

$$\| Tf_{t/2A_\infty}\|_\infty \leqslant A_\infty \|f_{t/2A_\infty}\|_\infty \leqslant \frac{t}{2}. \tag{10}$$

It follows from the subadditivity of T and (i) that

$$|Tf| \leqslant |Tf_{t/2A_\infty}| + |Tf^{t/2A_\infty}|,$$

and then from (10) that

$$\begin{aligned} \lambda_{Tf}(t) &\leqslant \nu(\{y: |Tf_{t/2A_\infty}(y)| > t/2\}) + \nu(\{y: |Tf^{t/2A_\infty}(y)| > t/2\}) \\ &= \nu(\{y: |Tf^{t/2A_\infty}(y)| > t/2\}). \end{aligned} \tag{11}$$

By combining (11) and (2), we see that

$$\begin{aligned} \lambda_{Tf}(t) &\leqslant A_r^r (t/2)^{-r} \|f^{t/2A_\infty}\|_r^r \\ &= 2^r A_r^r t^{-r} \int_{\{x: |f(x)| > t/2A_\infty\}} |f|^r \, d\mu. \end{aligned}$$

Using once more the formula (6), we deduce that

$$\begin{aligned} \frac{1}{p} \|Tf\|_p^p &\leqslant \int_0^\infty t^{p-1} \left\{ 2^r A_r^r t^{-r} \int_{\{x: |f(x)| > t/2A_\infty\}} |f|^r \, d\mu \right\} dt \\ &= 2^r A_r^r \int_0^\infty t^{p-r-1} \left\{ \int_{\{x: |f(x)| > t/2A_\infty\}} |f|^r \, d\mu \right\} dt. \end{aligned} \tag{12}$$

Now apply a Fubini-type argument again to (12), as in the final stages of the proof of A.2.1. The conclusion is that

$$\begin{aligned}\frac{1}{p}\|Tf\|_p^p &\leqslant 2^r A_r^r \int_M |f(x)|^r \int_0^{2A_\infty|f(x)|} t^{p-r-1}\,dt\,d\mu(x)\\ &= \frac{2^r A_r^r (2A_\infty)^{p-r}}{(p-r)} \int_M |f(x)|^p\,d\mu(x)\\ &= \frac{2^p A_r^r A_\infty^{p-r}}{(p-r)} \|f\|_p^p. \quad \square\end{aligned}$$

A.3. Vector-Valued Functions

The ideas and results of A.1 and A.2 apply with only notational changes to the cases in which complex-valued or extended-real-valued functions are replaced by vector-valued functions of the type discussed in Chapter 3. In these cases, T is assumed to map (suitably restricted) functions with values in one Hilbert space into functions or equivalence classes of functions with values in a second Hilbert space. In formulating the concepts of weak type and strong type, absolute values are replaced by the appropriate Hilbert space norms. We leave the reader to write down the translations of the definitions and the theorems and in particular to check that the proofs of the Marcinkiewicz theorems given in A.2.1 and A.2.2 go through for vector-valued functions with no more than the obvious notational changes.

Appendix B. The Homomorphism Theorem for Multipliers

Let G and H be LCA groups with duals X and Y respectively. Suppose that π is a continuous homomorphism of Y into X and $\hat{\pi}$ its *dual homomorphism* from G into H, defined by the requirement that

$$\hat{\pi}(x)(\gamma) = x(\pi(\gamma))$$

for γ in Y and x in G. Notice that $(\hat{\pi})^\wedge = \pi$.

Our aim in this appendix is to give a self-contained proof of what we have called the homomorphism theorem for multipliers. While elementary in the strict sense of the word, the proof is nonetheless quite intricate.

Theorem. *Suppose that* $1 \leqslant p \leqslant \infty$, $\phi \in M_p(X)$ *and* ϕ *is continuous. Then* $\phi \circ \pi \in M_p(Y)$ *and*

$$\|\phi \circ \pi\|_{p,p} \leqslant \|\phi\|_{p,p}. \tag{1}$$

B.1. The Key Lemmas

The first two lemmas are purely technical, involve only standard procedures, and amount to showing that it is enough to prove (1) for very "good" functions ϕ.

B.1.1. Lemma. *In order to prove the theorem it suffices to show that if* $\psi \in M_p \cap C_c(X)$, *then* $\psi \circ \pi \in M_p(Y)$ *and*

$$\|\psi \circ \pi\|_{p,p} \leqslant \|\psi\|_{p,p}.$$

Proof. Suppose that $\phi \in M_p \cap C(X)$ and let (k_α) be a net of functions on G with the following properties:

(i) $k_\alpha \in L^1(G)$ and $\|k_\alpha\|_1 \leqslant 1$;

(ii) $\hat{k}_\alpha \in C_c(X)$;

(iii) $\hat{k}_\alpha \to 1$ locally uniformly on X.

Since $\phi \in M_p(X)$, it is easily seen that

$$\left| \int_X \phi(\chi)\hat{k}_\alpha(\chi)\hat{f}(\chi)\hat{g}(\chi)\, d\chi \right| \leqslant \|\phi\|_{p,p}\|f\|_p\|g\|_{p'}$$

whenever f and g are integrable on G and have compactly supported transforms. So by 1.2.2(iii), $\phi\hat{k}_\alpha \in M_p(X)$ and

$$\|\phi\hat{k}_\alpha\|_{p,p} \leqslant \|\phi\|_{p,p}.$$

If the hypotheses of the lemma are satisfied, then by (ii),

$$(\phi\hat{k}_\alpha) \circ \pi \in M_p(Y) \quad \text{and} \quad \|(\phi\hat{k}_\alpha) \circ \pi\|_{p,p} \leqslant \|\phi\|_{p,p}.$$

Therefore, if h and k are in $L^1(H)$ and have transforms in $C_c(Y)$,

$$\left|\int_Y (\phi\hat{k}_\alpha) \circ \pi(\gamma)\hat{h}(\gamma)\hat{k}(\gamma)\, d\gamma\right| \leqslant \|\phi\|_{p,p}\|h\|_p\|k\|_{p'}. \tag{2}$$

Since π is continuous, and $\phi\hat{k}_\alpha \to \phi$ locally uniformly (see (iii)), it follows that $(\phi\hat{k}_\alpha) \circ \pi \to \phi \circ \pi$ locally uniformly. Moreover,

$$\|(\phi\hat{k}_\alpha) \circ \pi\|_\infty \leqslant \|\phi\|_\infty\|\hat{k}_\alpha\|_\infty \leqslant \|\phi\|_\infty,$$

by (i); and $\hat{h}\hat{k} \in L^1(Y)$. So the left side of (2) tends to

$$\left|\int_Y \phi \circ \pi(\gamma)\hat{h}(\gamma)\hat{k}(\gamma)\, d\gamma\right|$$

as $\alpha \to \infty$; therefore the result follows from 1.2.2 if we take the limit on α in (2). □

B.1.2. Lemma. *The theorem will be established if it is shown that*

$$\|\phi \circ \pi\|_{p,p} \leqslant \|\phi\|_{p,p}$$

whenever $\phi \in \mathfrak{F}L^1 \cap L^1(X)$.

Proof. Notice that, in any case, if $\phi \in \mathfrak{F}L^1(X)$, then $\phi \circ \pi \in \mathfrak{F}M(Y) = M_1(Y) \subseteq M_p(Y)$. We prove the present lemma by using B.1.1.

To this end, suppose $\psi \in M_p \cap C_c(X)$ and let (F_α) be an approximate identity on X consisting of functions in $L^1 \cap L^\infty(X)$ for which $\|F_\alpha\|_1 \leqslant 1$. Then

(i) $\psi * F_\alpha \in \mathfrak{F}L^1 \cap L^1(X)$;

and

(ii) $\psi * F_\alpha \to \psi$ uniformly on X.

We claim that, furthermore,

(iii) $\|\psi * F_\alpha\|_{p,p} \leqslant \|\psi\|_{p,p}$.

To see this, observe that by 1.2.2, it suffices to prove that

$$\left|\int_X \psi * F_\alpha(\chi)\hat{f}(\chi)\hat{g}(\chi)\, d\chi\right| \leqslant \|\psi\|_{p,p}\|f\|_p\|g\|_{p'} \tag{3}$$

whenever f and g are integrable and have compactly supported Fourier transforms.

By 1.2.2(iii) again, however, since $\psi \in M_p(X)$,

$$\left| \int_X \psi(\chi)\hat{f}(\chi)\hat{g}(\chi)\, d\chi \right| \leqslant \|\psi\|_{p,p}\|f\|_p\|g\|_{p'}.$$

More importantly,

$$\left| \int_X \psi(\chi - \chi_0)\hat{f}(\chi)\hat{g}(\chi)\, d\chi \right| \leqslant \|\psi\|_{p,p}\|f\|_p\|g\|_{p'}$$

for all χ_0 in X. Hence

$$\left| \int_X F_\alpha(\chi_0)\left\{ \int_X \psi(\chi - \chi_0)\hat{f}(\chi)\hat{g}(\chi)\, d\chi \right\} d\chi_0 \right| \leqslant \|\psi\|_{p,p}\|f\|_p\|g\|_{p'};$$

and by using Fubini's theorem, we conclude finally that

$$\left| \int_X \left\{ \int_X \psi(\chi - \chi_0)F_\alpha(\chi_0)\, d\chi_0 \right\} \hat{f}(\chi)\hat{g}(\chi)\, d\chi \right| \leqslant \|\psi\|_{p,p}\|f\|_p\|g\|_{p'}.$$

This is just (3); so (iii) holds.

To finish off the proof, suppose that (1) holds for all ϕ in $\mathfrak{F}L^1 \cap L^1(X)$. By (i) and (iii),

$$\|(\psi * F_\alpha) \circ \pi\|_{p,p} \leqslant \|\psi * F_\alpha\|_{p,p} \leqslant \|\psi\|_{p,p}.$$

Consequently, if h and k are integrable on H and have transforms in $C_c(Y)$, 1.2.2(iii) shows that

$$\left| \int_Y (\psi * F_\alpha) \circ \pi(\gamma)\hat{h}(\gamma)\hat{k}(\gamma)\, d\gamma \right| \leqslant \|\psi\|_{p,p}\|h\|_p\|k\|_{p'}. \tag{4}$$

Now notice that, by (ii) and the continuity of π, $(\psi * F_\alpha) \circ \pi \to \psi \circ \pi$ uniformly on Y; moreover, $\hat{h}\hat{k} \in C_c(Y)$. So we can deduce, by taking the limit on α in (4), that

$$\left| \int_X \psi \circ \pi(\gamma)\hat{h}(\gamma)\hat{k}(\gamma)\, d\gamma \right| \leqslant \|\psi\|_{p,p}\|h\|_p\|k\|_{p'}.$$

By 1.2.2(iii), this completes the proof. □

The next lemma establishes a generalised Parseval formula.

B.1.3. Lemma. *Suppose that* $\phi \in \mathfrak{F}L^1 \cap L^1(X)$, *and* $E \in \mathfrak{F}L^1 \cap L^1(H)$. *Then*

$$\int_G E \circ \hat{\pi}(x)\hat{\phi}(x)\, dx = \int_Y \phi \circ \pi(\gamma)\hat{E}(\gamma)\, d\gamma. \tag{5}$$

Proof. Let μ be the bounded measure on X such that

$$\mu(S) = \int_{\pi^{-1}(S)} \hat{E}(-\gamma)\, d\gamma.$$

It is a simple matter to check that

$$\int_X f(\chi)\, d\mu(\chi) = \int_Y f \circ \pi(\gamma)\hat{E}(-\gamma)\, d\gamma \tag{6}$$

for say all bounded Borel functions f on X. In particular if $x \in G$ and $f(\chi) = \overline{\chi(x)}$, we deduce from (6) that

$$\begin{aligned}\hat{\mu}(x) &= \int_Y \overline{\pi(\gamma)(x)}\, \hat{E}(-\gamma)\, d\gamma \\ &= \int_Y \overline{\gamma(\hat{\pi}(x))}\, \hat{E}(-\gamma)\, d\gamma \\ &= E(\hat{\pi}(x))\end{aligned}$$

by the definition of $\hat{\pi}$ and the inversion formula. So the left side of (5) is just

$$\int_G \hat{\mu}(x)\, \hat{\phi}(x)\, dx,$$

which, by the inversion formula again, is equal to

$$\int_X \phi(-\chi)\, d\mu(\chi).$$

Yet, by (6),

$$\begin{aligned}\int_X \phi(-\chi)\, d\mu(\chi) &= \int_Y \phi \circ \pi(-\gamma)\hat{E}(-\gamma)\, d\gamma \\ &= \int_Y \phi \circ \pi(\gamma)\hat{E}(\gamma)\, d\gamma\end{aligned}$$

by the reflection-invariance of Haar measure. This establishes (5). □

B.2. The Homomorphism Theorem

B.2.1. Theorem. *Let X and Y be LCA groups and π a continuous homomorphism from Y to X. Then if $1 \leqslant p \leqslant \infty$ and $\phi \in M_p \cap C(X)$, it follows that $\phi \circ \pi \in M_p(Y)$*

and

$$\|\phi \circ \pi\|_{p,p} \leqslant \|\phi\|_{p,p}. \tag{1}$$

Proof. By Lemma B.1.2, it will suffice to prove the inequality (1) for functions ϕ which are integrable and have integrable transforms. Let ϕ be such a function. Then by 1.2.2(iii), (1) is equivalent to the inequality

$$\left|\int_Y \phi \circ \pi(\gamma)\hat{h}(\gamma)\hat{k}(\gamma)\, d\gamma\right| \leqslant \|\phi\|_{p,p}\|h\|_p\|k\|_{p'} \tag{2}$$

for h and k integrable functions on H with compactly supported Fourier transforms. We proceed to establish (2).

By Lemma B.1.3,

$$\int_Y \phi \circ \pi(\gamma)\hat{h}(\gamma)\hat{k}(\gamma)\, d\gamma = \int_G (h * k) \circ \hat{\pi}(x)\hat{\phi}(x)\, dx. \tag{3}$$

Since $\hat{\phi} \in L^1(G)$, we may, given $\varepsilon > 0$, choose a compact set K in G so that

$$\int_{G \setminus K} |(h * k) \circ \hat{\pi}(x)\hat{\phi}(x)|\, dx \leqslant \varepsilon\|\phi\|_{p,p}\|h\|_p\|k\|_{p'}. \tag{4}$$

An examination of the proof of Theorem (31.37) of [20] shows that we can choose functions f and g in $L^1 \cap L^\infty(G)$ so that

(i) $1 \geqslant f * g \geqslant 0$ and $f * g = 1$ on K;

(ii) $\|f\|_p\|g\|_{p'} \leqslant 1 + \varepsilon$.

In order to prove (2), it will suffice to prove that

$$\left|\int_G (h * k) \circ \hat{\pi}(x) f * g(x)\hat{\phi}(x)\, dx\right| \leqslant (1 + \varepsilon)\|\phi\|_{p,p}\|h\|_p\|k\|_{p'}. \tag{5}$$

For if (5) holds, we conclude from (4) and (i) that

$$\left|\int_G (h * k) \circ \hat{\pi}(x)\hat{\phi}(x)\, dx\right| \leqslant (1 + 2\varepsilon)\|\phi\|_{p,p}\|h\|_p\|k\|_{p'}. \tag{6}$$

Since ε is arbitrary, (2) follows from (3) and (6).

In establishing (5), the key step is the following integral expression for the pointwise product $(h * k) \circ \hat{\pi}.f * g$

$$(h * k) \circ \hat{\pi}(x) f * g(x) = \int_H [(\tau_u h) \circ \hat{\pi}.f] * [(\tau_{-u} k) \circ \hat{\pi}.\, g](x)\, dx. \tag{7}$$

In (7), the function $\tau_u h$ is the u-translate of h defined by the formula

$$\tau_u h(x) = h(x - u).$$

The equality (7) is established as follows. By the translation-invariance of the Haar measure on H,

$$h * k(\hat{\pi}(x))f * g(x) = \int_H h(\hat{\pi}(x) - u)k(u)\,du \int_G f(x - y)g(y)\,dy$$

$$= \int_G \int_H h(\hat{\pi}(x) - \hat{\pi}(y) - u)k(u + \hat{\pi}(y))f(x - y)g(y)\,dy\,du. \quad (8)$$

The right side of (8) can be rewritten

$$\int_G \int_H [(\tau_u h) \circ \hat{\pi}.f](x - y)[(\tau_{-u}k) \circ \hat{\pi}.g](y)\,du\,dy$$

and, by the Fubini-Tonelli theorem, this is the same as

$$\int_H \int_G [(\tau_u h) \circ \hat{\pi}.f](x - y)[(\tau_{-u}k) \circ \hat{\pi}.g](y)\,dy\,du$$

$$= \int_H [(\tau_u h) \circ \hat{\pi}.f] * [(\tau_{-u}k) \circ \hat{\pi}.g](x)\,du.$$

So (7) is established.

Return now to the left side of (5) and write it

$$\left| \int_G \hat{\phi}(x) \int_H [(\tau_u h) \circ \hat{\pi}.f] * [(\tau_{-u}k) \circ \hat{\pi}.g](x)\,du\,dx \right|,$$

which, by the Fubini-Tonelli theorem, is

$$\left| \int_H \int_G \hat{\phi}(x)[(\tau_u h) \circ \hat{\pi}.f] * [(\tau_{-u}k) \circ \hat{\pi}.g](x)\,dx\,du \right|.$$

By using the Parseval formula, this can be written

$$\left| \int_H \int_X \phi(\chi)[(\tau_u h) \circ \hat{\pi}.f]^\wedge(\chi)[(\tau_{-u}k) \circ \hat{\pi}.g]^\wedge(\chi)\,d\chi\,du \right|,$$

which is at most

$$\int_H \left| \int_X \phi(\chi)[(\tau_u h) \circ \hat{\pi}.f]^\wedge(\chi)[(\tau_{-u}k) \circ \hat{\pi}.g]^\wedge(\chi)\,d\chi \right| du.$$

By the definition of $\|\phi\|_{p,p}$, this last expression is at most

$$\int_H \|\phi\|_{p,p}\|(\tau_u h) \circ \hat{\pi}.f\|_p\|(\tau_{-u}k) \circ \hat{\pi}.g\|_{p'}\,du,$$

which, by Hölder's inequality, is bounded above by

$$\|\phi\|_{p,p}\left(\int_H \|(\tau_u h)\circ\hat{\pi}.f\|_p^p\,du\right)^{1/p}\left(\int_H \|(\tau_{-u}k)\circ\hat{\pi}.\,g\|_{p'}^{p'}\,du\right)^{1/p'}$$
$$= \|\phi\|_{p,p}\left(\int_H\int_G |h(\hat{\pi}(x)-u)f(x)|^p\,dx\,du\right)^{1/p}\left(\int_H\int_G |k(\pi(x)+u)g(x)|^{p'}\,dx\,du\right)^{1/p'}$$
$$= \|\phi\|_{p,p}\left(\int_G\int_H |h(\hat{\pi}(x)-u)f(x)|^p\,du\,dx\right)^{1/p}\left(\int_G\int_H |k(\pi(x)+u)g(x)|^{p'}\,du\,dx\right)^{1/p'}$$
$$= \|\phi\|_{p,p}\|f\|_p\|h\|_p\|k\|_{p'}\|g\|_{p'}$$
$$\leqslant (1+\varepsilon)\|\phi\|_{p,p}\|h\|_p\|k\|_{p'}$$

by Fubini's theorem and (ii), provided of course that $1 < p < \infty$. In case $p = 1$ or $p = \infty$ the reasoning is quite analogous and leads to the estimate (5) again. □

B.2.2. Corollary. *Let X_1 and X_2 be LCA groups and ϕ an element of $M_p(X_1)$, where $1 \leqslant p \leqslant \infty$. Then the function Φ on $X_1 \times X_2$ defined by the formula*

$$\Phi(\chi_1, \chi_2) = \phi(\chi_1)$$

belongs to $M_p(X_1 \times X_2)$, and

$$\|\Phi\|_{p,p} \leqslant \|\phi\|_{p,p}.$$

Proof. Notice that the crux of the matter is that ϕ need not be continuous. The proof consists in combining Theorem B.2.1 with a standard regularisation procedure, as follows.

Let π_1 be the canonical projection of $X_1 \times X_2$ onto X_1, so that $\Phi = \phi\circ\pi_1$. Denote by G_i the dual group of X_i ($i = 1, 2$). We wish to prove that if h and k are functions on $G_1 \times G_2$ with compactly supported Fourier transforms, then

$$\left|\int_{X_1\times X_2} \Phi\hat{h}\hat{k}\,dm(\chi_1,\chi_2)\right| \leqslant \|\phi\|_{p,p}\|h\|_p\|k\|_{p'}. \tag{9}$$

Let K be the compact support of $\hat{h}\hat{k}$ and write $K_1 = \pi_1(K)$. It follows from (20.15) of [20] that, for every $\varepsilon > 0$, there exists a function f in $L^1(X_1)$ such that

$$\|f\|_1 = 1 \tag{10}$$

and

$$\int_{K_1} |\phi * f - \phi|\,dm(\chi_1) \leqslant \varepsilon. \tag{11}$$

Now by virtue of (10) it is also the case that

$$\phi * f \in M_p(X_1)$$

and

$$\|\phi * f\|_{p,p} \leqslant \|\phi\|_{p,p}. \tag{12}$$

For if u and v are integrable functions on G_1 whose Fourier transforms have compact supports, then

$$\begin{aligned}
&\left|\int_{X_1} \phi * f(\chi)\hat{u}(\chi)\hat{v}(\chi)\,d\chi\right| \\
&\quad = \left|\int_{X_1} f(\chi')\int_{X_1} \phi(\chi - \chi')\hat{u}(\chi)\hat{v}(\chi)\,d\chi\,d\chi'\right| \\
&\quad \leqslant \int_{X_1} |f(\chi')|\left|\int_{X_1} \phi(\chi)\hat{u}(\chi + \chi')\hat{v}(\chi + \chi')\,d\chi\right| d\chi' \\
&\quad \leqslant \|\phi\|_{p,p}\int_{X_1} |f(\chi')|\,\|\bar{\chi}'u\|_p\|\bar{\chi}'v\|_{p'}\,d\chi' \\
&\quad = \|\phi\|_{p,p}\|u\|_p\|v\|_{p'}
\end{aligned}$$

because of the translation-invariance of Haar measure and (10). Since (10) implies that $\|\phi * f\|_\infty \leqslant \|\phi\|_\infty$, we can assert, thanks to (11) and (12), that there is a sequence $(\phi_j)_1^\infty$ of functions on G such that

(i) each ϕ_j is continuous;

(ii) $\phi_j \in M_p(X_1)$ and

$$\|\phi_j\|_{p,p} \leqslant \|\phi\|_{p,p}; \tag{13}$$

(iii) $\|\phi_j\|_\infty \leqslant \|\phi\|_\infty$; (14)

and

(iv) $\phi_j \to \phi$ a.e. on K_1.

Returning now to the proof of (9) we notice that, by Fubini's theorem and (iv),

$$\Phi_j = \phi_j \circ \pi_1 \to \Phi = \phi \circ \pi_1 \text{ a.e.}$$

on K. Yet by Theorem B.2.1 (which applies here because of (i)) and (13)

$$\left|\int_{X_1 \times X_2} \Phi_j \hat{h}\hat{k}\,dm(\chi_1, \chi_2)\right| \leqslant \|\phi\|_{p,p}\|h\|_p\|k\|_{p'}. \tag{15}$$

Now apply the dominated convergence theorem to (15), taking note of (14). □

Remarks. The idea of using the integral representation (7) for the pointwise product of convolutions in establishing results about multipliers was first introduced by Herz [19]. It has been subsequently highly developed and used to great effect by many authors. See, for instance, [27], [28] and [29], [14] and [15], and most recently [8]. All of these authors use systematically the properties of the space A_p introduced by Figà-Talamanca [13] (and its variants); in particular, the fact that the dual of A_p is M_p. We have deliberately avoided introducing the space A_p since our aim here has been more modest. While the framework of the duality between the spaces A_p and M_p is undeniably useful and suggestive of ideas, it seemed valuable, and in the present context more appropriate, to give a proof of the homomorphism theorem which relies on little more than the Parseval formula and the Fubini-Tonelli theorem.

Appendix C. Harmonic Analysis on $\mathbb{D}_2$ and Walsh Series on [0, 1]

The fact that harmonic analysis on the Cantor group $\mathbb{D}_2$ is "the same as" the theory of Walsh series on [0, 1] is well known to all practising harmonic analysts. Since, however, it is difficult to cite a reference where the appropriate identifications are carried out in detail, it seems worthwhile to carry out some of the details for the sake of beginners unfamiliar with this piece of folklore.

The *Rademacher functions* $r_0, r_1, \ldots$ on [0, 1] are defined by the formulas

$$r_0(t) \equiv 1;$$

and for $j > 0$ and t not a dyadic rational,

$$r_j(t) = \operatorname{sgn} \sin(2^j \pi t);$$

r_j is extended to all of [0, 1] by requiring that it be right-continuous at each dyadic rational in [0, 1] and left-continuous at 1. The set of functions $\{r_j\}$ is orthonormal on [0, 1] with Lebesgue measure. For clearly

$$\int_{[0,1]} |r_j|^2 = 1;$$

and if $j > k$, then on each of the dyadic intervals where r_k is constant, r_j takes the value 1 on half the set and -1 on the other half (measurewise). So it is evident that

$$\int_{[0,1]} r_j r_k \, dt = \delta_{jk}.$$

However, the Rademacher system is not complete. The easiest way to see this is to check that the function $r_1 r_2$ is orthogonal to all the Rademacher functions.

The characters of $\mathbb{D}_2$. The *Cantor group* $\mathbb{D}_2$ is defined in the introduction to Chapter 4. To each character χ of $\mathbb{D}_2$ corresponds a unique element

$$\boldsymbol{a} = (a_j)_1^\infty$$

of the weak direct product

$$\prod_1^{\infty}{}^* \mathbb{Z}(2)_j.$$

The value of χ at the point $\boldsymbol{x}$ of $\mathbb{D}_2$ is given by the formula

$$\chi(\boldsymbol{x}) = (-1)^{\sum_{j=1}^{\infty} a_j x_j}. \tag{1}$$

Notice that the series appearing in the exponent in (1) converges since it is a finite series: at most finitely many of the entries a_j are nonzero.

Conversely, each element $\boldsymbol{a}$ of $\prod_1^{*\infty} \mathbb{Z}(2)$ determines a character of $\mathbb{D}_2$ via the formula (1).

Notice that the group $\hat{\mathbb{D}}_2 = \prod_1^{*\infty} \mathbb{Z}(2)$ is generated by the elements

$$\rho_0 = (0, 0, \ldots)$$

and

$$\rho_j = (\delta_{ij})_{i=1}^{\infty} \qquad (j \geqslant 1). \tag{2}$$

If we agree to write $\boldsymbol{a}(\boldsymbol{x})$ in place of the left side of (1) and $\rho_j(\boldsymbol{x})$ for the value at $\boldsymbol{x}$ of the character determined by ρ_j, then

$$\boldsymbol{a}(\boldsymbol{x}) = \rho_1(\boldsymbol{x})^{a_1} \rho_2(\boldsymbol{x})^{a_2} \cdots \rho_N(\boldsymbol{x})^{a_N} \tag{3}$$

where a_N is the last nonzero entry in $\boldsymbol{a}$.

Integration on $\mathbb{D}_2$. We have not yet given an explicit construction of the Haar measure on $\mathbb{D}_2$. This we now do. In the course of the construction we show that $\mathbb{D}_2$ is essentially Borel isomorphic to [0, 1]. The group $\mathbb{D}_2$ is of course not homeomorphic to [0, 1].

Let S be the countable set of points (x_j) in $\mathbb{D}_2$ with the property that $x_j = 1$ from a certain stage on. Denote by ψ the following mapping of $\mathbb{D}_2$ into [0, 1]:

$$\psi(\boldsymbol{x}) = \sum_{j=1}^{\infty} x_j/2^j.$$

Then ψ is a continuous mapping of $\mathbb{D}_2$ onto [0, 1], but it is not one-one. In fact, every dyadic rational in (0, 1) has two pre-images under ψ, one corresponding to the terminating dyadic expansion, the other to the repeating expansion. To get over this minor problem, remove from $\mathbb{D}_2$ all the points of S. Then ψ, restricted to $\mathbb{D}_2 \backslash S$, is a one-to-one continuous mapping of $\mathbb{D}_2 \backslash S$ onto [0, 1).

For each positive integer N, denote by G_N the closed subgroup of $\mathbb{D}_2$ consisting of those elements having 0 in the first N places. The mapping ψ carries the 2^N cosets of G_N onto the dyadic intervals of length 2^{-N}: $[0, 1/2^N]$, $[1/2^N, 2/2^N]$, ... $[(2^N - 1)/2^N, 1]$. If U is a coset of G_N in $\mathbb{D}_2$, ψ maps $U \backslash S$ to an interval of the form $[r/2^N, (r+1)/2^N)$, $0 \leqslant r < 2^N$; since every open set in $\mathbb{D}_2$ is a countable union of cosets of the groups G_N $(N \geqslant 1)$, it follows that ψ maps open sets of $\mathbb{D}_2 \backslash S$ onto Borel sets in [0, 1). The inverse mapping ψ^{-1} carries open sets to open sets. Therefore, by a standard argument ([34], Theorem 1.12), ψ is a Borel isomorphism of $\mathbb{D}_2 \backslash S$ and [0, 1).

Let $\mathscr{B}(\mathbb{D}_2\backslash S)$ be the Borel σ-algebra on $\mathbb{D}_2\backslash S$, and m the ordinary Lebesgue measure on [0, 1]. Define the measure μ' on $\mathscr{B}(\mathbb{D}_2\backslash S)$ by the rule

$$\mu'(E) = m(\psi(E)).$$

Now extend μ' to a function on $\mathscr{B}(\mathbb{D}_2)$ by agreeing that S is null, viz. set

$$\mu(E) = \mu'(E\backslash S).$$

Since $E\backslash S$ is Borel in $\mathbb{D}_2\backslash S$ whenever E is Borel in $\mathbb{D}_2$, μ is well-defined. It is routine to check that μ is a Borel measure on $\mathbb{D}_2$. Since ψ carries S to a countable and hence null subset of [0, 1], we see that

$$\mu(E) = m(\psi(E)) \tag{4}$$

for all Borel sets E in $\mathbb{D}_2$. Now complete $(\mu, \mathscr{B})$ ([34], Theorem 1.36) and check that the relation (4) holds true for all E in the completion. We continue to use the letter μ to denote the complete measure.

We claim that μ is the normalised Haar measure on $\mathbb{D}_2$. To verify this claim, we have to show that μ is regular and translation-invariant. The measure μ assigns finite mass to every Borel set; every open set in $\mathbb{D}_2$ is a countable union of compact sets. A general theorem of measure theory ([34], Theorem 2.18) shows that μ is regular both on $\mathscr{B}$ and on the completion. As to translation-invariance, if E is an open set in $\mathbb{D}_2$, E is a countable, pairwise disjoint union of cosets of the groups $(G_N)_1^\infty$; and for a coset $\boldsymbol{x} + G_N$,

$$\mu(\boldsymbol{x} + G_N) = \mu(G_N) = 1/2^N.$$

Consequently, $\mu(E + \boldsymbol{x}) = \mu(E)$ whenever E is open, and by regularity, this relation continues to hold for arbitrary measurable E.

We have therefore set up an "identification" of $(\mathbb{D}_2, \mu)$ and $([0, 1], m)$. As a consequence the Lebesgue spaces $L^p(\mathbb{D}_2)$ and $L^p([0, 1])$ are identified.

Characters on $\mathbb{D}_2$ and Walsh functions. If $(\rho_n)_0^\infty$ are the characters of $\mathbb{D}_2$ defined in (2) and which generate $\hat{\mathbb{D}}_2$, then the n-th Rademacher function r_n and the character ρ_n are related by the formula

$$\rho_n(\boldsymbol{x}) = r_n \circ \psi(\boldsymbol{x}) \qquad (\boldsymbol{x} \in \mathbb{D}_2\backslash S). \tag{5}$$

If $\boldsymbol{a} = (a_j)$ is a character of $\mathbb{D}_2$, then (3) and (5) show that $\boldsymbol{a}$ is identified, by ψ, with the function

$$r_1^{a_1} \cdots r_N^{a_N}$$

on [0, 1). It is clear therefore that the set of characters of $\mathbb{D}_2$ can be identified, by using the mapping ψ, with the set of all finite products of Rademacher functions on [0, 1]. This latter set of functions is called the *Walsh system* on [0, 1]. It is a complete orthonormal system in $L^2[0, 1]$.

By now it should have been made clear that questions of a measure-theoretic or integration-theoretic character concerning Walsh series on [0, 1] are "the same as" the corresponding questions about Fourier series on $\mathbb{D}_2$. One must of course always keep in mind that there may be a world of difference between the two set-ups when the question is of a topological character.

Appendix D. Bernstein's Inequality

In Chapter 7, we use the L^1-norm version of Bernstein's inequality for the groups $\mathbb{R}$, $\mathbb{T}$ and $\mathbb{Z}$. The precise statements and proofs of the inequality in the first two cases are set down in Sections D.1 and D.2 respectively. In Section D.3, a statement is given of a more general result, applicable to any LCA group; from this the Bernstein inequality for $G = \mathbb{Z}$ is easily deduced.

D.1. Bernstein's Inequality for $\mathbb{R}$

D.1.1. Theorem. *There is a number $A > 0$ such that*

$$\int_{\mathbb{R}} |f(x - a) - f(x)|\, dx \leqslant A\lambda|a| \int_{\mathbb{R}} |f(x)|\, dx$$

for every $\lambda > 0$, every a in $\mathbb{R}$ and every integrable function f such that support $(\hat{f}) \subseteq [-\lambda, \lambda]$.

Proof. Choose and fix κ in $C_c^2(\mathbb{R})$ such that $\kappa(y) = 1$ for $|y| \leqslant 1$, and define

$$k(x) = \frac{1}{2\pi} \int_{\mathbb{R}} \kappa(y) e^{iyx}\, dy.$$

Then k has bounded and continuous derivatives of all orders, all of which are integrable. Also, by the Fourier inversion formula, $\hat{k} = \kappa$. Put

$$k_\lambda(x) = \lambda k(\lambda x);$$

then k_λ is integrable, and

$$\hat{k}_\lambda(y) = \hat{k}(y/\lambda) = 1 \quad \text{for} \quad |y| \leqslant \lambda. \tag{1}$$

From (1) it follows that

$$\begin{aligned} \tau_a f - f &= k_\lambda * (\tau_a f - f) \\ &= (\tau_a k_\lambda - k_\lambda) * f, \end{aligned}$$

where, as usual, $\tau_a f(x) = f(x - a)$. Consequently,

$$\|\tau_a f - f\|_1 \leqslant \|\tau_a k_\lambda - k_\lambda\|_1 \|f\|_1. \tag{2}$$

On the other hand,

$$\begin{aligned}\|\tau_a k_\lambda - k_\lambda\|_1 &= \int_{\mathbb{R}} |k_\lambda(x - a) - k_\lambda(x)|\, dx \\ &= \int_{\mathbb{R}} |k(x' - \lambda a) - k(x')|\, dx' \\ &= I(\lambda a), \text{ say.} \end{aligned} \tag{3}$$

To majorise $I(\lambda a)$, let g be an arbitrary measurable function with bounded support such that $\|g\|_\infty \leqslant 1$. Consider

$$F(s) = \int_{\mathbb{R}} k(x - s) g(x)\, dx.$$

Since k has a bounded, continuous derivative, it follows at once that F is differentiable, and

$$F'(s) = -\int_{\mathbb{R}} k'(x - s) g(x)\, dx.$$

Hence

$$|F'(s)| \leqslant \|k'\|_1.$$

By the mean value theorem, there is a real number s in $\mathbb{R}$ such that

$$\begin{aligned}\left|\int (k(x - \lambda a) - k(x)) g(x)\, dx\right| &= |F(\lambda a) - F(0)| \\ &\leqslant |\lambda a F'(s)| \\ &\leqslant \lambda |a| \|k'\|_1. \end{aligned} \tag{4}$$

If we take the supremum of the left side of (4) over the set of functions g of the kind specified earlier, we conclude that

$$I(\lambda a) \leqslant \lambda |a|\ \|k'\|_1. \tag{5}$$

On collecting (2), (3) and (5) together, we obtain the desired result with $A = \|k'\|_1$. □

D.2. Bernstein's Inequality for $\mathbb{T}$

D.2.1. Theorem. *There is a number $A > 0$ such that*

$$\int_{\mathbb{T}} |f(xe^{-ia}) - f(x)|\, dm(x) \leqslant AN|a| \int_{\mathbb{T}} |f(x)|\, dm(x)$$

for every positive integer N, every real number a, and every trigonometric polynomial with spectrum in $[-N, N]$.

Proof. If we define

$$g = \sum_{|n| \leqslant N} \chi_n,$$
$$h = \sum_{|n| \leqslant 2N} \chi_n,$$

and

$$k = \frac{1}{2N+1} gh,$$

where $\chi_n(e^{it}) = e^{int}$, then it is easily verified that $\hat{k}(n) = 1$ for $|n| \leqslant N$. Hence

$$\tau_b f - f = k * (\tau_b f - f) = (\tau_b k - k) * f$$

when $b \in \mathbb{T}$; so

$$\|\tau_b f - f\|_1 \leqslant \|\tau_b k - k\|_1 \|f\|_1. \tag{1}$$

On the other hand, Hölder's inequality shows that

$$\begin{aligned}\|\tau_b k - k\|_1 &= \frac{1}{2N+1} \|(\tau_b h - h)g + (\tau_b g - g)\tau_b h\|_1 \\ &\leqslant \frac{1}{2N+1} (\|\tau_b h - h\|_2 \|g\|_2 + \|\tau_b g - g\|_2 \|\tau_b h\|_2),\end{aligned} \tag{2}$$

while Parseval's formula shows that

$$\|g\|_2 = \|\hat{g}\|_2 = (2N+1)^{1/2}; \tag{3}$$
$$\|\tau_b h\|_2 = \|h\|_2 = \|\hat{h}\|_2 = (4N+1)^{1/2}; \tag{4}$$

and

$$\begin{aligned}\|\tau_b g - g\|_2^2 &= \|(\tau_b g)^\wedge - \hat{g}\|_2^2 = \sum_{|n| \leqslant N} |b^n - 1|^2 \\ &\leqslant (2N+1) \sup_{|n| \leqslant N} |b^n - 1|^2 \\ &\leqslant (2N+1) N^2 |b - 1|^2,\end{aligned} \tag{5}$$

by the mean value theorem. Similarly,

$$\|\tau_b h - h\|_2^2 \leqslant (4N+1)4N^2|b-1|^2.$$

It follows from (1)–(6) that

$$\begin{aligned}\|\tau_b f - f\|_1 &\leqslant \frac{1}{2N+1}\|f\|_1 N|b-1|\{(2N+1)^{1/2}2(4N+1)^{1/2} \\ &\quad + (2N+1)^{1/2}(4N+1)^{1/2}\} \\ &= 3\frac{(4N+1)^{1/2}}{(2N+1)^{1/2}}\|f\|_1 N|b-1| \\ &\leqslant 3\sqrt{2}N|b-1|\,\|f\|_1 \\ &\leqslant 3\sqrt{2}N|a|\,\|f\|_1,\end{aligned}$$

by the mean value theorem, if $b = e^{-ia}$. This completes the proof and shows that A can be taken to be $3\sqrt{2}$. □

D.3. Bernstein's Inequality for LCA Groups

The method we used to prove Theorem D.2.1 can be modified to yield a form of Bernstein's inequality for any LCA group G.

D.3.1. Theorem. *Let G be an LCA group with character group X; let K be a relatively compact subset of X, and M a relatively compact open neighbourhood of* 0 *in X. Then*

$$\|\tau_a f - f\|_1 \leqslant 3\left(\frac{m_X(K+M-M)}{m_X(M)}\right)^{1/2}\omega_{K+M-M}(a)\|f\|_1$$

for all integrable functions f on G with $\hat{f}$ supported in K, and all a in G.

Here

$$\omega_D(a) = \sup\{|\chi(a)-1| : \chi \in D\}$$

whenever $a \in G$ and $D \subseteq X$; m_X denotes the Haar measure on X.

Remarks. (a) When $G = \mathbb{R}$, we may take $K = [-\lambda, \lambda]$, $M = (-\lambda, \lambda)$ and deduce Theorem D.1.1 by observing that

$$|e^{iya} - 1| \leqslant |ya|$$

when y and a are real.

(b) Similarly, Theorem D.2.1 can be deduced by taking $K = M = [-N, N]$.

(c) When $G = \mathbb{Z}$, take $K = \{e^{it}\colon |t| \leqslant \lambda\}$ and $M = \{e^{it}\colon |t| < \lambda\}$ where $\lambda > 0$.

(d) Theorem D.3.1 and other material on Bernstein's inequality for LCA groups can be found in [1].

Historical Notes

Chapter 2. The systematic study of singular integrals of the type considered in Chapter 2 has its origins in the fundamental paper of Calderón and Zygmund [5]. The subsequent literature is very extensive, and has given rise to many important generalisations and refinements of their methods and results. For a glimpse at this vast literature, see the references in [38] and [39].

Chapter 5. As far as we can determine, the martingale approach to Littlewood-Paley theory is due originally to D. L. Burkholder [4]. A little later, an alternative treatment of the martingale version of the Littlewood-Paley theorem was given by R. F. Gundy [17].

Chapter 6. (i) The conjugate function theorems (6.2.3, 6.3.3 and 6.4.3) and the Hilbert transform theorems (6.7.4 and 6.7.6) were discovered by M. Riesz.

Two approaches were used by Riesz, one based on complex variables (the Math. Z. paper), the other on his then-new convexity methods (the Acta paper); see the references below.

The case $p = 2$ of 6.7.4(ii), expressed bilinearly, is older, and due to Hilbert.

In the Acta paper, Riesz also discussed Fourier multipliers, referring back to earlier work by S. Sidon and W. H. Young, on $M_1(\mathbb{Z})$ and $M_\infty(\mathbb{Z})$, and M. Fekete. Fekete had dealt with the multipliers of various classes of functions: continuous functions, Riemann integrable functions, etc. The papers referred to are as follows.

Fekete, M.: Uber Faktorenfolgen welche die "Klasse" einer Fourierschen Reihe unverändert lassen. Acta Sci. Math. Szeged **1**, 148–166 (1923).

Riesz, M.: Sur les maxima des formes bilinéaires et sur les fonctionnelles linéaires. Acta Math. **49**, 465–497 (1926).

Riesz, M.: Sur les fonctions conjuguées. Math. Z. **27**, 218–244 (1927).

Sidon, S.: Reihentheoretische Sätze und ihre Anwendungen in der Theorie der Fourierschen Reihen. Math. Z. **10**, 121–127 (1921).

Young, W. H.: On Fourier series of functions of bounded variation. London Roy. Soc. Proc. **88**, 561–568 (1913).

Young, W. H.: On the Fourier series of bounded functions. Proc. London Math. Soc. **12**, 41–70 (1913).

(ii) The proof which we have given of Theorem 6.5.2 (the vector Riesz theorem) consists in applying the singular integral techniques of Chapters 2 and 3 to a particular vector-valued kernel. However, if the scalar M. Riesz theorem—equivalently, the continuity of the conjugate function operator T_c on L^p $(1 < p < \infty)$—is granted, then it is possible to deduce Theorem 6.5.2 (cf. 6.5.2(2)) immediately from an important principle due to J. Marcinkiewicz and A. Zygmund. The Marcinkiewicz-Zygmund principle states roughly that a linear operator T, continuous on $L^p(X)$ $(0 < p < \infty)$, has a natural extension which is continuous on $L^p(X, \mathscr{H})$, the space of p-th power integrable functions on X with values in the Hilbert space $\mathscr{H}$.

It seems worthwhile to give a precise statement and a proof of the Marcinkiewicz-Zygmund theorem.

Theorem (Marcinkiewicz-Zygmund). *Let $(X, \mathscr{M}, \mu)$ and $(Y, \mathscr{N}, \nu)$ be measure spaces, and assume that $0 < p < \infty$. If S is a vector subspace of $L^p(X, \mathscr{M}, \mu)$ and T is a linear mapping from S into $L^p(Y, \mathscr{N}, \nu)$ such that*

$$\|Tf\|_p \leqslant M\|f\|_p \tag{1}$$

for all f in S, then, for every positive integer N and every N-tuple $(f_1, \ldots, f_N)$ of elements of S, we have

$$\left\|\left(\sum_{i=1}^{N} |Tf_i|^2\right)^{1/2}\right\|_p \leqslant M \left\|\left(\sum_{i=1}^{N} |f_i|^2\right)^{1/2}\right\|_p. \tag{2}$$

Proof. Denote by Σ the unit sphere in $\mathbb{C}^N$ and by σ the normalised surface measure on Σ. (If $\mathscr{U}$ denotes the compact group of unitary transformations of $\mathbb{C}^N$, and m the normalised Haar measure on $\mathscr{U}$, then σ can be defined, in terms of m, by the requirement that

$$\int_\Sigma g\, d\sigma = \int_{\mathscr{U}} g(U(p_0))\, dm(U) \tag{3}$$

for all continuous functions g on Σ, where p_0 is an arbitrarily chosen point of Σ.) The main property of σ that we shall utilise is that σ is "unitarily invariant" in the sense that

$$\sigma(UE) = \sigma(E) \tag{4}$$

for (say) all Borel subsets E of Σ and all unitary transformations U of $\mathbb{C}^N$. The property (4) follows immediately from (3).

If ω_1 and ω_2 are points on Σ, there is a unitary transformation U of $\mathbb{C}^N$ such that

$$U\omega_1 = \omega_2. \tag{5}$$

Let (x, y) denote the standard inner product in $\mathbb{C}^N$. Then, thanks to (4) and (5),

$$\begin{aligned} c_p = \int_\Sigma |(s, \omega_2)|^p\, d\sigma(s) &= \int_\Sigma |(s, U\omega_1)|^p\, d\sigma(s) \\ &= \int_\Sigma |(U^*s, \omega_1)|^p\, d\sigma(s) \\ &= \int_\Sigma |(s, \omega_1)|^p\, d\sigma(s), \end{aligned} \tag{6}$$

U^* denoting the adjoint of U. Observe also that $c_p > 0$ since the function $s \to |(s, \omega_2)|$ is continuous, and positive at ω_2.

To come now to the proof proper, let $(f_1, \ldots, f_N)$ be as in the enunciation, and let $s = (s_1, \ldots, s_N)$ be an arbitrary point of Σ. Then by (1),

$$\int_Y \left| \sum_{i=1}^N s_i Tf_i(y) \right|^p dv(y) \leqslant M^p \int_X \left| \sum_{i=1}^N s_i f_i(x) \right|^p d\mu(x). \tag{7}$$

For each point y of Y for which $(Tf_1(y), \ldots, Tf_N(y)) \neq 0$, we can write

$$\psi(y) = (Tf_1(y), \ldots, Tf_N(y)) \Big/ \left(\sum_{i=1}^N |Tf_i(y)|^2 \right)^{1/2},$$

a unit vector in $\mathbb{C}^N$, i.e. an element of Σ. Similarly, for each point x of X for which $(f_1(x), \ldots, f_N(x)) \neq 0$, we can write $\phi(x)$ for the point

$$(f_1(x), \ldots, f_N(x)) \Big/ \left(\sum_{i=1}^N |f_i(x)|^2 \right)^{1/2}$$

of Σ. It then follows from (7) that

$$\begin{aligned} &\int_\Sigma \left\{ \int_{\{y:(Tf_1(y),\ldots,Tf_N(y)) \neq 0\}} |(s, \psi(y))|^p \left(\sum_{i=1}^N |Tf_i(y)|^2 \right)^{p/2} dv(y) \right\} d\sigma(s) \\ &\quad \leqslant M^p \int_\Sigma \left\{ \int_{\{x:(f_1(x),\ldots,f_N(x)) \neq 0\}} |(s, \phi(x))|^p \left(\sum_{i=1}^N |f_i(x)|^2 \right)^{p/2} d\mu(x) \right\} d\sigma(s). \end{aligned} \tag{8}$$

We deduce, by applying Fubini's theorem to (8), that

$$\begin{aligned} &\int_{\{y:(Tf_1(y),\ldots,Tf_N(y)) \neq 0\}} \left(\sum_{i=1}^N |Tf_i(y)|^2 \right)^{p/2} \left\{ \int_\Sigma |(s, \psi(y))|^p \, d\sigma(s) \right\} dv(y) \\ &\quad \leqslant M^p \int_{\{x:(f_1(x),\ldots,f_N(x)) \neq 0\}} \left(\sum_{i=1}^N |f_i(x)|^2 \right)^{p/2} \left\{ \int_\Sigma |(s, \phi(x))|^p \, d\sigma(s) \right\} d\mu(x). \end{aligned} \tag{9}$$

But, because of the invariance property (6), (9) reduces to the inequality

$$c_p^p \int_Y \left(\sum_{i=1}^N |Tf_i(y)|^2 \right)^{p/2} dv(y) \leqslant c_p^p M^p \int_X \left(\sum_{i=1}^N |f_i(x)|^2 \right)^{p/2} d\mu(x),$$

which, since $c_p > 0$, is what we had to prove. □

Remark. The Marcinkiewicz-Zygmund theorem is, in a sense, a cultural antecedent of later work of A. Grothendieck on extensions of continuous linear operators from domains of scalar-valued functions to domains of Hilbert-space-valued functions.

Marcinkiewicz, J., Zygmund, A.: Quelques inegalités pour les opérations linéaires. Fund. Math. **32**, 115–121 (1939).

(iii) The analogue for the circle group of (22) in 6.7.6 is due to S. B. Stečkin. It appears as Theorem 2 in the paper referred to below. Stečkin's proof rests essentially on the appropriate form of the Hilbert transform theorem.

Further developments of Stečkin's work have been given by I. I. Hirschman, Jr.. See also 16.4.7 in [9].

Hirschman, I. I., Jr.: On multiplier transformations. Duke Math. J. **26**, 221–242 (1959). MR **21** #3721.

Stečkin, S. B.: On bilinear forms. Doklady Akad. Nauk. SSSR (N.S.) **71**, 237–240 (1950) (Russian). MR **11** p. 504.

Chapter 8 (i) The (strong) Marcinkiewicz multiplier theorem is due to J. Marcinkiewicz.

Marcinkiewicz, J.: Sur les multiplicateurs des séries de Fourier. Studia Math. **8**, 78–91 (1939).

(ii) For the first versions of the Littlewood-Paley theorem, the reader should consult the following three fundamental papers.

Littlewood, J. E., Paley, R. E. A. C.: Theorems on Fourier series and power series (I). J. London Math. Soc. **6**, 230–233 (1931).

Littlewood, J. E., Paley, R. E. A. C.: Theorems on Fourier series and power series (II). Proc. London Math. Soc. **42**, 52–89 (1936).

Littlewood, J. E., Paley, R. E. A. C.: Theorems on Fourier series and power series (III). Proc. London Math. Soc. **43**, 105–126 (1937).

References

1. Bloom, W. R.: Bernstein's inequality for locally compact Abelian groups. J. Austral. Math. Soc. **XVII**, 88–101 (1974).
2. Bourbaki, N.: Eléments de mathématique. XIII. Première partie: Les structures fondamentales de l'analyse. Livre VI: Intégration. Actualités Sci. Ind. No. 1175. Paris: Hermann 1952. MR **14**, p. 960.
3. Brainerd, B., Edwards, R. E.: Linear operators which commute with translations. I. Representation theorems. J. Austral. Math. Soc. **VI**, 289–327 (1966). MR **34** #6542.
4. Burkholder, D. L.: Martingale transforms. Ann. Math. Statist. **37**, 1494–1504 (1966). MR **34** #8456.
5. Calderón, A. P., Zygmund, A.: On the existence of certain singular integrals. Acta Math. **88**, 85–139 (1952). MR **14**, p. 637.
6. Chung, K. L.: A course in probability theory, 2nd ed.. New York: Academic Press 1974.
7. Coifman, R. R., Weiss, G.: Analyse harmonique non-commutative sur certains espaces homogènes. Lecture Notes in Mathematics, No. 242. Berlin-Heidelberg-New York: Springer 1971.
8. Cowling, M. G.: Spaces A_p^q and L^p-L^q Fourier multipliers. Doctoral Dissertation, The Flinders University of South Australia 1974.
9. Edwards, R. E.: Fourier series: a modern introduction. Vols. I, II. New York: Holt, Rinehart and Winston 1967 and 1968. MR **35** #7062 and **36** #5588.
10. Edwards, R. E.: Functional analysis: theory and applications. New York: Holt, Rinehart and Winston 1965. MR **36** #4308.
11. Edwards, R. E.: Changing signs of Fourier coefficients. Pacific J. Math. **15**, 463–475 (1965). MR **34** #564.
12. Fefferman, C.: The multiplier problem for the ball. Ann. of Math. (2) **94**, 330–336 (1971). MR **45** #5661.
13. Figà-Talamanca, A.: Translation invariant operators in L^p. Duke Math. J. **32**, 495–501 (1965). MR **31** #6095.
14. Figà-Talamanca, A., Gaudry, G. I.: Multipliers of L^p which vanish at infinity. J. Functional Analysis **7**, 475–486 (1971). MR **43** #2429.
15. Figà-Talamanca, A., Gaudry, G. I.: Extensions of multipliers. Boll. Un. Mat. Ital. (4) **3**, 1003–1014 (1970). MR **43** #5255.
16. Garsia, A. M.: Martingale inequalities: Seminar notes on recent progress. Reading, Mass.: W. A. Benjamin 1973.
17. Gundy, R. F.: A decomposition for L^1-bounded martingales. Ann. Math. Statist. **39**, 134–138 (1968). MR **36** #4625.
18. Hardy, G. H., Littlewood, J. E., Pólya, G.: Inequalities. Cambridge: Cambridge University Press 1934.
19. Herz, C.: Remarques sur la note précédente de Varopoulos. C. R. Acad. Sci. Paris **260**, 6001–6004 (1965). MR **31** #6096.
20. Hewitt, E., Ross, K. A.: Abstract harmonic analysis. Vols. I, II. Berlin-Göttingen-Heidelberg 1963 and 1970. MR **28** #158 and **41** #7378.
21. Hörmander, L.: Estimates for translation invariant operators in L^p spaces. Acta Math. **104**, 93–140 (1960). MR **22** #12389.

22. Inglis, I. R.: Martingales, singular integrals and approximation theorems. Doctoral dissertation, The Flinders University of South Australia 1975.
23. Kahane, J.-P.: Some random series of functions. Lexington, Mass.: D. C. Heath 1968. MR **40** #8095.
24. Knapp, A. W., Stein, E. M.: Singular integrals and the principal series I. Proc. Nat. Acad. Sci. U.S.A. **63**, 281–284 (1969). MR **41** #8588.
25. Larsen, R.: An introduction to the theory of multipliers. Berlin-Heidelberg-New York: Springer 1970.
26. de Leeuw, K.: On L_p multipliers. Ann. of Math. (2) **81**, 364–379 (1965). MR **30** #5127.
27. Lohoué, N.: Algèbres $A_p(G)$ et convoluteurs de $L^p(G)$. Thèse de Doctorat ès Sciences Mathématiques. Orsay 1971.
28. Lohoué, N.: Sur certains ensembles de synthèse dans les algèbres $A_p(G)$. C. R. Acad. Sci. Paris Sér. A-B **270**, A589–A591 (1970). MR **41** #8933.
29. Lohoué, N.: Sur le critère de S. Bochner dans les algèbres $B_p(\mathbf{R}^n)$ et l'approximation des convoluteurs. C. R. Acad. Sci. Paris Sér. A-B **271**, A247–A250 (1970). MR **42** #8180.
30. Meyer, Y.: Endomorphismes des idéaux fermés de $L^1(G)$, classes de Hardy, et séries de Fourier lacunaires. Ann. Sci. Ecole Norm. Sup. (4) **1**, 499–580 (1968). MR **39** #1910.
31. Paley, R. E. A. C.: A remarkable system of orthogonal functions. Proc. London Math. Soc. **34**, 241–279 (1932).
32. Phillips, K., Taibleson, M.: Singular integrals in several variables over a local field. Pacific J. Math. **30**, 209–231 (1969). MR **40** #7886.
33. Rivière, N. M.: Singular integrals and multiplier operators. Ark. Mat. **9**, 243–278 (1971).
34. Rudin, W.: Real and complex analysis. New York: McGraw-Hill 1966. MR **35** #1420.
35. Rudin, W.: Fourier analysis on groups. New York: John Wiley and Sons 1962. MR **27** #2808.
36. Schwartz, L.: Sur les multiplicateurs de $\mathcal{F}L^p$. Kungl. Fysiografiska Sällskapets i Lund Förhandlingar, **22**, no. 21, 5 pp. (1953). MR **14**, p. 767.
37. Spector, R.: Sur la structure locale des groupes abéliens localement compacts. Bull. Soc. Math. France Suppl. Mém. **24** (1970). MR **44** #729.
38. Stein, E. M.: Singular integrals and differentiability properties of functions. Princeton Mathematical Series, No. 30. Princeton, N. J.: Princeton University Press 1970. MR **44** #7280.
39. Stein, E. M.: Topics in harmonic analysis related to the Littlewood-Paley theory. Annals of Mathematics Studies, No. 63. Princeton, N. J.: Princeton University Press 1970. MR **40** #6176.
40. Zygmund, A.: Trigonometric series, 2nd ed.. Vols. I, II. New York: Cambridge University Press 1969. MR **21** #6498.

Terminology

We list below the definitions of some terms which are used, but not defined, in the text. Those terms which are defined in the text are listed, with page references, in the Index of Authors and Subjects; any terms used without definition are considered by us to be so standard as to require no explanation.

Borel isomorphism. Let $(X, \mathscr{M})$ and $(Y, \mathscr{N})$ be measurable spaces. A *Borel isomorphism* ψ of $(X, \mathscr{M})$ and $(Y, \mathscr{N})$ is a one-one mapping of X onto Y such that $\psi(E) \in \mathscr{N}$ if and only if $E \in \mathscr{M}$.

Borel σ-algebra. Let X be a topological space. The *Borel σ-algebra on X* is the σ-algebra generated by the family of all open subsets of X.

First countable topological space. A topological space X is said to be *first countable* if each point x of X has a neighbourhood base which is countable.

Locally integrable function. A function f on a measure space $(X, \mathscr{M}, \mu)$ is said to be *locally integrable* if

(i) f is measurable for $\mathscr{M}$;

and

(ii) $f\xi_E$ is integrable with respect to μ for every set E in $\mathscr{M}$ which is of finite μ-measure.

Locally null set. A measurable subset E of a measure space $(X, \mathscr{M}, \mu)$ is said to be *locally null* if $\mu(E \cap F) = 0$ for all sets F in $\mathscr{M}$ which are of finite μ-measure.

Pseudomeasure. Let G be an LCA group, X its dual, and

$$A(G) = \mathfrak{F}L^1(G) = \{\hat{f} : f \in L^1(X)\}.$$

Define the norm on $A(G)$ by the rule

$$\|\hat{f}\|_{A(G)} = \|f\|_{L^1(X)}.$$

A *pseudomeasure σ on G* is a continuous linear functional on $A(G)$.

Relatively compact set. Let X be a topological space. A subset E of X is said to be *relatively compact* if its closure is compact.

Spectrum of a trigonometric polynomial. Let G be a compact Abelian group, and f a trigonometric polynomial on G. The *spectrum* of f is the support of the function $\hat{f}$.

Index of Notation

We list below those symbols which may not be in common use, or may be subject to more than one interpretation, and which are used systematically through the book. In a few cases, the symbols listed are not defined in the text, and we indicate briefly the interpretation intended. In all other cases, the definitions of the symbols are to be found on the page(s) indicated.

Index of Authors and Subjects

Ergebnisse der Mathematik und ihrer Grenzgebiete

1. Bachmann: Transfinite Zahlen
2. Miranda: Partial Differential Equations of Elliptic Type
4. Samuel: Méthodes d'algèbre abstraite en géométrie algébrique
5. Dieudonné: La géométrie des groupes classiques
7. Ostmann: Additive Zahlentheorie. 1. Teil: Allgemeine Untersuchungen
8. Wittich: Neuere Untersuchungen über eindeutige analytische Funktionen
10. Suzuki: Structure of a Group and the Structure of its Lattice of Subgroups. Second edition in preparation
11. Ostmann: Additive Zahlentheorie. 2. Teil: Spezielle Zahlenmengen
13. Segre: Some Properties of Differentiable Varieties and Transformations
14. Coxeter/Moser: Generators and Relations for Discrete Groups
15. Zeller/Beckmann: Theorie der Limitierungsverfahren
16. Cesari: Asymptotic Behavior and Stability Problems in Ordinary Differential Equations
17. Severi: Il teorema di Riemann-Roch per curve, superficie e varietà questioni collegate
18. Jenkins: Univalent Functions and Conformal Mapping
19. Boas/Buck: Polynomial Expansions of Analytic Functions
20. Bruck: A Survey of Binary Systems
21. Day: Normed Linear Spaces
23. Bergmann: Integral Operators in the Theory of Linear Partial Differential Equations
25. Sikorski: Boolean Algebras
26. Künzi: Quasikonforme Abbildungen
27. Schatten: Norm Ideals of Completely Continuous Operators
30. Beckenbach/Bellman: Inequalities
31. Wolfowitz: Coding Theorems of Information Theory
32. Constantinescu/Cornea: Ideale Ränder Riemannscher Flächen
33. Conner/Floyd: Differentiable Periodic Maps
34. Mumford: Geometric Invariant Theory
35. Gabriel/Zisman: Calculus of Fractions and Homotopy Theory
36. Putnam: Commutation Properties of Hilbert Space Operators and Related Topics
37. Neumann: Varieties of Groups
38. Boas: Integrability Theorems for Trigonometric Transforms
39. Sz.-Nagy: Spektraldarstellung linearer Transformationen des Hilbertschen Raumes
40. Seligman: Modular Lie Algebras
41. Deuring: Algebren
42. Schütte: Vollständige Systeme modaler und intuitionistischer Logik
43. Smullyan: First-Order Logic
44. Dembowski: Finite Geometries
45. Linnik: Ergodic Properties of Algebraic Fields
46. Krull: Idealtheorie
47. Nachbin: Topology on Spaces of Holomorphic Mappings
48. A. Ionescu Tulcea/C. Ionescu Tulcea: Topics in the Theory of Lifting
49. Hayes/Pauc: Derivation and Martingales
50. Kahane: Séries de Fourier absolument convergentes
51. Behnke/Thullen: Theorie der Funktionen mehrerer komplexer Veränderlichen
52. Wilf: Finite Sections of Some Classical Inequalities
53. Ramis: Sous-ensembles analytiques d'une variété banachique complexe
54. Busemann: Recent Synthetic Differential Geometry
55. Walter: Differential and Integral Inequalities
56. Monna: Analyse non-archimédienne
57. Alfsen: Compact Convex Sets and Boundary Integrals
58. Greco/Salmon: Topics in $\mathfrak{m}$-Adic Topologies
59. López de Medrano: Involutions on Manifolds
60. Sakai: C^*-Algebras and W^*-Algebras
61. Zariski: Algebraic Surfaces

62. Robinson: Finiteness Conditions and Generalized Soluble Groups, Part 1
63. Robinson: Finiteness Conditions and Generalized Soluble Groups, Part 2
64. Hakim: Topos annelés et schémas relatifs
65. Browder: Surgery on Simply-Connected Manifolds
66. Pietsch: Nuclear Locally Convex Spaces
67. Dellacherie: Capacités et processus stochastiques
68. Raghunathan: Discrete Subgroups of Lie Groups
69. Rourke/Sanderson: Introduction to Piecewise-Linear Topology
70. Kobayashi: Transformation Groups in Differential Geometry
71. Tougeron: Idéaux de fonctions différentiables
72. Gihman/Skorohod: Stochastic Differential Equations
73. Milnor/Husemoller: Symmetric Bilinear Forms
74. Fǫssum: The Divisor Class Group of a Krull Domain
75. Springer: Jordan Algebras and Algebraic Groups
76. Wehrfritz: Infinite Linear Groups
77. Radjavi/Rosenthal: Invariant Subspaces
78. Bognár: Indefinite Inner Product Spaces
79. Skorohod: Integration in Hilbert Space
80. Bonsall/Duncan: Complete Normed Algebras
81. Crossley/Nerode: Combinatorial Functors
82. Petrov: Sums of Independent Random Variables
83. Walker: The Stone-Čech Compactification
84. Wells/Williams: Embeddings and Extensions in Analysis
85. Hsiang: Cohomology Theory of Topological Transformation Groups
86. Olevskiĭ: Fourier Series with Respect to General Orthogonal Systems
87. Berg/Forst: Potential Theory on Locally Compact Abelian Groups
88. Weil: Elliptic Functions according to Eisenstein and Kronecker
89. Lyndon/Schupp: Combinatorial Group Theory
90. Edwards/Gaudry: Littlewood-Paley and Multiplier Theory

In action for Clydebank – using my RIGHT foot. That's got to be a collector's item.

Dreams are made of this … me with the European Cup.

Bertie Auld and I share the stage at my pal's pub, the Ranza Bar in Provanmill. What a player Bertie was for Celtic.

Oh, we do like to be beside the seaside! I am third from the right as my Partick Thistle team-mates – plus donkeys – fool around on the beach at Blackpool.

The jolly Jags. I am sitting second from the right in this Partick Thistle team photo. Manager John Lambie, as usual, takes centre stage.

Left: Happy days! Me with Partick Thistle colleagues Davie Craig, John Lambie, Brian Gallagher and Alan Dinnie.

Having a ball. John Lambie and Gerry Collins join me for a snap outside Firhill.

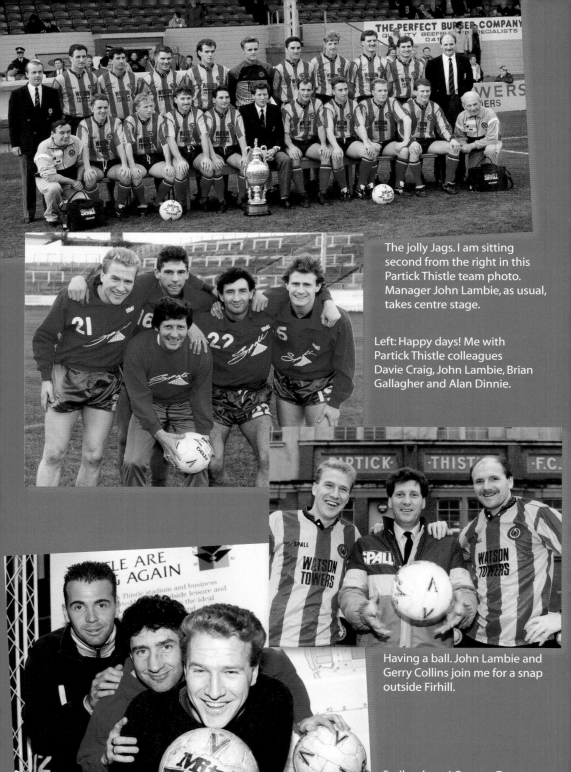

Smile, please! Cammy Duncan, Brian Gallagher and me grin for the camera.

© HERALD AND TIMES GROUP

Happy hat-trick! I am celebrating after scoring three goals against Alloa and picking up a £500 bonus from League Cup sponsors Coca-Cola.

Take that! I am sticking my tongue out at the Rangers fans at Ibrox. I was only joking – honest!

Celebrations! I am delighted as Partick Thistle are on their way to a 3-1 victory over Celtic at Hampden.

Ouch! Celtic winger Paul Byrne is on the receiving end of a crunching tackle as we both challenge for a loose ball.

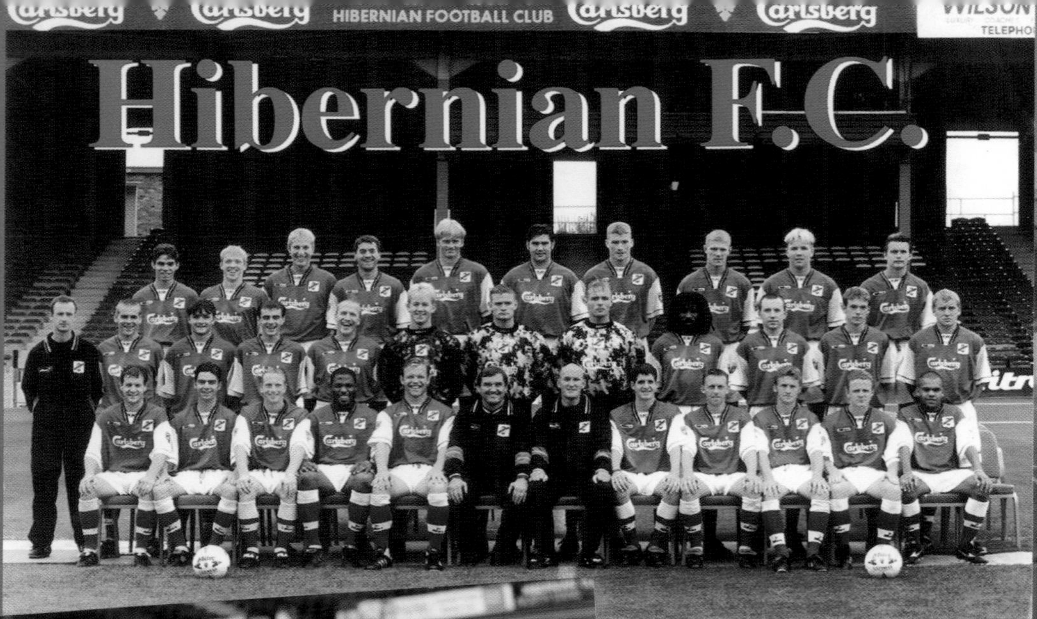

Hib, Hib, hooray! Here I am posing with my Easter Road colleagues. If you look carefully you will see that one half of the group is laughing – I've just cracked a joke.

Edinburgh buddies. Here I am with good pal David Elliott at Hibs. We were both together at St Mirren, too.

Down and out! I am hovering over Celtic's Andreas Thom. I got the winning goal in a 2-1 victory at Easter Road that day in 1997.

Bottom Left: What a belter! I am celebrating my winner against the Parkhead outfit.

In with a shout! John Hughes, a former Celtic player, joins in the fun.

The inspirational Ross Freeland, the Partick Thistle supporter who called me Uncle Chic.

Sign here, please. Boss Murdo MacLeod looks on as I put pen to paper to join Dumbarton.

Right: Me with the late, great Sir Bobby Robson as we take kids for training with the Coca-Cola-sponsored Maximise.

Cesar and me. Celtic great Billy McNeill shakes my hand. Why didn't he sign me?

Thanks, fans. I applaud the Celtic support for their remarkable backing after my one and only appearance in the hoops – a 3-1 win over Manchester United at Old Trafford.

Oh, no! Celtic have just conceded a goal as they go down 3-2 to Porto in the UEFA Cup Final in Seville in 2003.

There ain't nothing like a dame! John Lambie and I get ready for the panto season.

What a mover and shaker! My ma Isa goes through a *Strictly Come Dancing* routine.

Posing in front of the statue of Christ The Redeemer on top of the Corcovado mountain in Rio de Janiero. What a place!

Hail! Hail! Here I am with son Gary and daughter Danielle as we celebrate Celtic winning the Scottish Cup in 2004, beating Dunfermline 3-1 at Hampden.

Yes, I realise all the ifs and buts are part of the fascination of football. But, believe me, luck does play a major role. What was it Napoleon said? 'Don't give me a great general, give me a lucky one.' That is so true in our game. Reputations can be destroyed on the bounce of the ball or a deranged decision from the referee. Obviously, I am in Jim Duffy's corner when I make this observation. His knowledge of the game is second to none and he still possesses fabulous enthusiasm. Brechin City are the benefactors of his wide range of abilities at the moment, but I would still put a few quid on Duff returning to a big club, where he undoubtedly belongs.

There are a lot of things that go on behind the scenes at football clubs that would undoubtedly surprise – or shock – your average fan. Managers have got to put up with so much from certain individuals and, as you might expect, there can be clashes of personalities. I remember Jim Leighton, the former Scotland international goalkeeper, when we were both with Hibs. He was so dour he reminded me of Alex Miller in our brief spell together at St Mirren. Jim wasn't big on smiling, that's for sure. I doubt if he spoke more than two words me to me in all the time we were together. I always got the impression he wasn't too happy at Easter Road and no-one was surprised when he hot-footed it back to Aberdeen as soon as he could. Maybe he once applied for a personality and was turned down!

Possibly he was a bundle of laughs before his old Dons gaffer Sir Alex Ferguson signed him for Manchester United. Jim, rightly, had a growing reputation in the game at the time and his stock was high. I think the Old Trafford side had struggled for a consistent No.1 since the days of Alex Stepney, their 1968 European Cup-winning keeper. Gary Bailey settled the ship a little, but Sir Alex certainly didn't inherit any genuine class when he arrived. Chris Turner, a decent keeper from Sheffield Wednesday, made a few outings without impressing. So, it was inevitable that Sir

Alex would look again to Aberdeen because he knew the capabilities of Jim, who had won an avalanche of honours under Fergie's guidance at Pittodrie. He looked like a safe pair of hands just when Fergie needed them most.

For whatever reason, Leighton just didn't cut it at Old Trafford. He made a couple of blunders in his first few games and that was enough for the United fans to groan, 'Oh, here we go again!' Being Scottish didn't help, either, I suppose. Jimmy Greaves, the old England striker, used to lay into our shotstoppers every week on his Saturday lunchtime TV show, Saint and Greavsie when he shared a studio with Ian St John. If a Scottish keeper chucked one into the net it was shown in slow motion over and over again. It did our reputation no good at all. Leighton just couldn't get the supporters onside and it wouldn't be until Peter Schmeichel turned up that Fergie could breathe easily over his problem position.

It was all over for Leighton after his display against Crystal Palace in the 1990 English Cup Final. Fergie, of course, was desperate for his first trophy as United manager. He had gone four years without silverware and, believe it or not, there were cries from the United faithful for him to be sacked. So, you can imagine how he was feeling on that particular afternoon as the team bus pulled up at Wembley on a gloriously sunny afternoon. United were ninety minutes away from a trophy breakthrough and were odds-on favourites to beat Palace. Things didn't go according to plan. Leighton conceded three goals in a 3-3 draw and, although I have never played in goal and I realise it is almost an art form, I thought he might have done better with at least two of them. Fergie agreed. He dropped Leighton for the replay the following midweek and brought in veteran Les Sealey. United won 1-0 with a goal from Lee Martin, later to play for Celtic, and it was the beginning of the end for Leighton at United. One minute a hero, the next a zero. Football is littered with such stories.

Leighton's mood wouldn't have been enhanced when he went to Dundee and was criticised on a few occasions by their ex-boss Simon Stainrod. Apparently, the Dens Park gaffer wasn't impressed by Leighton's rather bizarre deadball kicking style. He seemed to slide along the ground as he took a goal-kick. If kicking was a problem, Leighton only had to ask me for advice. Mind you, that might have taken more than two words!

Anyway, Fergie had a momentous decision to make before that Wembley replay and, as ever, he wasn't afraid to take responsibility. Could I have played for Fergie? Why not? Could I have accepted the infamous 'hair-dryer' treatment? Have you ever been in a dressing room with John Lambie when he is not happy? If you can cope with that you can cope with anything. Fergie would have been a pussycat after that!

Speaking of the great man, I recall my old mate Frank McDougall telling me this little story and, frankly, I'm amazed he is still around to reveal this tale. Fergie was far from happy with his Aberdeen players after a particular game on a Saturday and ordered them back to Pittodrie for extra training the following day. Frank had made arrangements to see his daughter in Glasgow on the Sunday, anticipating a day off. That cut no ice with the Aberdeen manager. 'No exceptions,' he informed my pal. 'You'll be in with the rest of them tomorrow. Your daughter will have to wait.' Frank must have been in a defiant mood because he informed Fergie he was going ahead with his original plans. It was time for Fergie to get up close and personal. What happened next must have stunned the Pittodrie gaffer. Frank had been chewing gum and, as his boss got closer, he removed it from his mouth and slung it at Fergie. I think everything after that is unprintable!

Talking of bosses with magical qualities, I was saddened at the death of Sir Bobby Robson in August 2009. The former England team manager was a national treasure, a great ambassador for football and a wonderful human being. I met him on a couple of occasions although I doubt if he would have remembered the first time. I have to admit I was fairly drunk on that particular occasion!

I travelled down to Wembley in 1981 with my Uncle William and pals Bogey Redmond and Anthony Gaughan to support Scotland in the Home International match against England. I was only eighteen years old at the time and, naturally enough, we made a few pitstops on our way to London. My uncle got lumbered with the driving, so he wasn't allowed to touch a drop of alcohol, but his three fellow travellers made up for him, I can tell you that. Our original plan was to make our first stop at Carlisle for a light refreshment – aye, that'll be right! – but I don't think we actually got outside Glasgow before we stopped at a certain hostelry. That set the tone for the rest of the journey.

Goodness knows how many pubs we popped into on our way to England's capital, but it was probably in double figures. Well, it was Wembley weekend, after all, and, like most Scottish supporters, we had put aside a few bob each week and were determined to enjoy ourselves. We would show the English how to drink – none of that half-pint nonsense for us. Unless it was a half-pint of vodka or whisky! By the time we reached our destination three out of the four of us were more than just a little tipsy. OK, we were blitzed. As we walked to the famous old ground I spotted Bobby Robson signing autographs for some fans. I seized my opportunity. I staggered over to this football legend and threw my arms around him. I recall he looked more than just slightly taken aback as I cuddled him. So, there I am outside Wembley, a totally unknown drunken Scot embracing the great Bobby Robson!

It was a superb occasion and made even sweeter when John Robertson slotted a penalty-kick past Joe Corrigan to give us a 1-0 victory. Although I had a few pints beforehand I can still recall that goal vividly. Davie Provan, the stylish Celtic outside-right, released Steve Archibald inside the English penalty box. Steve was a fox. He must have seen their captain Bryan Robson coming in from a difficult angle to make a challenge. I think Stevie even invited the tackle. Crunch! And down went Stevie. Penalty kick.

As always there is a bit of kidology when there is a spot-kick awarded. Players will do all sorts of things to put off the kicker. They'll sling mud in front of him as he starts his run-up and there will always be a bit of banter. I was astonished to see an English League game a few years ago when an opponent actually stood on a balloon and burst it just as a spot-kick was about to be taken. That's got to be ungentlemanly conduct. I wish I had thought of it! As John Robertson prepared to take aim, Trevor Francis, his team-mate at Nottingham Forest at the time, raced forward to have a word with the English goalkeeper. It looked as though he was informing him of his Forest colleague's favoured spot for placing the ball. He was wasting his time. Our man was completely unfazed and rolled the ball into the corner of the net as Corrigan took off in the opposite direction. Thanks, Trev. Drinks all round. Happy days.

I caught up with Sir Bobby Robson again much later in life when I met him in Newcastle as I was invited down by Maximise to take a training session with about thirty local kids. It was an absolute honour and privilege to be in the company of this first-rate guy who, despite all his fame, had no airs or graces. Actually, he allowed me to take training with the youngsters and Charlie Mularvey, the Maximise boss, said, 'Obviously, he doesn't know you, Chic.' I didn't think it was the time or the place to remind Sir Bobby of the day he was given a bear hug by a boozed-up Scottish fan outside Wembley all those years ago!

The aforementioned Steve Archibald, of course, used to play for Barcelona and still stays in the Catalan capital. He tells a good story about Sir Bobby when he popped into the Nou Camp one morning. Sir Bobby was then in charge of Barca and was supervising a training session. The Spanish giants had all sorts of nationalities playing for them at the time. They were going through some free-kick routines and the manager was talking very slowly in some sort of pigeon English. He was speaking to Bulgarian international striker Hristo Stoichkov and saying, 'This . . . is . . . how . . . I . . . would . . . like . . . you . . . to . . . take . . . this . . . kick.' Steve had arranged a quick meeting with Sir Bobby and he stood on the sidelines while all this was going on. After about half an hour or so Sir Bobby came over to Steve. He offered him his hand and said, 'Hello . . . Steve . . . very . . . sorry . . . to . . . keep . . . you . . . waiting.' Sir Bobby was still speaking as though he was conversing with a foreigner. Steve let this go on for a few minutes before saying, 'Bobby, it's OK, you can speak normally. I am Scottish, you know!'

The guy oozed class. Would we have ever heard of the Special One, Jose Mourinho, if it hadn't been for Sir Bobby? I doubt it. When the Englishman was at Barcelona he employed Mourinho as his eyes and ears in the dressing room. He also acted as his interpreter because, I'm told, Sir Bobby's mixture of English and Spanish ended up a hilarious sort of Spanglish which seemed to baffle one and all. Remember, Mourinho had absolutely no pedigree as a player or a coach until he met Sir Bobby. However, he was taken under his wing, pushed up the ladder and did enough, of course, to return to Portugal and get a coaching post before taking over as manager of Porto. The boss who made Brian Clough look shy and retiring was in charge of Porto when they beat Celtic 3-2 to lift the UEFA Cup in 2003. I was there that unforgettable evening in Seville. I should hold that against Sir Bobby! Mourinho then became boss of Chelsea and the rest, as they say, is history. Oh, by the way, I was also told that Sir Bobby could sing like Sinatra. Were there no ends to this man's talent? He is missed by all who ever came in contact with him – a true gent.

Another unusual incident on a Scotland v. England weekend came at Hampden and I recall Kenny Dalglish coming out of Hampden with a signed ball under his arm. I was about twelve years old at the time and was there with my Uncle John on this occasion. Kenny met up with his wife Marina and made his way to his parked car. Can you imagine this legendary footballer's embarrassment when his vehicle wouldn't start? It coughed and spluttered and there was only one thing for it. My uncle and I rolled up our sleeves and gave it a push start. It eventually came to life and Kenny and his missus roared out of the car park with a cheery wave.

One manager I'd have loved to work with is Bertie Auld. He was a complete one-off as a player and was fairly unique as a manager. Hibs must have been an interesting place when he was there with John Lambie as his assistant. I thoroughly enjoyed watching Bertie as a player when I was a youngster. He and Bobby Murdoch were different class in the middle of the park. It was obvious someone with a footballing brain like Bertie's would go into management when he retired from playing and so it proved when he went to Partick Thistle. I never played under his guidance or influence, but I would have responded to the man, I'm sure of that.

Bertie was a character and stories about him are legendary. There's the one where he took off Jim Melrose and left Partick Thistle with only ten men because he had used all his substitutes. Melrose, completely fit, was removed from the action and must have been more than just a shade bewildered as he got the hook. Asked why afterwards, Bertie shrugged and said, 'He wasn't doing as he was told.' Just like that. It was either Bertie's way or no way. I'm delighted to say Bertie and I are still friendly today and he has some sharp observations on how the game should be played. He still has a good eye for a player and I respect his judgements. There is a flood of great Bertie Auld stories and I recall one he told me when he had taken over as manager at Dumbarton. He got his new players around him in the dressing room as he prepared to make his first team announcement. He was going through the list in numerical order – goalkeeper Smith No.1, right-back Bloggs No.2 and so on. Now the Boghead club possessed a guy called Gerry McCoy at the time and he wasn't a bad player. In fact, he was a consistent goalscorer, but lacked a bit of personality and charisma. Nice bloke, though.

Bertie got to the forward line – No.7 Jones and then he came to No.8. 'McCoy,' said Bertie. 'Wait a minute, boss,' came the cry from the player. 'I always play No.10, that's my shirt.' Obviously these were days before squad numbers were given out at the start of the season and they stuck with you all the way through the campaign. 'Well, you're No.8 today,' answered Bertie. McCoy wasn't having it. 'No, boss, you don't understand, I always play in No.10,' persisted McCoy. 'You can play No.10 next week,' said Bertie, 'you're No.8 today.' McCoy continued to protest. After a couple of minutes a clearly frustrated Dumbarton boss said, 'OK, you're No.12 now.' McCoy was flummoxed – he was on the substitutes' bench! The player should have known better than to cross swords with my old mate. I doubt if he ever did again.

14

PAMPERED PRIMA DONNAS – AND PERFECT PROS

Some of today's footballers have egos so large that they have their own postcode. A fair percentage of them are legends in their own lunchtime, as far as I am concerned. A lot of them wouldn't know a ball from a banana. Many are over-rated and overpaid and the money they are earning is absurd.

Okay, I realise this may come across as someone with a gripe about missing out on the gravy train, but I can assure you that is not the case. I watch a lot of football and I am continually amazed at mediocre players who can pick up £5,000-per-week. That appears to be the going rate for a distinctly average player in the SPL days. How on earth can some so-called superstars justify picking up wages of around £50,000-per-week? Good luck to them all, but, for me, the figures just don't stack up. I would go as far as to say these payments are just a joke. It's all Hollywood now, isn't it?

I don't want to come across as football's version of Methuselah, but a lot of these players wouldn't have got a kick of the ball in my day. They might have struggled to get a game in the reserves. In my opinion, the standards have dropped fairly dramatically. I read my good pal Bertie Auld's excellent autobiography *A Bhoy Called Bertie* recently and I have to agree one hundred per cent with the Lisbon legend when he stated many of today's footballers are pampered prima donnas. I couldn't have put it better, Bertie. Money appears to be their God and they are losing touch with the people who matter most in this game – the supporters who shell out hard-earned money each week to follow their team.

My jaw dropped the other day when I was reading a newspaper and some footballer I had never heard of and who played for some obscure outfit in the third tier of the English League had been in a car crash. His vehicle was a £150,000 Lamborghini! That just about says it all. My old Partick Thistle pal Brian Gallagher was joking when he said, 'Chic, can you imagine if we were playing today and earning five grand a week? We could work shifts. You could get sent off every second game, be suspended for a fortnight and still put £10,000-a-month in the bank. You could return and then I could get suspended for two weeks. We could sort out a rota, play only half a season each and still have loads of dosh in the bank.' Well, I think he was only joking.

However, that daft scenario just underlines how ridiculous the cash situation is in football these days. Of course, it is up to the clubs how they distribute the cash among the players, but you have to question the wisdom of some chief executives who agree to part with such astronomical sums. Agents are obviously doing a very good job for their clients. Yes, it would have been nice to have pocketed a small fortune in my day, but I am not jealous of today's performers. Simply put, I believe a lot of them aren't worth what they are picking up. Many of them are journeymen footballers at best.

There is another side of the coin, of course, and there are players you would travel miles to see in action. Roy Keane, for me, is one of those performers. What a player. Even when he was winding down his career and had that brief spell with Celtic you could still see he was something special. What I would have given to play alongside that guy. He was a genuine Manchester United icon and I use the word advisedly. I thought he had everything; unbelievable vision, immense strength, fabulous leadership qualities, inch-perfect passing ability and, just as important, a marvellous presence. I would have paid money to play alongside Roy Keane. What an experience that would have been. He was a textbook player that any young kid could learn from.

He wasn't afraid to ruffle a few feathers and made his views known to everyone. He must have upset a few of Manchester United's better-healed supporters when he called them the 'prawn sandwich brigade'. Somehow I get the feeling he didn't give a toss how they might react. He had an opinion and he expressed it, as is his right. He refused to be gagged and even had a go at his Old Trafford team-mates on Manchester United TV. The interview had to be pulled, but, as you might expect, some journalists got a hold of the tape and made some of his outspoken, and even controversial views, known to the public. He refused to back down and that may have even cost him his career at the club. However, United's loss was very much Celtic's gain.

As a fan of the club I was intrigued to see Roy Keane in those famous hoops. I realise he was carrying an injury that would ultimately end his career at the end of season 2006/07 after only six months in the east end of Glasgow, but he didn't disappoint. I recall an incident in a game against Rangers when he pointed out rather forcibly to full-back Mark Wilson that he had done something wrong. It wouldn't have been apparent to a lot of onlookers, but the defender had been sucked out of position and Roy, in that inimitable fashion of his, told him so. I don't think Mark Wilson put a foot wrong for the rest of the game. There was no way he was going to run the risk of incurring the wrath of the Irishman again!

Roy Keane was a powerhouse of a footballer and a rare talent. I wouldn't be surprised if he went onto become a first-rate manager. He obviously has a ferocious passion for the game and, with it, a massive understanding of how it should be played. Maybe his strong-minded attitude might have frightened off some chairmen when he left his first managerial position at Sunderland. Ipswich Town gave him another opportunity and, once again, to Roy Keane's credit he wasn't afraid to step down a division to start again. The man is a winner and I believe that, unlike so many others, earned every single penny of his wages. As a season ticket holder at Celtic, I can say I was happy to fork out to watch him play.

Another player I loved watching was Paul McStay, who, like Roy Keane, possessed all the talents required for a midfielder. I played against Paul a few times, of course, and was always impressed by his astonishing range of skills. A lot of people talked about his precision passing – and he did have an abundance of skill in that particular department – but I'll tell you something that may surprise you. Paul McStay was incredibly strong. He was as hard as nails yet no-one seemed to notice this part of his game. When you were on the receiving end of one of his tackles your bones used to rattle. I know because I was tackled a few times by the former Celtic and Scotland player. I'm sure I've still got the bruises to prove it. Don't get the idea Paul McStay was a dirty player. Far from it. He was tough, but always fair.

I am now about to go into print and say something that will probably upset a lot of my fellow-Celtic fans – I think Paul McStay should have quit the club and moved on to pastures new at some stage of his career. I believe he should have tried his luck in England or on the Continent and performed to an even bigger audience. I am positive he would have been a huge success wherever he went. Of course, it did look as though he was on his way at one stage in the early nineties. He was out of contract and there were stories popping up everywhere about him going to England, Spain, Italy or Germany. At the end of season 1992 he actually ran over to the fans in The Jungle at Parkhead and threw his shirt into the crowd. It looked like a farewell gesture to most of us.

However, Liam Brady was the Celtic manager at the time and obviously possessed a lot of natural Irish charm because he persuaded Paul to sign an extension to his contract much to the obvious joy of the Celtic support, me included. However, I can't help but wonder what might have happened to his career if he had left his boyhood idols. We'll never know now, but with his talent he would have been a success anywhere on this planet. Actually, I believe he stagnated a bit at Celtic and I don't mean that as undue criticism. I think he might have relished a new challenge in a different environment. It must have been a bit wearing for the player to come up against the same teams season in, season out.

I recall having a debate – probably more of an argument! – with my Partick Thistle pal Gerry Collins about the merits of Paul McStay. Gerry insisted he thought former Rangers captain Barry Ferguson was a better player than Paul. I wasn't having any of that and I told him so. Look, Barry Ferguson is a fantastic footballer, but I would not put him in Paul's class. No way. Don't accuse me of Celtic bias, either. I would like to think I, too, could judge a footballer and, for me, Paul McStay was a far superior footballer to Barry Ferguson. End of debate.

I've often been asked who was the best player I ever played alongside and I have to give the nod to Frank McAvennie. I didn't get to team up with him too often at St Mirren, but when I did I can tell you it was always a pleasurable experience. I recall a game when the Saints were playing Hibs at Easter Road. The home fans were singing 'Hail! Hail! The Hibs are here.' They were mimicking the Celtic song and if anyone had listened carefully, Frank and I – and Frank McDougall – were singing, 'Hail! Hail! The Celts are here!' There we were, getting ready to play for St Mirren singing Celtic songs. It could only happen when Macca was around.

I'm told he was nicknamed 'Swan Vesta' when he first turned up at Love Street as a young lad. Apparently, he was pale white, stick thin with a mop of red hair. It seemed natural that he would be named after a box of matches! What a player, though. Frank didn't think twice about putting his head where a lot of players would have hesitated putting their foot in a crowded penalty box. Mind you, any defender stupid enough to kick Frank on the napper probably ran the risk of breaking his foot!

Frank had a natural goalscorer's instinct and that is something you just cannot coach into a player. You can spend time trying to tell strikers to go to the near post, peel off a defender, time their run and so on. However, it is out on the pitch on matchday that guys like Frank come alive. Defenders always knew when this guy was around. Former Celtic manager Billy McNeill continually fined my mate for being late for training, but it didn't stop him playing in every game. Frank would fly down to London after a game on a Sunday to enjoy the nightlife with friends he had met during his first stint at West Ham United. Inevitably, he would miss the early flight back to Glasgow on Monday morning and once again Big Billy would want a word. That led to a fine, but it didn't stop Frank from doing the exact same thing the following week. That guy had more money than sense. Thankfully, Captain George and the Eye in the Sky came to the rescue every now and then.

He was a brilliant team player, making selfless runs all over the place. Opponents were wasting their time trying to intimidate Frank. He put himself about and would received some fairly fierce treatment in return, but he still came back for more. His goalscoring record speaks for itself and he is still a good mate to this day. I appreciate his friendship. He's a good man to have in your corner.

Talking of favourite players, I once asked Jackie McNamara Senior, Jim Duffy's assistant manager at Hibs, who was the best footballer he had worked with. Immediately, he answered, 'Kenny Dalglish.' I queried, 'Okay, who is your second best?' Jackie thought for a moment, looked me straight in the eye and replied, 'You, of course.' Mind you, his nose was growing Pinocchio-style as he made that observation!

15

ROSS FREELAND: AN INSPIRATION

The name Ross Freeland won't really mean too much to most football fans, but the lad had a profound and lasting effect on me. Ross was Partick Thistle crazy, a genuine supporter of the club. Sadly, Ross was diagnosed with cancer at a very early age and died in 2004. He was only twelve years old. I will always remember Ross and the world is a poorer place without his presence.

Football is not all about mansions, flash cars and bulging salaries. There is another level and that's where you meet the likes of Ross. I enjoyed his company many times, but it was distressing to witness the deterioration in his condition over the period. He fought on bravely and to start with I found it emotionally hard to be in his company. I soon discovered that he was, in fact, a delight to be around because of his wonderful enthusiasm for football in general and Partick Thistle in particular.

His mum Marion and dad Andy got in touch with the authorities at Firhill to tell them of their son's dreadful plight. George Carson, who runs the Junior Jags, and his partner Rosie, two extremely generous and genuine people, got to hear about Ross and they set about making his last days as memorable as possible. They arranged tours of Firhill and took him out onto the pitch where his heroes played on matchday. The lad was thrilled; his illness forgotten for a time.

John Lambie and I went out to see him one day at the family home in Knightswood. You should have seen his eyes light up when the Partick Thistle manager and a player came to his front door. It was heartwarming, believe me. A simple gesture, but it meant so much to Ross who was in a wheelchair by this time. We had picked up a Panini football sticker album for him and you would have thought we were handing him the Crown Jewels. He loved it when the boss slapped me on the back of the head and said, 'See that, Ross, that's how you treat this guy.' It was all an act and the kid thrived on it. Over the months I got to know Ross very well and he used to call me Uncle Chic. I was overcome to realise he felt so close to me. Honestly, it was just a bit overwhelming.

We knew it was getting near the end for this fabulous wee character when he became desperately ill. George Carson, Rosie, myself and a few others decided to organise a charity event to raise some much-needed money for the family. It wasn't difficult to get people to throw their weight into the project. We put together everything we could from Partick Thistle, Henrik Larsson donated a Celtic shirt and Barry Ferguson did likewise with a Rangers top. We arranged it for Firhill for a Sunday night and journalist Hugh Keevins very kindly agreed to be MC. Former Motherwell player John Gaghan, now a very funny comedian, got involved, too.

It was good to put together some money for such a worthy cause and we were satisfied the night had gone well. The following day I was at Motherwell's ground at Fir Park as I was involved in the Coca-Cola Sevens with a pile of kids. Murdo MacLeod appeared and pulled me aside to tell me the awful news that Ross had passed away earlier that day. We all realised there were no miracles around the corner for this gallant wee lad, but it was still devastating at the time to be told he had died. Memories of his beaming, bright smile came flooding back in an instant. I remembered his laughter and his infectious enthusiasm.

I immediately thought about his family. I attended Ross's funeral and his dad Andy told me his dying wish was to have his ashes scattered at Firhill, home of his team, Partick Thistle. Andy, mum Marion, his sister, the minister and myself made our way to Maryhill to take care of his last request. I was told the ashes couldn't be thrown on the playing surface by one of the groundsmen. We were in the middle of a pre-season, so I didn't think it mattered. He must have noted my expression. I turned to the parents and asked, 'Where specifically would Ross like to have his ashes placed?' They informed me he wished to have them scattered in the goalmouth he used to stand behind when he was growing up. 'Fair enough, let's go,' I said. Thankfully, no-one interfered during this extremely solemn and moving occasion.

Ross's ashes were then put in position in the resting place he had chosen. Experiences such as that leave you chastened. I'm telling the story of Ross Freeland simply because I believe it deserves to be told. It puts football in perspective. There are more important things than this game of ours. I can smile now at the memory of Ross and what he brought to me. He enhanced my life and I can only hope that I, too, gave him something back, however small.

It was a privilege to have Ross Freeland call me Uncle Chic.

16

SWEDE DREAMS

Former England international Alan Ball had an unmistakeable squeaky speaking voice. Jim Baxter called him the Clitheroe Kid after Jimmy Clitheroe, the veteran actor who played a schoolboy in a TV series in the Sixties. My phone rang one morning and on came Ball. I pretended not to recognise his dulcet tones. 'Is that Chic Charnley?' enquired Alan, who was manager of Exeter City at the time. 'Aye,' I replied, adding, 'Who's this please?' There was a moment's silence before Alan came back with, 'If you don't know a World Cup winner when he's talking to you, then you are a waste of time!'

I dissolved into laughter. There were two Alan Balls – his dad had the same name and had also been involved in football – but there was only one distinct voice like the former Blackpool, Everton and Arsenal footballer. He got straight to the point, 'How do you fancy playing for Exeter City?' It was 1992 and I was a free agent after leaving St Mirren. To be perfectly truthful, I didn't fancy playing for Exeter although I did take it as a compliment that Alan Ball wanted me. He had been Jock Wallace's assistant at Colchester at one stage and had seen me play. He thought I could do a good job for his club.

My only involvement in English football prior to that had been a three-game loan stint at Bolton Wanderers when I was with St Mirren. It was a painful experience, believe me. The contract was scrapped after I broke some ribs and had to return to Love Street. Phil Neal, the former England and Liverpool full back, enquired about me and arranged the deal with Saints boss Davie Hay. Ironically, I had the opportunity to sign for Swindon Town around the same time and they were managed by ex-Celtic player John Gorman, a great friend of Davie. However, the thought of playing for Bolton appealed to me and I agreed to go to Burnden Park.

Phil Neal actually telephoned Graeme Souness, then the Rangers boss, to enquire about me. I must have been given a clean bill of health because I was whisked south and I thought this could be a new and exciting chapter in my career. I have to admit I was made more than welcome at Bolton and I met up with former Celtic striker Andy Walker and also saw the emergence of a teenage talent that would eventually make the switch to Parkhead. Alan Stubbs was just a youngster coming through the ranks, but you could see immediately that he had what it took to make the grade. He could play in central defence or midfield and I recall one game when we were playing West Brom.

We were awarded a free kick about thirty yards out and Alan strode forward to take it. He hit it with unbelievable venom and the ball simply took off at a huge rate of knots and thundered into the net. It was one of the most ferocious deadball efforts I have ever witnessed. I was delighted when Tommy Burns signed him for Celtic. I thought he was a great player for the club. Anyway, I decided to give it my all at Bolton until I picked up my unfortunate injury. I recall I had to apologise to Sammy Lee, who later became manager of the club, one day after a reserve game against Newcastle United. We were awarded a penalty kick and I claimed the ball. 'I'll take this,' I said before Sammy could intervene. Obviously, I wasn't the regular penalty taker, but I fancied myself this particular day. Yes, you've guessed – I missed.

I ran up feeling confident and blasted the ball goalwards. Alas, the keeper guessed right and managed to block the effort and it was scrambled to safety. Sammy and my new team-mates weren't best pleased, I can tell you that. Poor Sammy got it in the neck afterwards when someone asked him, 'Why the hell did you let him take that penalty-kick?' I had to pull him aside and tell him how sorry I was. I returned to St Mirren and by the time I was fully fit Davie Hay had left the club and had been replaced with Jimmy Bone.

When I left St Mirren for a second time I was quite surprised to learn that a Swedish club, Djurgardens, were interested in me. Actually, Gordon Smith, who had been Davie Hay's assistant boss at Love Street, was instrumental in linking us together. I'll always be grateful to him for that. I telephoned Davie for some advice after he had had a successful spell in Norway with Lillestrom. I wanted to know how much I could ask the Swedes in wages. I reckoned the two nations would be fairly similar. Djurgardens had offered a basic weekly wage of £600 and Davie said, 'Ask for another £100. You'll probably get it.' And I did. I also got a £20,000 signing-on fee. I negotiated, as ever, on my own behalf and I had it written into my contract that I could leave on a free transfer at some stage.

Thankfully, my reputation didn't follow me to Sweden and, believe it or not, I don't think I was even booked once. I played twelve games in three months before their winter break and I returned to Scotland during the rest period. I thought I had played well and I scored three goals in the initial twelve-week period. I have to say I enjoyed myself in Sweden and, of course, wasn't hassled by card-happy referees. They were a club that made you feel welcome and that was important to me. They were well run with a good set of fans. So, I was dumbfounded to say the least when I received a telephone call from the club chairman one morning. He said, 'We want to sell you.' I wasn't interested and told him so. 'I'm not going,' I insisted.

He was quite a persuasive gentleman. He continued,

'Huddersfield Town want you and we are willing to sell.' I told him I still wasn't interested in leaving Sweden. The next thing he said did capture my attention though. 'We will give you £50,000 to go,' he said. 'Are you interested now?' I didn't have to spend too much time mulling that one over! 'Agreed,' I answered and went down to Yorkshire to meet Neil Warnock, who was their boss at the time. I don't know how much they were going to pay for me, but it must have been fairly sizeable if Djurgardens were happy enough to give me £50,000 when you consider they had agreed a £20,000 signing-on fee only a few months beforehand. I spent three days at Huddersfield and played one trial game. Honestly, I had a shocker; an absolute embarrassment of a performance. I couldn't do a thing right. I was perhaps too eager to please and the more I tried the worse I became.

I wasn't taken aback when Neil Warnock came to me and said, 'You're not what we're looking for at the moment, Chic. Thanks for your time.' Clearly he hadn't been impressed and I don't blame him. I was awful. Djurgardens still held my registration, so I thought I would be returning to play for them for the remainder of the season. However, fate stepped in once more and I found myself heading back to Partick Thistle for the second time.

17

LOOK ON THE MONEY SIDE

I must have picked up something in the region of £300,000 in signing-on fees during my twenty years in football and the most I ever earned in a basic wage was £750-per-week at Hibs. Dumbarton even promised to pay me what they took from a fruit machine in their social club!

I didn't realise how well off I was when Davie Hay signed me for St Mirren in 1991. We had a meeting to discuss terms and somehow I managed to lose £36,000 in the translation. Thankfully, as everyone in football is aware, Davie Hay is a fair man and he set the record straight. I didn't use an agent, so when we sat down I agreed an £18,000 signing-on fee and a basic wage of £400-per-week. At Partick Thistle I was used to getting a £15,000 signing-on fee spread over two years. When I got home I suddenly wondered if the £18,000 I had said yes to was to be apportioned over the three years of my contract. I was so desperate to work with Davie Hay that I hadn't taken it all in. I knew my basic wage would rise to £430-per-week for the second season and £460 for the third term. But what about the signing-on fee? Would I be getting £6,000-per-year while I could get £7,500 at Firhill?

I had to telephone the boss. I asked, 'Mr Hay, that £18,000 – is it spread over three years?' Davie sighed, 'Chic, have you looked carefully at the contract? You're going to get £18,000 for signing this year, another £18,000 up front for the following year and yet another £18,000 in a lump sum the year after that. You're getting a total of £54,000 in signing-on fees.' Delighted? You bet. I chanced my arm. 'Is there any chance of getting it all up front?' There wasn't even a pause. 'No chance,' said my new gaffer. However, it was a pleasure doing business with Davie Hay.

Then there was my transfer to Dumbarton from Partick Thistle in 1995. Actually, I had the opportunity to go St Johnstone at the same time and I admit it was an honour that a manager like Paul Sturrock, such a wonderful striker for Dundee United and Scotland in his playing days, was interested in me. I went up to Perth and actually played a trial match. Paul must have been impressed enough to offer me a contract. He put £10,000 on the table for signing and there would be a basic wage of £400-per-week. We would discuss the possibility of a car afterwards. I gave them my word I would give it due consideration, but I also told St Johnstone that Murdo MacLeod wanted to talk to me. Everything was done up front.

I went to meet Murdo and the Dumbarton board and, although they had just earned promotion to the First Division, they were willing to better anything St Johnstone had to offer. That surprised me because at the time the Saints had been well established in the top division for a few years. Dumbarton were willing to shell out £15,000 for my signature and a further basic of £450-per-week. They were also going to throw in a car. Okay, it may not have been something that would get a place in Cristiano Ronaldo's garage, but it was still a means of transport.

We were discussing my basic salary when I caught Murdo smiling. The directors were going to make up my wages with money from a puggy machine. One of them said, 'That machine there brought in £250 last week. It's just about guaranteed.' I asked, 'What if the fans stop playing the puggies?' There was silence from the directors. They were going to have to find the money from another source to sort out the difference. It sounded just a little dodgy, so I did a bit of negotiating. 'I'll sign if I am guaranteed a free transfer at the end of the season.' I said. 'If we stay up, I want an extra £50-per-week on my wages.' The directors agreed and I signed on with the Sons.

Looking back, I now realise I should have joined Paul Sturrock at St Johnstone. I believe that would have been better from a purely football-playing point of view. Sturrock had been a player I admired and, as I have said, it was marvellous that he thought enough of me that he wanted me to play for his club. I couldn't turn down Dumbarton, though, because their offer was far superior. As it happens, they were relegated at the end of the campaign, but neither Murdo nor myself were around to witness it. He quit as manager before I had a chance to play for him. I often rattled his cage by saying he got on his bike because he knew he couldn't handle me! Murdo was replaced by his assistant Jim Fallon. Jim Duffy came in for me around about Christmas time and I was off on my travels once more, this time to Dundee.

It was Duff, of course, who then took me to Hibs and gave me the biggest basic wage I had ever earned. He always made me feel wanted and I was closing in on my mid-thirties when I signed for him. When he offered me £750-per-week I knew I couldn't let him down. He also agreed to pay me £500-per-week appearance money. I really wanted to perform for him and repay his faith in me. Obviously, I was sick when he lost his job at the Edinburgh club. I knew it was time up for yours truly the day he left Hibs.

I thoroughly enjoyed myself at Hibs. We tried to be entertaining and that's probably what undid us in the end. We left the backdoor open too often and suffered the consequences. Former Celtic striker Andy Walker joined us from Bolton Wanderers and made an immediate impact by scoring two goals in a good win over Aberdeen. I set him up with both and Andy told everyone afterwards, 'I don't think I saw Chic for another four days.' Actually, he was wrong – it was five! Talking of Andy reminds me of turning out for Celtic in the Masters six-a-sides. Andy keeps reminding everyone that I have had seven years of wearing the hoops because of my one and only game for the club in the Testimonial Match against Manchester United. 'My God, Chic,' he said, 'you've made plenty out of that ninety minutes, haven't you?' You bet.

I should have realised my St Mirren career was coming to an end shortly after Jimmy Bone replaced Davie Hay as boss. Once again, I have no-one to blame but myself. I look back at what I did in my last game for the Saints and feel nothing but overwhelming shame. It's deeply embarrassing to even talk about the incident because I think what I did is despicable and disgusting. I spat at an opponent. Please believe me when I say how shocked I was at my own reaction. It's not something I would normally consider doing in any circumstance. It's totally unacceptable, but I snapped at the time.

I had been booted all over the place by some bloke in a game against Ayr United, who were managed at the time by present Scotland international boss George Burley. It was obvious the bloke was trying to get me to retaliate. I had gone into the game with a couple of red cards behind me and I was determined to resist the temptation. The referee, knowing my reputation, was keeping a close eye on me while ignoring the Ayr player flying into tackles all over the place. He went over the top and could have broken my ankle when I finally caved in. Instead of just kicking the guy I spat on him. Why I did that I will never know. I was brought up in Possil and you settled your differences with your fists. Spitting was a no-no. However, on this occasion, I just couldn't stop myself.

As luck would have it, the referee was right behind me. He didn't have to tell me I was about to be sent off. I was heading for the dressing room even before he flashed the red card under my nose. I deserved the punishment. It was the only course of action for the match official to take. I was raging at myself. I had risen to the bait once again and I had no-one to blame but myself. Jimmy Bone had been in my corner from day one and even made me captain of the club. However, I detected something in his tone when he telephoned me the following morning. 'Chic,' he said, 'I'm getting bad vibes from the directors. I think the board want to sack you.'

I couldn't put up much of an argument in the circumstances, but I believed they were seizing the opportunity to get me off the payroll. I had been given a good contract by Davie Hay, as I have said. Sure enough, the board decided to give me my P45 and refused to honour the rest of my contract which still had a year to run. I was due the third £18,000 of my agreed signing-on fee, but they didn't want to know. I decided to get the PFA involved and I took St Mirren to court. George Fulston, my old chairman at Hamilton, and Rangers director Campbell Ogilvie were on the tribunal committee. St Mirren director Stewart Gilmour, a practising lawyer, represented the Paisley side. Tony Higgins, of the PFA, was in my corner.

Before the case started, someone suggested we should have one last go at trying to settle out of court. The club offered me £2,000 which I immediately rejected. The directors had a quick confab and came back and put £5,000 on the table. I still wasn't entirely satisfied, but I took it. I just couldn't be bothered with all the hassle of a court case. My career with St Mirren was over and it was time to move on, but even I was surprised when my next transfer was to Djurgardens in Sweden.

By the way, there are an awful lot of players who believe they are financial experts who can negotiate unbeatable contracts. That certainly wasn't the case for an old St Mirren team-mate when he was trying to get more cash out of Davie Hay. The Saints gaffer offered the player – I couldn't possibly name him – an

£80,000 signing-on fee covering three years. It was good money, but the player was convinced Rangers wanted to sign him and stalled on signing the contract. He asked the boss for time to mull it over. He should have snatched at the deal as he discovered a couple of weeks later when he met again with Davie Hay. 'Okay, boss, I'm ready to sign,' he said. 'Fine,' said the gaffer and put the contract in front of the player. The £80,000 which had originally been on offer had shrunk to £20,000. 'That's the best we can do,' said Davie. My mate was so dumbfounded that he put pen to paper. In the space of a fortnight he had lost £60,000!

Speaking of my old pals reminds me of the time I was robbed of a perfectly good goal against, would you believe, Rangers. I was playing in Campbell Money's Testimonial Match and the Ibrox side provided the opposition. I came on as a substitute and I am still convinced I scored one of the best efforts of my career that day. I fired in a curling 30-yard free kick and it completely bamboozled their keeper, Chris Woods. It smacked off the underside of the crossbar and came down. I'm sure it was over the line, but the referee was having none of it and waved play on. Maybe if my name had been Geoff Hurst and we had been playing West Germany at Wembley and there was a Russian linesman it might have stood!

Footballers are no different from anyone when it comes to money. We've got mortgages to pay, everyday bills that need to be sorted and so on. That's why it's always nice to pick up a little extra when you can. When Coca-Cola sponsored the League Cup they had an incentive package and one was to shell out a £500 bonus to anyone scoring a hat-trick. That sounded quite enticing to yours truly although I was hardly a goal machine. One day I picked up the telephone to call my Ma and simply tell her that I loved her. She answered back, 'And I love you, son.' Now that may not seem too unusual for other people, but it was something I rarely did although, of course, I did love her. I just suddenly got the urge to put in the call. While I was on, I asked her, 'We're playing Alloa in the Coca-Cola Cup tonight. Do you fancy coming along?' She readily agreed and was in her place in the stand by kick-off time.

Somehow I sensed something good was in the air. I scored one, then another and suddenly I realised I was on a hat-trick. If any of my Thistle team-mates thought I would be passing the ball to them when I had a chance to have a pop at goal they were wasting their time. I duly notched my third goal and, shortly afterwards, got a call from the League Cup sponsors to come and collect my £500. I turned up for my bonus and bumped into Ally McCoist. He was in his usual bubbly mood. 'See that £500, Chic,' he said, 'that and all the other hat-tricks I score help to pay for my holidays every year!'

18

GUNSHOTS IN RIO

The strangest things seem to happen to me. I appear to be a magnet for odd happenings and, for the life of me, I can't figure out why. I can't even escape the bizzare twists and turns that follow me even when I am on another continent. How about this little story from Brazil?

Bobby McCulley, who was a coach at Partick Thistle, and I were on the world famous Copacabana Beach in Rio in 1994. Our hotel was nearby and we decided to go down and watch some kids play football on the beach. Their skills were fantastic even at a very early age. Their technique would put some of today's SPL players to shame. Anyway, Bobby and I were enjoying the entertainment when suddenly we heard something that sounded like a car backfiring. The kids immediately hit the sand as they had a better idea of what was going on. This was no car backfire – it was a shoot-out between the police and a bank robber.

'Bang!' Then another bang. Bobby and I hit the deck behind a parked car. It was like a scene from a movie. I asked Bobby, 'Are we in Bridgeton or Possil?' He replied, 'I don't know, but I can tell you I'm going nowhere. I'm staying here.' A full-blown siege was going on around us and bullets seemed to be flying all over the place. And then there was silence. After a while Bobby and I decided to take a peek to see what was going on.

There were cops everywhere and there was the bank robber lying dead in a pool of blood inside his car. Just another day in downtown Rio, we were told.

Actually, it really is a wonderful city. It's a bewildering mix of posh and poverty. In some quarters it is genuine upmarket quality, but you can turn a corner and you are in some shanty town. I've got to say they make rather bizarre bedfellows. I went with wee Bobby to get a close-up view of that iconic statue of Jesus Christ – known as Christ the Redeemer – that sits atop Corcovado mountain. The views from that vantage point are simply breathtaking. This is a city of 11 million inhabitants, double the population of Scotland, and it is utterly fascinating. I loved the place, despite people getting blown away by the local constabulary!

Then there was the time I was ordered to go into the crowd to retrieve my shirt. Partick Thistle became the last team to win the Glasgow Cup in its old format when we beat Clyde at Firhill. It may only have been the Glasgow Cup, but it was silverware and the Thistle players decided to do a lap of honour. I was so elated I stripped off my shirt and threw it to the fans. Jackie Husband, our trainer and a true Jags legend, was far from pleased when I returned to the dressing room. 'Where's your top?' 'I tossed it into crowd,' I replied. 'Go and get it back – it's part of a set.' I was speechless. 'Go back out there and try to find my shirt? You're having a laugh.' But Jackie was deadly serious. He genuinely wanted me to go into the crowd and try to get my shirt. Like that was going to happen! All the time, John Lambie was standing behind Jackie doing his best to keep a straight face. Needless to say, that jersey was long gone and some happy Thistle fan probably has it under his bed to this day.

Did you know that I once appeared on the hit television detective show *Taggart*, set, of course, in Glasgow? They were celebrating their fiftieth episode and they had Paul Barber, the guy who played Denzil in *Only Fools and Horses*, as one of the stars. The STV producers came up with the idea of using real footballers for the drama that was all about a referee being murdered. I have to say it wasn't the first time I wanted a referee murdered!

It was all quite amusing. We didn't get to speak, naturally enough, but the TV people thought all footballers should drive big flash cars like Jags and Mercedes. There weren't too many of them in the Firhill car park, that's for sure. So, they took us down to a local car showroom and we were only allowed to drive them about six yards because of insurance problems. In real life, you were lucky to get a Transcard at Thistle!

I used to swim quite a lot as I think it is great for all-round fitness. In fact, I still swim today to try and keep in shape. There was one day I was minding my own business doing a few laps at Bishopbriggs baths. I realised I had been spotted by a few obvious Rangers supporters. They were teenagers and were wearing Rangers shorts. One of them thought I would be a good idea to dive bomb yours truly. I saw him getting into position and I anticipated what was coming next. Sure enough, the young lad threw himself in my general direction and sent the water cascading all over the place. As he sank in the pool I kicked him on the leg. The wee pest got to the surface and shouted at his mates, 'The bastard kicks under water as well!' Aye, and don't you forget it!

Mind you, that was all very innocent compared to what I experienced when I was playing for Rockerfellers in a game in Castlemilk. A confrontation between a team from Possil and one from Castlemilk is always likely to be tasty. There aren't too many fainthearts or shrinking violets in either of those two housing schemes. However, even I couldn't have predicted how this one would have ended. I took my little brother Frank to the game. He was only about seven years old at the time.

As you might imagine, there was plenty of welly going in that day. We kicked them and they kicked us back. It was a typical Sunday morning in Glasgow. My mate Jas Carbon took umbrage when one of our opponents scythed him down. Jas didn't hold back. He whacked the guy and the bloke was knocked stone cold. He went down like a collapsing tree. The rest of the game was a bit frantic and, to be honest, I was happy to hear the full-time whistle. There were no dressing rooms, so we had to change at the Labour Club just up the road. We made our way there but found the doors locked. We hung around for about fifteen minutes or so and then a van drove up and stopped. About seven guys came out the back and, as they say in quaint Glasgow parlance, they were tooled-up. Basically, they were all carrying blades. One of these gorillas walked up to us and growled, 'OK, who wants to fuckin' fight now?'

Believe me, my mates could more than handle themselves in tough situations, but on this occasion they said nothing. I stepped forward and tried to reason with our aggressors. 'Lads, we've got a wee lad here,' and I pointed to Frank. 'We don't want any trouble, mate.' This guy glanced in my direction and muttered, 'Don't call me mate. I'm not your fuckin' mate. Understand?' I remained silent. Then, remarkably, the Castlemilk guys looked at each other and nodded. They weren't going to attack us while Frank was in the vicinity. 'Fuck off, Possil shitbags,' they said. 'Next time we'll do you – with or without the kid.' We didn't hang around to debate the matter. Without getting changed and still caked in mud, we piled into our cars and made our getaway. Thank God for wee Frank!

19

FAREWELL TO FIRHILL

There is a quaint old Glasgow expression that goes along the lines of 'putting yourself in the poorhouse.' Basically, it means you are skint. That's exactly the situation I found myself in when I left Partick Thistle in 2004. How did I put myself between the rock and the hard stuff? Well, I was stunned when I walked into Firhill one Monday morning to be told that manager Gerry Collins and his assistant Bobby McCulley had been sacked. I had been part of the backroom staff and we were a team. We all got on so well with each other and I was genuinely choked when I was given this stunning news. I was then informed by chief executive Alan Dick that my job as a coach was safe.

Partick Thistle were bringing in Gerry Britton and Derek Whyte as joint managers. I heard one director describing them as 'intellectuals'. What did that make Gerry and Bobby? And me, for that matter? I was far from happy about this turn of events and I mean that as absolutely no disrespect to Gerry and Derek, two guys I just happen to like. I didn't hold any grudges against them because that's just the way it is in football. To get a job in the game you have to wait until someone moves out, voluntarily or otherwise. There are just so many managers you can squeeze into the one dug-out. However, whoever made the decision at Firhill that I should remain while Gerry and Bobby were booted out the door obviously didn't know Chic Charnley.

I was having none of it and quit there and then. I put myself in the poorhouse.

Do I regret it? Would I do the same again today? Even after the years that have passed I have no hesitation in saying I would do exactly the same again. OK, that doesn't make me Businessman of the Year material, but that never entered into it. I felt for my colleagues, I knew how much those jobs meant to them and I simply couldn't carry on without them.

Thistle had lost on the Saturday to Celtic and the board had a meeting the following day. I believe the vote was 4-3 in favour of sacking my pals. It seemed a rash act to me. The club's next two games were at home to Aberdeen and Motherwell and were matches Thistle could have won. Gerry and Bobby could have turned things around if they had been given the opportunity and time, I have no doubt about that. John Lambie phoned me when he heard I had walked out. 'Get back there and get your job back,' he said. Jim Duffy gave me a call, too. 'What are you doing, Chic? Don't go. They want you to stay – you've still got a future at the club.' Lisbon Lion John Clark, now part of the Celtic backroom team, also got on the blower. He said, 'Don't chuck it, Chic. You've still got a lot to offer.' I wasn't listening. I had made up my mind.

I was still simmering when I talked to Gerry later that evening. He asked casually, 'What sort of compensation package did you get, Chic?' I replied, 'What compensation?' Gerry fixed me in his gaze and replied, 'Bobby and I got a few quid to leave. They sacked us while we were under contract. We were due money. What did you get?' I was silent for a moment while this little gem of knowledge had been imparted. 'Nothing,' I said lamely. 'I resigned.' My big mate said, 'You quit? You silly beggar, they won't have to pay you a penny.' And they didn't. My mates left with a wedge in their pockets and I went out with zilch. I was never going to make it as a Philadelphia lawyer.

I loved that club – still do – but I just couldn't envisage me taking training without Gerry and Bobby around. We weren't just colleagues, we were good mates. I was given advice by everyone and it all went the same way. I think wee Bobby, one of the best coaches in the business, came close to falling out with me because I refused point blank to change my mind about quitting. However, I know he also admired me afterwards for the stance I took.

I wonder if wee Bobby would have hung around Maryhill if Gerry and I had been sacked. Or Gerry, for that matter, if Bobby and I had been given our P45s. I know the answers and so, too, do my mates. We were a team and we all had Thistle nearest to our hearts. We were striving to bring stability and then, hopefully, success to the club. I can assure everyone that we worked hard for the Jags. We took in a lot of games, looked at a lot of players who were within our price range. We all knew what was required and we were working to that end. Unfortunately, there were those at Firhill who didn't understand that. It takes time – and that's what we didn't get. It was all very unfortunate.

To my way of thinking we were all in it together. I thoroughly enjoyed working with Gerry and Bobby and I don't think my heart would have been in it if they weren't around. The best thing to do was leave because I wouldn't have taken Thistle's money under false pretences. However, it was an extremely sad day when I left the club for the last time. I said my goodbyes and walked out into the street and wondered what football held in store for me. That was my last connection with mainstream football and that has, rather obviously, left me fairly sad. Let me say here and now how much I loved being part of Partick Thistle. The fans were brilliant to me and still are any time I turn up at a Thistle match or function. Let me take this opportunity to say a massive 'thank you' to them all. I will never forget them and I hope they will never forget me.

20

IF THE CAP FITS

I was as amazed as anyone when there were calls for me to get capped by Scotland when I was heading for my mid-thirties. To be honest, I was utterly flattered to be even mentioned with the international set-up. To be equally truthful, it didn't seem to matter too much at the time. Now that I've stopped playing it's an entirely different matter. I would have loved to have played for Scotland. It now really annoys me that I didn't.

Andy Roxburgh and Craig Brown were the managers who were around at the time I was peaking, but neither seemed too impressed with yours truly. I was too busy just getting on with business with my club sides to even think about moving onto another level, but I wish I had got the nod from either – or both – of them. Could I have done the business for the international team? With all humility I have to say I think I could have achieved something for my country. When I was playing for Jim Duffy at Hibs in 1998 I thought I was doing well and that was when there was a major push from some quarters for me to at least get in the squad.

Possibly neither Roxburgh nor Brown thought I was worth the risk because of my so-called bad-boy image. I couldn't argue with that, but they did select other players around that time whose temperaments could be more than questionable. I look back now and, as I have said often enough, I know I could have done so much better. I'm not being big headed, but the potential was there and I just didn't make the most of it. I would have been thrilled to have played for my country even just once to say I had represented my nation at the highest level.

In fact, I might have made it onto the international stage – with the Republic of Ireland! Jackie Charlton was the Irish manager at the time and he telephoned John Lambie at Partick Thistle to see if I was eligible. Back then, the joke was that you only had to drink a pint of Guinness to be allowed to play for Big Jack. He loaded up his team with Glaswegians such as Gorbals-born Owen Coyle and Castlemilk-born Ray Houghton, who were available because of the birthplace of their parents or grandparents. English-born players were brought in, too, as the manager transformed his adopted nation. I suppose the English World Cup-winning centre-half had heard about my allegiance to Celtic and possibly presumed there was a bit of Irish in me. I was told of his interest, but I didn't bother pursuing the matter.

That reminds me of a story of a former team-mate of mine and I will spare his blushes by not naming him. He was invited into a Scotland Under-21 squad for a friendly match. He fretted a bit because he had set his heart on playing for the Republic of Ireland although he had been born in Glasgow. He picked up a 'mysterious' injury after playing for his club on Saturday and had to withdraw from the Scotland squad for their midweek game. He believed his chances of pulling on the green jersey of the Republic of Ireland would have been scuppered if he had turned out for Scotland, even at Under-21 level. As it happens, he never got another chance to play for the nation of his birth and there was no call from the Irish, either, for one simple reason – he was ineligible. There was not even the hint of an Irish background in his family's history and even Big Jack couldn't persuade FIFA, the world governing football body, to allow him to field players in these circumstances!

Celtic's Aiden McGeady, of course, elected to play for the Republic of Ireland despite being born in Scotland and that is his right. I don't think he should be criticised for wanting to play for the birthplace of his grandparents. He had a choice to make and I'm sure he had very good reasons for making the one that he did. I have to admit it would have been extremely difficult to reject the opportunity of playing alongside Roy Keane when he was at his best.

It would have been intriguing, also, to have played for a manager such as Jackie Charlton. I still laugh at a story I heard about him and his assistant manager Maurice Setters. I think they were with Sheffield Wednesday at the time. It is alleged they went fishing on the morning of a Saturday match and completely lost track of time as they sat and dozed off by the banks. Anyway, the story, true or not, goes that they awoke with a jolt and realised they only had about half an hour or so to get to Hillsborough for the 3pm kick-off. I'm told the crowd were treated to the sight of their manager and his assistant racing into the dug-out in the nick of time. Did I mention they were also said to be eating fish suppers out of newspapers? And I thought John Lambie was a character!

So, unfortunately for me, there was no call from my country to perform in the dark blue. Chic Charnley in a blue strip? Doesn't seem right, does it? No, I would have thrived on getting recognition for my country. It would have been a major thrill to play for Scotland at Hampden. Imagine lining up against England? I had travelled with the fans to Wembley and Hampden on umpteen occasions to witness these spectacles and to actually get the opportunity to play against the Auld Enemy would have been something else altogether. I wouldn't have needed any extra motivation for a confrontation like that. I wouldn't have kicked a single Englishman. Honest!

Having said all that, I'm not too sure how I would have fitted into any international set-up. Andy Roxburgh and Craig Brown expected the players to have their socks pulled up even for training. Why? Does looking impeccable going through all sorts of routines make you a better player? I would have questioned that. In fact, I'm in Richard Gough's corner. It was headline news that the former Rangers skipper had fallen out with the hierarchy because he refused to get involved in a team bonding thing when he was on a plane with the rest of the squad. I don't know what game the SFA chiefs wanted him to play on that particular flight, but I'm told he was reading a book at the time and was quite happy to give it a miss. I would have done exactly the same thing and that would probably have been noted by the bosses.

My God, I've heard tales of the players getting involved in handball games as part of the bonding process. Leave it out. Only goalkeepers are allowed to use their hands, so what's the point of the outfield players being expert in this field? None. Once again, I might have been tempted to tell the gaffers to sling it. As I have said, I have absolutely no doubt I could have played in the international side back then, but I'm not so confident I could have handled all the other things entailed with being a Scotland player. I've always just wanted to prove what I could do where it mattered most – out on the pitch. All the bonding nonsense would probably have done my nut in!

Don Revie used to have pre-match rituals for his Leeds United players back in the late Sixties and early Seventies. Apparently, he was an extremely superstitious bloke and would turn up on matchday with the same suit, shirt and tie all the time if they won when he first wore it. He also ordered his players to nights at the bingo. I bet Big Jackie Charlton and wee Billy Bremner must have revelled in that. They played indoor bowling and were allowed to play cards, but only for matchsticks. That would have been laughed out of court by some of my former team-mates, I can tell you. How many pints can you get for a fistful of matches?

I'll always remember a meeting with Andy Roxburgh, who was the international gaffer at the time. His uncle was Jack Steedman, the Clydebank chairman, and Roxburgh had popped into the dressing room before a game. I was with the Bankies at the time and shouted, 'Hey, Andy, how about a Scotland cap?' He didn't even break stride as he replied, 'Why? Is the sun in your fuckin' eyes?' My heart sank.

The SFA have made more than a few mistakes over the years. I don't think I am alone in believing that. They made the wrong decision is giving the manager's job to Berti Vogts. George Burley wasn't a good appointment, either, if you ask me. What was the Berti Vogts thing all about? Who thought that was a great idea? OK, he was a superb defender in a fabulous German team and he had won the European Championship with a now-defunct Golden Goal against the Czech Republic in the final at Wembley in 1996. Mickey Mouse could probably have taken care of that talented line-up. However, I believe he was actually sacked by the German FA after a string of terrible results and he was the first national coach to be axed in the nation's history. So it's only right and fitting that he turns up as Scotland's boss, isn't it? I'm not being smart after the event because I said so at the time. I thought he would be a disaster in the post and obviously I knew something the SFA didn't.

He didn't actually get off to the best start, did he? We were hammered 5-0 by France in Paris in his debut game and, for me, that said it all. I realise the French were world and European champions at the time and were a formidable outfit, but our resistance that night was next to nil. We were 4-0 down by the interval and, thankfully, the French took their foot off the gas after the turnaround. It was one of those encounters where the opposition could have scored at will. If they had needed ten to win on aggregate I would have put money on them achieving their target. Wee Berti had hardly settled into his managerial chair when we heard stories about a stand-up argument with Rab Douglas at the airport on the way home. Apparently, Big Rab, Celtic's goalkeeper at the time, had been promised a run-out in Paris, but was overlooked. It got worse when it emerged that Neil Sullivan, the keeper who played the entire ninety minutes, was carrying an injury. Berti, apparently, got confused when he put on Kilmarnock's Gary Holt as a substitute and then, minutes later, took him off and put on another sub. Gary, I must add, was perfectly fit. My hopes of ever seeing my country lift the World Cup were dwindling fast!

And who on earth made the decision to appoint George Burley as the Scotland boss instead of either Tommy Burns or Graeme Souness? I'm sure Burley is a lovely chap, but did he really have the pedigree for that job? I don't think so. Yes, I realise he had a good couple of months at Hearts at the start of one season, but any club can go on these sort of runs. They hadn't encountered the nitty-gritty of the campaign at that stage. Burley went to Southampton and they were struggling in the wrong end of the league by the time the SFA were looking for a successor for Alex McLeish. What a farce that was. McLeish actually went with the SFA big-wigs to South Africa for the World Cup qualifying draw and then joined Birmingham City a couple of days later. Astonishing. Terry Butcher as Scotland's second-in-command? Give me a break. I've nothing against the former England captain, but that's not the post for him. Really, he shouldn't be there. It's marvellous that he has kept up a friendship with Burley that started in their playing days at Ipswich Town, but, as far as I am concerned, he is in the wrong place at the wrong time.

Tommy Burns, who had been Vogts's assistant, was told he would have to apply for the manager's post if he wanted it.

Why? Didn't the SFA know what he had contributed during the reign of Vogts? Did they not have a clue what was happening right under their noses? Tommy Burns, I hasten to say, is one of the most distinguished characters I have ever had the pleasure to meet. The guy was class and extremely knowledgeable about the game. Maybe he knew too much. But the SFA should be ashamed of themselves for even daring to think he should be interviewed for the post. Simply put, that was nonsensical. He should have got the job the minute McLeish defected to the Midlands. I will always remember Tommy saying, 'No' bad for a wee lad from the Calton, eh?' He loved Celtic, of course, but I know he would have leapt at the opportunity of taking over Scotland. It wasn't to be, alas.

Do you think for a single minute that Barry Ferguson and Allan McGregor would have got involved in that bevvy session at Cameron House on Loch Lomond after that 3-0 World Cup hammering from Holland in Amsterdam if Graeme Souness had been around? Not a chance. Obviously, it was wrong for the blokes who rule our football to allow it to happen in the first place. What on earth were they celebrating, anyway? We've just been humped with the world looking on. I might just have crept under my bed and hid for a few hours in the darkness. Souness would have had plenty to say about the situation and you better believe he would have taken swift action. Burley had them sitting on the substitutes' bench for the following midweek game against Iceland at Hampden. I'll never be able to fathom that one out.

Was our international manager the only person in Scotland to be surprised when the Tartan Army booed the names of the Rangers double-act when they were announced? Did they actually have any chance of coming on that night? I doubt it. So, what on earth were they doing there in the first place because there were other subs who could have taken their place? And, of course, Ferguson and McGregor took the opportunity to give the V-sign to all and sundry. I'll wager they would have been nowhere near our national stadium that evening if Souness had been in charge. The SFA wouldn't have had to wait a few days before coming to a decision on how to reprimand this pair because Souness would have done it there and then. It would be goodnight and farewell to both of them.

By the way, I hope Ferguson is not simply going to be remembered for his silly actions over those few days. It wouldn't be fair if Boozegate was now going to follow him around for the rest of his life. I think he is a first-rate player and was a great servant to Rangers. It would be a real shame if that's all we now believe he contributed to football. He did so much more and it would be out of order if he is only going to be thought of as a boozer who likes to flick V-signs at people.

The fact that Souness wouldn't have hung around mulling over making a decision about Ferguson and McGregor might be the very reason he is not the international manager. Were the SFA afraid of appointing such a strong-willed, single-minded individual? He was never one for tugging on the old forelock and I don't believe time has mellowed him too much. I'm sure he would still say exactly what he means and he would still stand on people's toes when he thought it necessary. I read that he did not get the job when he admitted he would not come across the border to live in Scotland. I'm informed he was happy enough to spend two or three days a week here, but the SFA wanted him to come home lock, stock and barrel and be on call twenty-four hours a day. Like that was ever going to happen. Apparently, he would have been asked to do school visits and so on and I'm sure Souness, as Scotland's boss, would have been only too pleased to carry out these duties, but only on his time. Quite right, too. Why have we got so many community coaches? Isn't that a job for them? I think so.

In my opinion George Peat, the Scottish FA President, shouldn't be there, either. What possessed him saying what he did about Chris Iwelumo on the eve of an important World Cup-tie against Macedonia at Hampden in September 2009? To blame the Wolves striker for our position in the qualifying group was nothing short of ridiculous. OK, the big frontman did contrive to miss an easy chance against Norway in an earlier qualifier and the game ended goalless. It wasn't his finest moment, but he missed because he is only human. What gave Peat the right to criticise the player? What pedigree does he have? He was on the board at both Airdrie and Stenhousemuir. Big deal. So, he will be used to football when the stakes are really high, then? He'll be aware of what it is like to perform when the pressure is on? Has he ever kicked a ball in his life? I thought what he said about a good professional, trying his best at international level, was scandalous. And you better believe this – the other players will have taken note of his words. They won't be impressed, either.

So, how was George Burley treated for his failure to get Scotland to the World Cup Finals in South Africa in 2010? Yes, of course, he was given a vote of confidence and remained in his job. And did someone high up in the SFA echelons really say, 'We are moving in the right direction'? Am I missing something? Is this the same Scotland who last qualified for a major tournament when they reached the World Cup Finals in France in 1998? Is it the same Scotland who were knocked back by former skipper Gary McAllister, out of a job at the time, who was offered a managerial post alongside Burley and Butcher? It's all very mind-boggling.

Maybe international football wasn't for me, after all.

21

EUROPEAN FRIGHTS

My good friend Jim Duffy knows more about football than just about anyone I know. So, when he talks about the plight of Scotland in Europe I tend to listen. Well, most of the time! I agree entirely with him when he says we have got to get things right at grassroots level and I love his take on the predicament.

He says, 'Barcelona's top-class players Xavi and Iniesta would not have made it through the system in Scotland. On the deck they wouldn't have had enough power in their legs to kick the ball from one end of the pitch to the other. So, their coach would have found other lads who could.'

Xavi and Iniesta, of course, are on the slight side and I know precisely what Jim means when he says that. It is a spot-on observation. Now the nearest I ever got to playing in Europe was on the Spanish beaches with my mates in kickabouts. I almost got into the St Mirren travelling squad for a European tie against Feyenoord, but, as I disclosed in an earlier chapter, I put the kybosh on that possibility by decking team-mate Gardner Speirs in training.

So, what gives me the right to give my verdict on the lamentable showings from Scottish club sides at this level? I think I've got a right to my opinion because you don't have to be a horse to pick the Grand National winner, do you? You don't have to be a mountaineer to realise Mount Everest is a rather large mountain. Need I go on? Plus I've had two decades in the game. You can have knowledge of a subject without actually participating in it and I believe our performances in Europe have been nothing short of disgraceful.

You only have to look at season 2009/10 to see what I am talking about. Aberdeen thrashed at home 5-1 by a Czechoslovakian team called Sigma Olomouc. Who? It could never have happened in the Alex Ferguson era, that's for sure, when Pittodrie was a fortress. Falkirk went out to Vaduz, a Liechtenstein bunch of nonentities. Hearts? Gubbed 4-0 by Dinamo Zagreb and it was all over after ninety minutes of the first leg. It didn't get much better for Motherwell, either. Another first game hiding – this time 3-0 – from Steaua Bucharest. In short, these results are simply disastrous. They are also unacceptable for the good of Scottish football.

It was painful to see Celtic beaten home and away by Arsenal although it must be said the only luck they got in those two ties was bad luck. However, there can be no excuses for their display in Tel Aviv against a fairly ordinary side in Hapoel. Even I can't defend that 2-1 Europa League result. Tony Mowbray's first six games on the European front brought four defeats, one draw and one success, the 2-1 last-gasp victory over Dinamo Moscow in Russia. That sort of record isn't designed to get you a rousing round of applause. My God, it seems such a long time ago that Glasgow hosted two European semi-finals on one night. It was back in 1972 that Celtic faced Inter Milan at Parkhead in a European Cup-tie and, after two successive goalless draws, only went out 5-4 on penalty-kicks when Dixie Deans fired his spot-kick over the bar. Across the Clyde at Ibrox, Rangers were beating Bayern Munich 2-0 on their way to lifting the now-defunct European Cup-Winners' Cup. What on earth has happened since those heady days? It would be too easy to say expectation levels are too high for Scottish football. Have we gone backwards? Or have other nations improved dramatically?

I think we have definitely stagnated. The calibre of youngsters coming through just isn't good enough and there must be a reason for that. And that's where I think Jim Duffy has a valid point. I realise it has been said before, but I think we have got to concentrate on youth at a very early stage and it might not be a bad idea to copy what they do in Holland. Their kids come through the ranks, but they don't play for trophies or medals. They are simply encouraged to hone their skills and to go out there and enjoy themselves. The pressure is lifted off them and that certainly helps them develop. What do we do in Scotland? You get dads – and mums, too – shouting and bawling from the sidelines as ten-year-olds play on full-size pitches. What a lot of nonsense. I've always felt sorry for the wee goalkeeper. What chance has he got? How does he develop his skills when he needs a ladder to touch the crossbar? Being lobbed twenty times in a game is hardly going to inspire confidence, is it?

The Dutch have scaled pitches for all ages and the kids also play with balls designed for them. It gives them a feel for the ball. Our wee lads have got to try to kick things the weight of cannonballs. I had to laugh the other day when I was watching a schoolboys' game at a local pitch and one skinny tiny tot, who looked in dire need of a good fish supper, went to take a corner-kick. His dad shouted, 'Put it to the back stick, son.' I had to smile. The poor wee soul looked as though he would have difficulty getting the ball out of the arc never mind into the oppositions' penalty area. That hardly helps confidence in your ability, too, when you continually fail to achieve what is asked of you. Honestly, you want to take these daft dads and give them a good shake at times.

But what has happened to a nation that used to bring through kids such as Jimmy Johnstone, Denis Law, Jim Baxter, Kenny Dalglish, Graeme Souness, Billy Bremner, oh, I could go on forever? The conveyor belt has come shuddering to a halt and

I can recall that every successful English team used to have Scots at the core. That's not the case these days. Maybe kids today prefer to go home and play with their computers rather than dig out a ball and practise their skills.

Another excuse – not a reason – that constantly crops up is people saying that we are only a nation of 5.5 million or so and we should remember that fact when we take on bigger countries. That's rubbish. Uruguay have won the World Cup twice and their population is just over 3 million. I accept the tournament might have been in its infancy at the time, but it was still two magnificent achievements. End of that argument, I think. No, I believe the youngsters need guidance from folk who are dedicated to the cause and not merely going through the motions. I'm sure that happens. People turn up, take a course, show little interest and go home. That gets through to kids. These are their formative years and are so important to their progress. They aren't given these years back to start again and rectify the earlier mistakes. 'Too late' was the cry.

Believe me, I know what I am talking about. I've seen some of my talented Possil pals lose their way because there is no guidance. With a little application from someone they could have had a good profession in football. I was one of the lucky ones although, as I have said, I realise I should have done so much better. What is it they say – that youth is wasted on the young? It's so true. If I knew back then what I know now then my career would have been vastly different. No point in crying about it now, though, is there?

I've heard all sorts of nonsense coming from all sorts of angles about our demise in Europe. Yes, Rangers deserve credit for reaching the UEFA Cup Final in 2008/09. However, and I realise I will be accused of Celtic bias here, it was hardly the most exhilarating voyage to Manchester, was it? However, at least they did get there. In the main, though, foreign clubs come to

Scotland and look as though they have the ability to play us off the park.

They have poise, technique and the ability to kill the ball in an instant. Most Scottish footballers' first touch is a tackle. The foreigners caress the ball; we try to clatter it. I have also heard it said that our clubs have lost their fire, fight and passion because we have too many foreigners in our teams. Try telling that to Celtic fans who witnessed Henrik Larsson playing for them. Sorry, I'm not having that.

Europe used to be so important to Scottish clubs. Celtic, Rangers, Aberdeen, Dundee United, Hibs and Hearts have all had their moments on this platform. Jock Stein was a visionary when he took over at Parkhead in 1965. He seemed to have the ability to see the bigger picture. Winning things in Scotland was all well and good, but Europe was where you really made your name. Big Jock, I know from talking to players such as Bertie Auld, thought it was absolutely vital and totally crucial to perform in that arena while all eyes were on you. What he did at Celtic was truly remarkable and to win the European Cup in 1967 was an exceptional breakthrough. Will another Scottish club conquer Europe? You're having a laugh. Well, at least, not until they start to get their act together, think positively, believe in themselves and, most importantly, encourage and develop youth.

I almost fainted when I heard a Scottish manager – I'm not going to embarrass him by naming him – answering a question shortly after his side had gone out of the Inter-Toto Cup. 'Are you disappointed?' he was asked. 'Not really,' he said, 'we can now concentrate on the domestic competitions.' What marvellous ambition. What would Big Jock have said?

I am making no apologies for getting on my soapbox here because I love to see Scottish clubs performing well in Europe. I went to Seville in 2003 to watch You-Know-Who against Porto in the UEFA Cup Final and I was so proud of Martin O'Neill's side. Wasn't it great to see them put it over on English duo Liverpool and Blackburn Rovers on their way to that final? And the way they played against the Portuguese in sweltering heat was wonderful. They may have lost 3-2 in extra-time, but it could so easily have gone the other way with a little bit of that much-needed commodity called fortune. It didn't favour the brave that night, unfortunately. But Celtic gave our nation a real lift with their exploits that season. Neutrals would surely agree with that.

Looking back on that run it now seems like a hundred years away. And the way results have been going lately it might be a hundred years before we witness again. We must encourage and nurture youth and be able to spot slight little kids such as Xavi and Iniesta – or Harvey or Innes, if you prefer – and point them in the right direction. I can always dream.

22

TAKE A CHANCE ON ME

After two decades of being involved in football as a player I have to admit that being on the outside these days is sore. Chic Charnley, football manager? Chic Charnley, football coach? I can hear people scoffing at the very notion. My reputation goes before me and there is nothing I can say or do that will change the past. I made mistakes and I paid for them. However, haven't some people heard the one about poacher turned gamekeeper?

It would be impossible to pull the wool over my eyes. I have been there and seen it all. Nothing would surprise me and you better believe I have picked up more than a few hints by working with guys such as John Lambie, Davie Hay and Jim Duffy, three of the best in my estimation. There was also Bobby McCulley at Partick Thistle and I really rated him as a top coach although he never got anything like the praise he deserved for all the work he did outside the glare of the spotlight. Gerry Collins and Jackie McNamara Senior are two other top football people I have studied.

As I have already stated elsewhere in this book, I watch a lot of football and would dearly love to have a role to play at a club somewhere. My main involvement at the moment is with Maximise when we take kids for training. I love coaching these enthusiastic youngsters, but I don't mess with their heads with all the technical jargon. At this stage of their development it should all be about pleasure. Give them some hints here and there, but don't bog them down with too much information. I get a lot of joy out of seeing these hopefuls going through their paces.

Actually, some of these kids can be quite funny. I recall a day when Murdo MacLeod and I were putting some of them through a simple routine. Murdo wanted the wannabes to practise with their weaker foot. He was explaining the importance of being two-footed and emphasising that they were at a stage when they should develop their skills. He asked if they could keep the ball up ten times with their unfavoured foot. This wee boy stepped forward and exclaimed, 'I can.' Murdo was impressed. 'You can keep the ball in the air ten times with your bad foot?' he queried. 'Sure,' said the wee boy. 'Nae problem.' He promptly threw the ball into the air and kept it up only twice. He asked for another shot and Murdo gave him a second chance. Again, the ball hit the deck after only two touches. Murdo turned to the class and asked, 'What does that tell you?' Back came the reply from one of these little urchins, 'That he's a fuckin' liar!'

In time, I would like to step up and, obviously, would like to put some of my own ideas in motion. I have always been a realist and I do understand that some chairmen wouldn't touch me with the proverbial bargepole. They are looking at my history of sendings-off and probably believe I will never change. I would ask them one question, 'If I was that wayward as a player how come so many managers bought me?' Yes, I did get up to some high jinks as I have already described in detail, some of it painful, but that was then and this is now. I have changed. Fairly dramatically, too.

I am no different from just about every footballer on the planet. You never think your playing career will one day come to an end. You don't believe for a minute that you will lose that ability, that your legs will ache for days after a ninety-minute game. Old Father Time takes no prisoners, does he? Once you know that it is time to go there is nothing else for it. Quit while you still have a fair reputation. Of course, I left as coach at Partick Thistle when Gerry Collins and Bobby McCulley were sacked otherwise I might still be at Firhill. Who knows?

People may even perceive me as a troublemaker, but folk who really know me will tell you quite a different story. I got into bother as a footballer – that's a matter of record – but I have already explained the reasons. I find watching football compared to actually playing it a lot less manic. I am not as hyper, for a start. I can watch a game at Senior or Junior level and take a note of things that, hopefully, could be of use in the future. I even take in kids' games, too, and I have to laugh at some of the comments from the proud parents on the touchline. They are well intended, of course, but they are quite ludicrous at the same time.

That's why Maximise is so important to me. I've got the opportunity to try to get the youngsters into good habits. I believe we should have an open mind about how to coach kids. As I have said, we should look at the Dutch. Their schoolboys are brought up playing the game on football pitches that are tailored to their age group. They are more compact and the ball is lighter and smaller. And would-be goalkeepers aren't asked to perform in goals that are the same size as the ones professionals use. It gives them confidence. The outfield players are happier with a ball they can kick without fear of breaking a toe.

I love football and had twenty memorable years in the sport. It put a lot into my life and I would now like to put something back into it in another capacity. All I need now is for someone to take a chance on me.

23

AWAY FROM IT ALL

I think I've got my career in reverse. I am doing more running now than I ever did as a player! Thankfully, I have never been one who has had to keep a daily check on their weight. As well as running, I do a lot of swimming and play a bit of tennis. I am forty-six years old at the time of writing and I still feel fairly fit. I play for Celtic in the National Six-a-Sides, of course, and I thoroughly enjoy being out there kicking the ball around again. It's a reasonable standard and there's a lot of fun before, during and after these games. Well, there's got to be with Macca around.

The Sixes used to be run by a former West Ham player called Paul Allen, who was a mate of Macca. The English Professional Footballers' Association were involved back then as the tournament came to life. I'll be forever grateful for that ninety minutes in a Celtic shirt against Manchester United at Old Trafford because that qualified me to play in the hoops in these games. I recall a time when we were in Dubai and we were mixing with the former players of Liverpool, Manchester United, Chelsea and Rangers among others. Macca, as you might expect, was scheming and dreaming. There was loads of hooky gear around the market places and he wanted to make a few extra quid. He said, 'Chic, how about you and me loading up with some fake stuff and selling it when we get back to Glasgow?' To be honest, I wasn't that bothered. The thought of carting half of Dubai's dodgy goods home to Scotland didn't exactly appeal.

If you know my old pal, you will also know that he is not one to accept 'No' for an answer too readily. 'Look, Chic,' he said, 'the stuff's really good. It's virtually real.' I told him that it was like saying something was semi-invisible. That didn't wash with Macca. 'It's fabulous. They've got watches, belts, handbags, denims – you name it, they're selling it. We could make a fortune.' I still wasn't too sure. We sitting around the bar and Macca turned to ex-Liverpool and England player John Barnes, who had a short stint as manager of Celtic, and former Manchester United and Leeds United winger Lee Sharpe. 'How about you guys?' asked Macca. 'Are you in? The stuff's good quality, I can assure you.' Neither of the former England internationals looked two excited at getting involved in Macca's 'Get Rich Quick' scheme. It must have escaped my mate that Barnes and Sharpe were wearing genuine Rolex wristwatches and I was willing to have a flutter on the gold chains around their necks being authentic, too. They probably lived in mansions with tennis courts and inside swimming pools. Politely, they declined Macca's offer. Macca went ahead with the scheme, anyway, and came home loaded down with all this fake stuff.

I had to laugh when Chris Kamara, the ex-Leeds United defender who is now a Sky sports presenter, talked a few of us into trying to ride camels. There weren't too many of them around when I was growing up in Possil, I can tell you that. We didn't have a lot of practice clambering onto these beasts. I can now reveal that I was not Peter O'Toole's stunt double in *Lawrence of Arabia*. The out-takes would have lasted longer than the entire film. However, I'm always up for a laugh and I agreed to give it a try. Once I had settled in my saddle – after about an hour or so – we took off. It's a bit of an experience. I think I'll stick to riding bikes in future, though. I was walking around like John Wayne for the remainder of the trip!

I'm beginning to think Kamara is a bad influence on me. On another evening out we went along to see some belly dancers. That's all well and fine, but, after a drink or two, you are invited onto the stage to see if you can emulate these supple young girls. Of course, anyone in their right mind would reject it out of hand. Ex-footballers on a night out? There was a race to see who could make the biggest fool of themselves!

There are so many good memories from these jaunts abroad. David May, the old Manchester United and Blackburn Rovers player, appeared at a few of these outings and I have to say if he hadn't made it as a footballer he could have cracked it as a professional singer. Honestly, the guy had an outstanding voice. If I had heard him on the radio or on a record I would have believed he was the real thing. He would croon away merrily after a few beers and we would put our feet up and sing along.

Fans are allowed to travel with the veterans to these functions and a lot pay good money to see us old codgers try to roll back the years. Or, in some cases, roll back the fat! I bumped into former Liverpool and England international defender Alan Kennedy one day. Naturally enough, I recognised him straight away, but he didn't return the compliment. I don't blame him. Anyway, he asked me, 'Which team are you supporting?' He must have wondered what he had said that had me blushing so furiously. I made my excuses and left. He must have been more than slightly surprised and bemused when he saw me turning out for Celtic later that evening.

And don't tell me we don't take these games seriously. Anyone who witnessed our game against Tranmere Rovers in the Masters Final at the Echo Stadium in Liverpool in September 2009 will know what I mean. That confrontation almost ended in a riot! We had the likes of Alan Stubbs, Tommy Boyd, Tommy Johnson, Darren Jackson, Mark McNally, Andy Walker and yours truly ready to go. Stewart Kerr, who would undoubtedly have made the grade at the highest level if it hadn't been for a severe back injury, was in goal. We were desperate to be crowned British champions. We had won the Scottish version and that got us into the finals against the regional winners. We met Tranmere in the knock-out quarter-final stages. There was an early flashpoint when Andy was mauled by an opponent. I went on shortly afterwards as a sub and let the guy know I had arrived!

They scored first, but Andy equalised. They netted a second with under a minute to go, completely against the run of play. We were fuming at that but we were raging seconds later when they gave away a free-kick and one of their players prevented us from taking it. The expected thing for the referee would be to stop the game. Not John Underhill. He waved play on and that took up vital seconds. The countdown began. Five . . . four . . . three . . . two . . . one . . . full-time! The ref hadn't added a single second. I was first to the ref to complain. Big Stubbsy was next. And then Boydy. And then Andy. Suddenly the match official was surrounded by angry Celtic players. You would have thought we had just gone out of the World Cup Finals. We were still arguing in the bar well into the night.

I get involved in charity football matches, too, and they are always good for a laugh meeting up with old mates and former adversaries. I recall one match a few years back when I was involved in a Celtic line-up that was playing against Rangers for the Bhoys Against Bigotry campaign. Tosh McKinlay and Pat McGinlay were among the ex-Celts and Richard Gough, John McDonald and Colin Jackson were among the former Ibrox performers. Alex Willoughby, who, sadly, has since passed away, was manager of the Rangers Old Crocks. Alex was a smashing wee guy. He was always extremely dapper and I don't think he ever had a hair out of place. Anyway, we played the first-half and someone thought it would be a good idea for some of us to swap teams for the second period. I was asked to play for the opposition. Me? In a Rangers strip? Help!

A lot of my pals were at the game and they would have invaded the pitch and strung me up if they had seen me playing for Rangers. Now I know that was the spirit of the occasion – let's wipe out bigotry and so on – but I just couldn't do it. I apologise for this admission, I really do. I am not a bigot, believe me. Anyone will tell you that. Yes, I get involved in the light-hearted banter, but a lot of my mates are bluenoses. John Lambie is a good pal and we all know which foot he kicks with, don't we? For the life of me, I just couldn't put on that Rangers top. What was I to do? I hid! No-one could find me and they had to restart the match without me. When it was safe to reappear, I did so and made up some lame excuse for my disappearance. I met up with Alex Willoughby at the reception afterwards. I had known him for years and went over to him at the bar. 'Can I buy you a drink, wee man?' I asked. Back came the reply, 'Fuck off you fenian bastard. I was only your pal when we were in public.' And then he burst out laughing. We had a good night, as I recall.

Actually, that particular game was a bit surreal. For a start, there was a busload of Tibetan Monks banging on all sorts of drums and chanting throughout the match. It wasn't quite the Hampden Roar. We were playing on a public pitch in the West End of Glasgow and they were going through one of their usual routines. No, they hadn't travelled over from Tibet to see Chic Charnley playing football. The Dalai Lama, the Buddhist religious leader, was in Glasgow at the same time and was appearing at the SECC the following night. Around 10,000 people turned up and, out of curiosity, I went along, too. It was an interesting night.

Then there was another evening when I was playing for a Partick Thistle Select against their Celtic counterparts in the Testimonial Match for our goalkeeper Kenny Arthur. Although these occasions are sometimes known as 'no-tackling games' there can be a little bit of needle and no-one wants to lose. Anyway, we were trailing 2-1 with about a minute or so to go. It was time for me to make my presence felt. I noted the Celtic keeper had come off his line and, when the ball dropped at my feet about thirty-five yards out, I lobbed it goalwards. It just kept on going and we were worthy of our 2-2 draw. I read in a football magazine not so long ago that Kenny rates it as one of the best goals he has ever seen. Thanks, mate.

I've still got my pals in Possil, of course. We have never lost touch and I would never have had it any other way. As I said right at the start of the book, some may not be angels, but I look at it from a different angle. Believe me, a lot of these guys have a lot of good in them. They'll go out of their way to do someone a favour.

Some of them are slightly off the wall. I recall one of my schoolboy chums, Joe Sutherland, had a pet tortoise and he named it Shuggy after the former Celtic player Johannes Edvaldsson. The versatile Icelandic player, who answered to Shuggy, was not the quickest around the field. Even he would probably have admitted that. Anyway, one day my wee pal thought it would be a smashing idea to paint his tortoise green, white and gold. His pet died of paint inhalation shortly afterwards!

I've always been a boxing fan – no jokes, please – and I thoroughly enjoy watching old films of Muhammad Ali in action. What a performer and he took boxing onto a whole new stratosphere. He brought a glamour to the sport that we had never witnessed before. Wee Jinky Johnstone is my all-time sporting great, but Ali is up there somewhere. Another boxer I also admired was Chris Eubank and I went along with my pals James Rodden and Jimmy Maxwell to watch him work out before one of his fights in Glasgow. There was one of those question-and-answer promotions at a hotel after training one day, so I popped in after training at Firhill. For reasons known only to myself, I wore a pink jacket that day. Chic Charnley in pink? What would my Possil pals have said if they had seen me? 'Big Jessie' would have been the best I could have hoped for. What had I been drinking before I bought that clobber? I must have had a few crème de menthes that day!

Anyway, Eubank was taking questions from the floor and I had read somewhere that he didn't actually like fighting. I found that strange for a guy who would go into the ring and try to cement his opponents. The pink jacket must have attracted his attention. He looked over at me. He said, 'Have you got a question?' I was on. Straightaway I asked, 'Why do you box if you don't like fighting?' Good question, I thought. He didn't blink as he stared at me and replied, 'I'll get back to you on that one, Pink Jacket.' I detected that I might have hit a raw nerve. After a while he swung round and looked in my direction again. 'OK, Pink Jacket, do you still want to know why I fight?' he asked. I nodded. 'Money,' he replied reasonably. 'I have a family to feed and this is what I do best. Simple as that. I do it for the money.' Question asked and answered.

I went along with Paul Lambert, Campbell Money, Kenny McDowall and Norrie McWhirter to see Eubank in action a couple of nights later and, as usual, he got in about it. His opponent must have been delighted that Eubank didn't really like what he was doing. God knows what would have happened to him if his rival was actually enjoying himself as he knocked him all over the place!

I am not a gambler, but I will have a wee punt every now and again. I had a mate called Bobo McAuley who used to own a greyhound. Like my schoolboy chum before him with the tortoise, he also called his dog Shuggy, acknowledging the legend that was Edvaldsson. Actually, as I recall, that greyhound could shift. It won a few races and I'm sure I picked up a bob or two along the way. There was a funny incident when I was out walking over the hills one day when I saw Bobo coming towards me carrying Shuggy – the dog, not the player – in his arms. 'What's happened here?' I enquired. My mate, completely deadpan, replied, 'Oh, nothing, it's just that I'm in a rush!'

It was only later that I discovered Bobo was only joking. In fact, the poor animal had cut its paw and he was on the way to the vet. He had me going for a moment, though.

24

BOOZE BROTHERS

Drink Canada Dry? I've got a few mates who would give it a right good try. I've got to say I have been in the company of some world class boozers and, just for a laugh, I've compiled a team of bevvy merchants who would guarantee Britain a gold medal in the Olympics if drinking ever became a sporting event.

I think it's only fair to start with yours truly. I once missed an important Old Firm encounter, but I had a very good excuse – I was in jail! I was at St Mirren at the time and, would you believe it, I was suspended. I had tickets for the Celtic v. Rangers game and decided it would be a good idea to take in the action. The law had other ideas. I was out with some of my old mates in a pub called Benson's in Possil on a Friday night and, because I wasn't playing at the weekend, I had a few beers. Around nine o'clock one of my pals decided to call it a day, but I was alarmed to see he had every intention of driving. I was tipsy, but he was well and truly out of the game. I got the keys off him and, knowing he stayed about five minutes up the road, I thought it was a good idea to drive him home. Well, it was a good idea at the time.

I got him home safely and the bright thing would have been to ditch his car outside his house. However, I decided it would be better to return to the pub and leave the car there. It would be fair to say I wasn't quite thinking straight. I drove off and, horror of horrors, I spotted a patrol car behind me. I steadied my nerves and kept well within the speed limit. I thought I had got away with it until I saw the flashing of their headlights. They wanted me to pull over. I kept going until I found a secluded spot and I suddenly stopped the car, pulled the keys out of the ignition and took off on foot. The cop car came to a halt, too, and one of the boys in blue came after me in hot pursuit. I was pounding down the pavement and this guy was chasing after me like his life depended on it. What to do next? I slung the keys into a bush and kept on running. The cop didn't hesitate – he stepped up the pace. Trust me to confront a Glasgow bobby who was superfit!

Eventually, he caught up with me and, as they say in the newspapers, I was apprehended and taken to the cop shop at Saracen Street. Ironically, I recognised one of the inspectors and, in fact, I had presented football medals to schoolboys on his behalf in Paisley shortly beforehand. There was no way he was going to return the favour. I was nicked. He came over to me and said, 'Chic, if you don't tell us where the car keys are you will be kept in until Monday.' I thought I would be dropping my mate in it if I told him he had allowed me to drive his car while I was over the limit. I decided to keep quiet. It was then I was informed I might be charged with stealing the car. Oh, great. It wasn't enough that I was about to be done for drunk driving, but now I was going to be hammered for car theft. I realised I was not in a nice place.

I refused to take the breathalyser and I wasn't going to give them a urine or blood sample, either. I wanted time to clear my head. They stuck me in a cell for an overnight stay. The following day was much the same. I refused to co-operate and I was still thinking of ways to get out of this jam. And, just as importantly, I was wondering if I would be out in time for the kick-off in the Old Firm game. I was told I was being just a tad optimistic. I had four tickets in the possessions the cops had taken from me the night before. There was no point in letting them go to waste so I got a message to my brother John to come down to the cop shop and pick them up. I thought he might pop in and see me. He might even have brought in the day's newspapers, too. Nope. John picked up the tickets, ignored me completely and took three of his mates to the game. He still owes me the money for those tickets!

I remained in the cells until Monday morning when we were moved to Baird Street police station before going to court. Needless to say, the cop shop was packed with fans who had been at the Old Firm game on the Saturday. They were coming in by the dozen for all sorts of offences and I was handcuffed to one little hooligan as they shoved about twenty of us into the back of a police van to take us to Baird Street. The holding cell was bulging and I was wondering what on earth I was doing there. At least the cops were not pursuing the theft of the car and seemed to accept my explanation of how I had got myself into this mess. It was a straight drink driving charge. While we were waiting to be called into court, I was still manacled to this wee nutter. A cop opened the cell door and was about to shout out the name of this next bloke to be sentenced. He had his hand on the doorway frame when the hoodlum beside me kicked out and slammed the heavy iron door on the policeman's hand. I was sitting right beside the door and the cop looked me straight in the eye and said, 'Fuckin' animals!' I felt like shouting, 'It wasn't me, officer.'

Eventually, I was carted off and was fined £400 and banned from driving for a couple of years. You could say it was a sobering occasion all round. Remarkably, it wasn't reported in the newspapers and I can only think that was because the court reporter didn't recognise the name James Charnley. If it had been Charles or Chic I'm sure I would have hit the headlines. Either way, it wasn't too clever. I wasn't proud of myself and I wouldn't recommend what I did to anyone. Looking back, I should have got my mate a taxi or even walked him home. Like I said, it was a good idea at the time, but my act of kindness backfired big-style.

My old mate Frank McAvennie was lethal behind the wheel of a car. I recall he had a Ford Capri and there was a day he was taking myself and Frank McDougall to training at St Mirren. It was a horrible winter's day, the snow was falling and the roads were icy. That didn't prevent Macca from doing his best impression of a Formula One driver. Frank McDougall was sitting in the passenger's seat and I was in the back as Macca sped along. He was nattering away, as usual, when I saw a couple of cars had been in a collision about forty yards ahead of us. 'Macca, slow down,' I implored. Too late. Macca hit the anchors and his car slid straight into the side of one of the other vehicles. Crunch! What a mess.

The door was hanging off at the passenger's side and Macca said, 'We better hurry – we've still to get to training.' We decided to try to tie the door and Macca and myself got out of the car with McDougall staying put. We were trying our best to apply some First Aid on the Capri when we saw another car coming in our direction. Like Macca, the driver of that car seemed to be a frustrated Jackie Stewart. He was going far too fast and we knew what was about to happen next. Macca put his finger to his lips and said, 'Shhh'. McDougall, unaware of what was happening behind him, was still sitting in his seat as we saw this car careering towards us, completely out of control. Macca and I got out of the way and, sure enough, the other car smashed right into the back of Macca's one-time pride and joy. 'At least, I'll be able to claim the insurance now,' said Macca. What McDougall said is unprintable! That was after he stopped screaming like a banshee.

McDougall was known as Luther to his friends. It was an

unusual nickname and had nothing at all to do with Superman's long-time arch-enemy Lex Luther. Frank won a Golden Boot trophy for being top scorer in the Scottish League while he was at Aberdeen. Luther Blisset, the former Watford and AC Milan striker, was England's main hitman that season. Before the start of that campaign, the prolific frontmen met each other during a pre-season tournament in the Isle of Man and, for reasons known only to himself, Frank decided to call him Luther Blizzard all night. Footballers, having a keen wit, then decided to call Frank by his new name of Luther. It could so easily have been Blizzard! Anyway, my mate concocted a lie saying he had travelled to Paris to get his trophy. According to Luther, he was so well known that even Europe's finest recognised him by his nickname.

The way Luther would tell it, Michel Platini, the French legend who is now a top man at FIFA, was in Paris, too, to receive some award for his services to football. Of course, Michel spotted the Aberdeen 'icon' and immediately pushed through the crowd to offer him his hand. 'Luther,' said Michel, 'it eez a pleasure to welcome you to zee French capital. We would not have started without you. Thank you so much.' Luther was unfazed. 'Aye, good to see you, too, Michel.' Then it was the turn of Gary Lineker to force his way through his fans to get to my pal. 'Luther,' said the England international. 'Great that you could make it. You've made my night.' Luther, again, took it in his stride, but replied in typical fashion, 'That's OK, Gazza.'

My good buddies Alan Dinnie and Steve Pittman enjoyed a pint or twelve when they were in the mood. There was a time when they were both at Dundee that they decided to give the booze a right good hammering. They drank their way round the clock and then went to training at Dens Park the following day. They were so pissed they couldn't find their gear. They were hurling hampers all around the dressing room as they searched for their boots. Manager Jim Duffy, one of the most patient guys

I have ever met, wasn't too happy or impressed. In fact, it got so bad that he sent them both home to sleep it off. I think they were both sold shortly afterwards. I wonder why. By the way, Pitts' wife Izzy would have to be in my drinking team, too. I know she'll take that as a compliment. That girl can shift her alcohol.

Ray Farningham was another who would get into any boozer's top line-up. There was a day when he commandeered Dundee's minibus for a day out at Perth races. I don't think the club was aware that their coach was being used that day, but the officials were not best pleased when the vehicle was returned. For a start, the windows had been kicked out. Possibly the boys hadn't enjoyed the best of fortunes at the horses and took it out on Dundee's property. Ray was put forward as the main culprit and I believe Roddie Manley might have had a foot in it, too.

Billy Abercromby and Jimmy Rooney were two others who liked a pint. They fell foul of Alex Miller at St Mirren which, to be honest, wasn't a difficult thing to achieve. I could do that on an hourly basis. Anyway, they were ordered to come back to the ground for extra training in the afternoons and one particular day they thought it was okay to go to one of the local pubs after the morning session. They downed a few beers and went for an Indian curry. Yes, it's not ideal preparation for another training stint, is it? They stumbled back to Love Street and they were to be supervised by Miller's assistant, Drew Jarvie. Now wee Drew was a big fan of the Bobby Charlton-style comb-over. If he thought the boys hadn't spotted his disguise he was put right that afternoon. Aber and Jimmy – known as Casper to everyone – were told to do a few laps. Aber was having none of it. He walked over to the dug-out where Jarvie was sitting and told him simply, 'Fuck off, baldie!' With that, he parked his backside beside him as Casper did a couple of laps before deciding to call it a day.

Lex Richardson was another I spent a few evenings with as

we tried to lower a vodka lake. I had to laugh when St Mirren were due to play Celtic and Lex was more excited than usual. He was a massive fan of the Parkhead side and he had just bought Roy Aitken's autobiography. 'I wonder if he'll sign it for me,' said Lex. He popped into the Celtic dressing room and came back beaming like a Cheshire cat. He had got Big Roy's autograph and, boy, was he pleased. I can't recall the score that afternoon, but I think Celtic won and Roy, unhampered, played well!

Brian Gallagher and Jimmy Rooney liked a wee half as much as Victor Meldrew liked to moan. At St Mirren, we used to be paid on a Wednesday, so, as you might expect, we were all skint on a Tuesday. Gal and Casper decided to pool their cash and discovered they had enough for three pints each. They duly went into Glasgow city centre for a mini-bevvy session. After a couple of pints – and realising they fancied a few more – they decided to put the remainder of their cash on a dog. They looked up the race card in the newspaper and Gal said, 'The dog in trap three – that's a winner. Let's go for it.' Casper agreed. Gal nicked round to the nearest bookie and put their wordly possessions on the dog. Needless to say, it lost. It's probably still running to this day. Gal stayed in the north side of Glasgow and Casper in the south. 'There's nothing for it, but to walk home,' said Casper. They left the pub and Casper was less than pleased to see his mate hop on a bus. 'I kept my fare home,' shouted Gal. 'See you tomorrow. Enjoy your walk.' And with that he gave his pal a cheery wave and took his seat on the bus.

Okay, that's a line-up to warm the heart of any publican, but here's a team I would pay good money to see performing out on the football pitch. I've played alongside so many talented footballers I have had to think long and hard about the ones who would get into this particular side. If I have overlooked anyone, please accept my apologies, but after two decades in the game, I have lined up with hundreds of players and some just have the edge over others.

Right, let's start with the goalkeeper. That was a tough one, but I have given the nod to my old St Mirren team-mate Billy Thomson. He was one of the most athletic and agile shotstoppers I have ever seen. Billy, who also played for Scotland, Dundee United and Rangers, had everything in his locker you would expect from a top-class keeper. He had a presence, too, and I think that is important to the guy between the sticks. I've seen footage of the likes of Gordon Banks, England's World Cup-winning keeper in 1966, and he always looked poised and assured. Same with the Russian legend Lev Yashin and Northern Ireland's Pat Jennings. Artur Boruc, too, always looks on top of his game even after he has thrown one into the net. It's a talent in itself. A goalkeeper must never look rattled. Once you see them falling to pieces everyone just piles in to keep the pressure on and that's when a lot of them fold. A lot of good scouts insist they actually like to see the keeper get involved in a howler when they are looking at them. They reason that they can then see how well he recovers. If he doesn't, then he is no use to them. Makes sense, I suppose.

Billy was a confident guy and that, too, is so vital for your last line of defence. So he gets the gloves in the Chic Charnley Select. Mind you, it could just as easily gone to another former Paisley team-mate, Campbell Money. Dibble was a solid performer, totally reliable in everything he did. He wasn't quite as flash as Billy, but he got the job done. I admired him in many ways, but Billy just shades it.

Roy Aitken walks into my team. There's absolutely no way I could leave him out. The bloke was so versatile he could probably have played in goal. He had all the ability in the world and he was a ferocious competitor. He wouldn't think twice about launching into a tackle and I know a lot of opponents who gave him a wide berth for ninety minutes on matchday. Sensible chaps. As I have said elsewhere, Big Roy – known as The Bear – was the ultimate professional. He trained like a beast and never held back. His leadership qualities were obvious from a very early age. If my memory serves correctly, I think Jock Stein gave him his Celtic debut when he was only sixteen-years-old and I believe it was against Aberdeen at Pittodrie. That must have been a baptism of fire, but I'm sure The Bear took it in his stride.

Jim Duffy would be an absolute must, too. As you will know by now, I really rated him as a classy player. His use of the ball from the back was simply outstanding and his timing in the tackle was excellent. He made his Celtic debut on the same night as Charlie Nicholas when they both played against Clyde in a Glasgow Cup-tie at Parkhead. Nick scored the first of many for the Hoops that evening – I think it finished 3-0 – and Duff was just as impressive. Celtic might just have got rid of him too early, but he went on to have a good playing career elsewhere. Another former Parkhead performer who would have to come into the reckoning in my squad from heaven would be John Hughes, the current Hibs manager. Affectionately known as Yogi after the bear in the TV cartoon, he is another wholehearted professional. I never witnessed him give anything less than 100 per cent for the cause. I enjoyed our time together at Hibs and I picked up a hint or two from him.

Alan Dinnie and Steve Pittman, my pals at Thistle, were two colourful characters off the field, but they could do the business on the pitch, too. I would play Alan at right-back with Steve over on the left. Any winger coming up against these two would know they were in a game. They liked a laugh away from the game, but when it came to putting in a shift on matchday you could count on these two guys.

Paul Lambert was a magnificent student of the game. Even when he was just a kid, it was obvious that he would go all the way. I recall Steve Archibald raving about him during his time at St Mirren. The former Barcelona and Scotland player – who had almost as many clubs as me! – was unstinting in his praise of Lambo. I know he was greatly impressed that Lambo always came back in the afternoons for extra training. He had a marvellous hunger for knowledge, too, and grilled Archibald on what life was like at Barcelona. That was the level where he wanted to perform and it was superb that he won a European Cup winners medal with Borussia Dortmund in 1997. Like Big Roy, he was a leader of men and it seemed fitting that he would captain both Celtic and Scotland. He deserved those accolades.

How about a double-act of the two Franks – McAvennie and McDougall – up front or with Macca sitting in just behind in midfield? What a partnership that would be. I'm not just including them because they are still my mates and would never talk to me again if I didn't select them. They are in there on merit and any manager would be delighted with the sort of firepower they would provide. Both were natural-born goalscorers and were huge presences in the penalty area. Everything Macca has done – on and off the field! – has been well chronicled over the years. He was lethal in that box and his timing was uncanny. Watching him in action as a predator, you almost felt sorry for the opposition. There was little point in putting a player on Macca. He was impossible to pin down and would wriggle this way and that way out of anyone's clutches.

I was as delighted as him when he joined Celtic. It's a strange thing to say, but he was a Celtic-type player. He was an exciting, colourful individual and, of course, the Parkhead fans took to him immediately. Macca was afraid of no-one. An opponent setting out to intimidate him was wasting his time. Macca would just keep coming back for more and the defender would be left wondering what he had to do to keep him in check. A cage, lock and key might have come in handy. Frank McDougall might not have hit the headlines in the same way Macca did, but he was a genuine threat to any opponent. He had mighty wallop and made good use of it, too. He wasn't as mobile as Macca, but he still got his fair share of goals. Alex Ferguson bought him for Aberdeen and he rarely got it wrong in the transfer market. Fergie, of course, had been an old fashioned rummel-'em-up centre forward in his playing days at Rangers, Dunfermline and Falkirk, among others, and he could spot a frontman who possessed menace. Frank didn't let him down.

I would give Andy Walker a starting place, too. There wasn't much of Andy, but he was as brave as anyone I have played alongside. He was at Bolton, of course, when I had my short stint on loan down there and we teamed up again at Hibs. He was as good a finisher as I have ever seen. He could strike the ball cleanly, but he was one of those sorts of frontmen who was quite happy to get the ball over the line by any manner or means. He was just as delighted to see the ball hit the net off his backside as he was with a great hit. Like Macca and McDougall, his record speaks for itself. He scored goals wherever he went and I believe he and Macca were two of the main reasons Celtic won the League and Cup double in the club's centenary season in 1988. Celtic actually defended from the front in that particular campaign and Macca and Andy certainly put themselves about. No defender could take a breather with those two around.

Brian Gallagher never got the recognition he deserved, but I would have him in my starting line-up. Okay, he's a friend, but he was also a top-class player who could have gone all the way if he had applied himself. Like myself, his head was full of nonsense, but he was a magnificent player when he got into his stride. He is a rarity in that he once scored a hat-trick away from home in a UEFA Cup-tie. How many Scots can claim to have achieved that? He claimed three in a match against Hammerby in Sweden in season 1985/86 and I think that feat underlined his capabilities. He would be a cert to start up front with Andy Walker and Frank McDougall with Macca sitting in behind them in midfield.

I think I can safely say Kenny McDowall is still a mate despite his defection to Rangers. I can't be accused of being biased and he, too, must get a mention in my dream team. I would have him on the substitutes' bench with the promise he would get a game at some point. Again, you knew what you were getting when Kenny was around – a hard-working guy up front who put himself about for the team. Maybe he wasn't one to stand out, but his fellow-professionals realised only too well what he brought to the table. He was totally unselfish and went in where it hurt. A smashing guy to have in your team and a great bloke, as well. If only he had stayed at Celtic! I once got him tickets for a Partick Thistle game against Morton for his dad and I remind him of that good deed to this day. Especially when I'm looking for tickets. It was a sad day for me when he left Celtic for Rangers, I can tell you.

David Elliot would be in the squad, too. Well, I've got to keep in with him because he's a cop these days. Handy guys to know. Where was he that night in Possil when I needed him? Alan McInally would be there or thereabouts, too. I played alongside Rambo at Ayr United and he was a powerful, strong-running frontman. He wouldn't have got into my boozers' eleven, though. Two pints and Alan's anyone's. Three pints and he's everyone's. Four pints and he's no-one's. Only joking, Rambo! Jimmy Hughes and John Hynd, from my days in the Juniors, would have to get a mention. Maybe they could come on for the last few seconds! John Flood, at Thistle, was another good pro who never got the headlines he deserved.

Okay, here's my line-up in a 4-3-3 formation: Billy Thomson; Alan Dinnie, Jim Duffy, Roy Aitken and Steve Pittman; Paul Lambert, Frank McAvennie and yours truly (well, I've got to pick myself, haven't I?); Andy Walker, Frank McDougall and Brian Gallagher. Subs: Jimmy Hughes, John Hynd, Alan McInally, David Elliot, Kenny McDowall and John Flood. I know I will have forgotten someone, so, as I said before, please accept my apologies if I have overlooked you. So, in an ideal world I could line up alongside Roy Aitken and Co on matchday and then go for a pint with Gal and Co afterwards. Heaven!

25

AGENTS: WHO NEEDS THEM?

Willie McKay is a big-time football agent and now lives in Monte Carlo. That's a bit of a difference from the days when he was just starting out in the business. He got in touch with me to see if I was interested on going onto his books. I had never used an agent before and did my own wheeling and dealing. I wondered if Willie would be good for me. He wanted to meet me the following day to discuss things. Where did he want to see me? Would you believe an ice cream parlour in Maryhill called Jaconelli's! I suppose he takes would-be clients to slightly more upbeat establishments these days. After mulling it over in my mind, I decided to give the meeting a miss. I was never that fond of ice cream, anyway. Looking back, though, that might be another of my regrets when you see what McKay has gone on to achieve in the game.

There was another agent who tried to sign me up, but I won't name him to save his blushes. Along with Derek McInnes, the former Rangers player and current St Johnstone manager, and ex-Partick Thistle team-mate Ian McDonald I was invited along to meet this guy in a hotel in Glasgow. Now I never dressed as a Beau Brummell, but this bloke looked as though he had been dragged through a hedge backwards. Sartorial elegance and this particular agent were strangers. He looked dishevelled and scruffy. His shoes were battered and scuffed. He might even have been mistaken for a dosser. He might have been great at his job, but, to be honest, I wasn't going to let this guy represent me. So, that was a no-go even before we sat down to talk terms. Michael Robinson, the former Liverpool and Brighton player who is now based in Spain, also wanted me on his books, but that didn't materialise, either.

In my opinion, a lot of agents are merely in the game to line their own pockets. I believe they take far too much out of football and put in so little. You can see where most of them are coming from and, if I was still playing today, I wouldn't want to know them. Possibly they are a necessary evil for players at the top end of the game, but I don't think they do a good job for players at the other end of the scale. Some of them are toiling to pay their mortgage, but the agent is still there with his hand out looking for his ten per cent or his twenty per cent every month.

Don't get me wrong, some of them really do a fantastic job of looking after their clients. Take Gary Mackay, for example. As I have said elsewhere, there was no love lost between Gary and me when we were in opposing camps. But I have got to know him very well through coaching classes with Maximise and, trust me on this, he works overtime in his capacity as an agent. I've been in his car when he has been doing all sorts of favours for players. I'm talking about simple little things about how to pay an electricity bill, or what's the best travel agency to use, where to get the best price for a car. All sorts of stuff. Naturally, there is no percentage in that for Gary, but he does it for all his players. He runs about daft after them. And I am not just talking about the big names, either. He runs around after the youngsters, too. Often I've said, 'How can you be bothered with all that nonsense?' Or 'Tell him to sling his hook.' Gary takes no notice and simply steams ahead giving out free advice.

Garry O'Connor, the Scotland international forward, was one

of his clients, but Gary missed out when he came back to Britain after his short stint in Moscow. Gary was the agent who got him his £1million move from Hibs to Lokomotiv Moscow. That deal allowed O'Connor to buy flash cars and the like. However, when it was evident that he wanted back to the UK he dropped Gary as his representative. It shows you the trust Gary has in his stable of clients that he didn't have the player on a written contract. A handshake was good enough for Gary and, in the end, it cost him a few bob when O'Connor ditched him and signed for Birmingham City. I believe it was the same situation when Christophe Berra left Hearts to go to Wolves. Gary isn't bitter, though, and I have never heard him complain.

Could I become an agent? No. However, what I would say is that I would look after a player, give him advice and try to point him in the right direction. But I would never be Mr Ten Per Cent. No chance. If I did a deal and the player was satisfied that I had done a good job then he could pay me what he thinks is fair. But I wouldn't be holding him to ransom or pointing out the small print in a contract. That's not my style. I would leave it entirely to their discretion. OK, that may not make me Businessman of the Year material, but it is the way I would work.

Some agents are worth their weight in gold, I'll admit, and Celtic surely will be forever thankful to the guys involved in bringing the likes of Henrik Larsson and Lubo Moravick to Parkhead. Rangers, too, owe a debt of gratitude to the agents involved in bringing in Brian Laudrup and Paul Gascoigne to Ibrox. Those individuals were good for the Scottish game and were genuine personalities. I accept Henrik might have been lured to Celtic by the presence of Wim Jansen with their Feyenoord connection and Lubo to the east end of Glasgow because his former Czech Republic international boss Jo Venglos was there. However, let's face it, top-notch continental players don't really want to come to Scotland.

Do they want to perform against Falkirk, Motherwell, St Johnstone and so on? And I mean absolutely no disrespect to those clubs by saying that. I just happen to believe it to be true. A lot of them have arrived on these shores merely to top up their pension. For most of them, their futures are behind them. I'm told Jonas Thern was picking up something in the region of £40,000-per-week during his stay at Rangers. That's remarkable money. How much would Tore Andre Flo have been earning if the Ibrox side thought enough of him to shell out £12million to buy him from Chelsea? He wouldn't have been on a pittance, that's for sure.

The cash has dried up now and clubs are being a little more careful on transfer matters. They are taking a bit more time in their deliberations and they are not prepared to splash out just for the sake of it. I don't blame the players one bit for raking in as much as they can. It is a short career, after all. Who am I to speak? I once knocked back St Johnstone to sign for Dumbarton because they put a few extra bob on the table. It should have been a no-brainer to go to the Perth club. Of course, some players come out of the profession, are confronted by the real world and have to start looking elsewhere for gainful employment. Football can turn its back on people and there are only so many coaching jobs to go around.

Some agents take care of their clients by making sure they have pensions well in place, but I think there are others who simply want their percentage of the deal and don't care what happens in the future to the player as they move on to someone else. Aye, it's a real rat race, isn't it?

26

IT MAKES ME SO ANGRY!

Do you ever get the impression that many of today's footballers would be more at home on the stage involved in amateur dramatics? I ask because we seem to have an awful lot of wannabe actors in our game at the moment and I have to admit it makes me so angry.

OK, my disciplinary record speaks for itself, but I don't recall even once getting a red or yellow card for acting as a conman. And that's exactly what some of our performers these days are all about, make absolutely no mistake about it. I have no time for these blokes. These are the guys who could quite easily get blokes like me into trouble. I would react to a rival going through the dying swan routine and, as is so often the case, I would be in the sharp focus of the referee. The injustice of it all would get to me. Meanwhile, the player who has kicked off the trouble is giving it the innocent-me act. What also annoys the hell out of me is that most match officials will fall for it. That's what makes it even more frustrating.

Where I come from you don't point out villains; you have your own set of rules as far as justice is concerned. There are some ridiculously well-paid footballers out there who don't think twice about trying to get a fellow-professional into trouble. You would think that with the cash they are earning they might actually concentrate on playing the game it was meant to be played. I think they are also cheating the fans. They pay hard-earned cash to watch their team on a matchday and they deserve to be entertained. I'm sure if they wanted to witness some wonderful acting they could go along to the local cinema. Frankly, I don't know how these guys live with themselves.

How often have you seen a player go down as though he has been struck by an invisible express train and bounce right back to his feet the moment an opponent has been booked or, even worse, banished from the field? We used to look at the antics of, say, the South Americans and shake our heads. Players over there thought nothing of going over, rolling around for about ten minutes, giving you the impression they were near-fatal cases in dire need of some urgent hospital treatment and then bouncing back as though nothing had happened. I remember as a kid I thought it was all very funny. It was slapstick soccer. I don't laugh now, that's for sure. I think it is disgusting.

What about the treatment my all-time sporting hero Jinky Johnstone received in the European Cup semi-final against the hatchetmen of Atletico Madrid at Parkhead in 1974? I was only nine years old at the time, but even I realised what the Spaniards, with a mix of Argentines, were up to that night. They set out to intimidate Celtic right from the kick-off and they weren't even remotely interested in making it a fair contest. I saw a picture in a football book recently that portrayed an Atletico defender, his face all screwed up in anger, pointing a finger under Jinky's nose and you wouldn't have to be Oxford-educated to realise he was not asking to swap shirts at the end of the game.

The point, though, is that these thugs kept up a steady assault on my wee pal and when the referee – an inexperienced Turk, if I remember correctly – called them over they would look contrite, hands clasped behind their backs, standing to attention and bow their heads. It was an orchestrated display and it was sickening then and it is just as nauseating today. How many match officials swallow this act? Far too many, if you ask me. The clogger boots his opponent then immediately develops acting skills that would be the envy of Robert De Niro to prevent themselves getting into trouble.

I was too young to recall much about the 1966 World Cup Finals in England, but I have seen a lot of footage since. Remember the big Argentine captain Antonio Rattin? I saw him in highlights as his country reached the quarter-finals against the host nation that year. He was no neanderthal and, in fact, looked an awesome player with such a fine touch for a big guy. He was called over by the referee in the first-half of that turbulent encounter at Wembley and immediately adopted the hands-behind-the-back, butter-wouldn't-melt-in-my-mouth routine. On this occasion, the West German match official didn't buy the ticket. He signaled he was going off and the look on Rattin's face was priceless. He protested his innocence and maybe that worked every week in South America, but he wasn't going to con a European referee and eventually, with police threatening to come on to forcibly march him off the field, he reluctantly made his way back to the dressing room. Nice one, ref. And you won't hear me saying that too often!

There now appears to be a public outcry about these fraudsters as if it was new. Take it from me, diving hasn't just arrived in our game. It's been around for years. There's the old joke about a Rangers player who could earn his side a penalty kick after being tripped on the halfway line. Well, I think it's a joke! The culprits are being caught more and more because of the better television facilities. Simple as that. Look at Eduardo's nonsensical dive that earned Arsenal a penalty-kick against Celtic in their Champions League play-off in season 2009/10. It was clear goalkeeper Artur Boruc did not touch the Gunners' frontman, but the referee didn't hesitate in pointing to the spot. A decade or so ago there would have been no controversy about the award because in real time it did look like a penalty-kick. However, images in slow motion from several different angles clearly proved otherwise. UEFA weren't impressed, of course, and banned the Croatian for a couple of games. So what? How did that help Celtic? If the referee had been doing his job properly he could have booked or sent off the Arsenal man for simulation, as it is called by the powers-that-be. Then we could have had an entirely different outcome to the tie. UEFA, of course, later overturned Eduardo's ban for diving. Go figure, as they say.

It seems to have become trendy for managers in recent times to come out after a game and start labeling opposition players as divers. It's happening almost every week and I confess it is a little worrying. For a start, kids are copycats. They witness what their heroes are doing and if it's all right for their idols then surely it is OK for them. That's just what we need, isn't it? The next generation of Scottish footballers going down clutching ankles, holding their heads, crawling around in pain – and that's just at the warm-up! I shudder to think.

Also, it is too easy for bad habits to creep into the game and it's not so easy to eradicate them. You'll probably find this amusing coming from me, but I think the authorities should hammer the cheats. Hit them in the pocket and put them on the sidelines for weeks. Name them and shame them. Strip them of their credibility and status. Then they might just think twice about indulging in something that has nothing whatsoever to do with sport. It's not just in football, either. What about that ropey rugby player who bit into a capsule and it gave the appearance he had blood streaming from his mouth? Silly sod then winked at his coach as he was coming off the field. Thankfully, he got his comeuppance, too. So, there are some unacceptable traits coming to the fore and we have to stop these disgusting 'tactics' in their tracks before the game is infested with cheats and drama queens

I made mistakes on the pitch and I'll hold up my hands and repeat that I wasn't too proud of some of them. I would blow up and pay the price. But I never deliberately went out with any thoughts of keeling over when a gust of wind came anywhere near me in an effort to get an opponent into trouble. I was never that calculating or cute. Mind you, if I had been I might have finished a few more games!

27

PLAYING FAVOURITES

Going to the movies for most people is a pleasurable experience; something to look forward to. For me, it's torture. You've got to sit on your backside for about two hours and keep still. That's a real test for me, believe me. I start to fidget after about ten minutes and, around the half-hour mark, I'm ready to leave. However, there have been a few films that I have been happy to sit through when they have caught my attention.

One of them is *The Godfather 2*. In fact, I have enjoyed the entire trilogy, but I really enjoyed the second one. When the original Godfather, starring Marlon Brando in the title role, of course, was released in Britain in the Seventies it was amid some incredible publicity. Even I had to go to see this movie that everyone was talking about. I had no option. I have to say I thoroughly enjoyed it and couldn't wait for the follow-up. That is most unlike me. I think it was around then that I took a liking to Robert De Niro – not in the Biblical sense, of course! He was excellent in the second movie and I admire the way he can switch from drama to comedy with effortless ease. *The Deer Hunter* was a barrel of laughs, wasn't it! Talking of *The Godfather*, I've often been told I look a little bit like James Caan who also starred in the first two movies. Handsome devil, isn't he? However, I have to say I rate De Niro as my favourite actor today. Mind you, he has some fierce competition from some of the guys who masquerade as footballers at the moment. Some of them deserve Oscars for their dying swan acts as soon as someone comes within a foot or two of their vicinity.

Another movie that kept me parked on my backside was the wonderful *One Flew Over The Cuckoo's Nest*. Jack Nicholson was absolutely wonderful and actually had you believing he was a headcase. I think I recognised some of my old mates from Possil in that film! I'm sure there were a few in *Midnight Express*, too. That was the rather harrowing film shot in a Turkish jail and, like *The Godfather*, was a must-see. It was relentlessly dark all the way through until it had a happy ending. It was good to leave the theatre that night with a smile on my face after two hours or so of despair and five minutes of delight. Al Pacino is another favourite film star and I really like his style. I thought he was immense in *Scarface*. You might get the notion by now that I have never seen a Doris Day film! I enjoy gritty stuff, real edge-of-the-seat stuff.

Music? Well, Rod Stewart has got to be in there, hasn't he? And, no, it's not just because he is a Celtic fan. Rod's got a unique singing voice and I like just about everything he has done. He first came to the fore with The Faces in the early Seventies before branching out on his own. I've spent many an evening with my feet up while Rod has gone through his routines on CD. I remember my Ma bought me a Roxy Music LP when I was a kid for my Christmas. Bryan Ferry, like Rod, had his own inimitable style of singing. It was the good old days of vinyl and I'm sure the plastic was worn out by the end of the weekend it was played and replayed so much. Anyone want to hear my rendering of 'More Than This'? No, I didn't think so!

Marc Bolan and T Rex always get my toes tapping. I enjoyed that brand of music, too, but I'm not too sure about the make-up. I didn't quite see me covered in glitter or mascara, do you? 'Ride A White Swan' was wonderful and very different during that period in pop music. There was a lot of innovative stuff as the industry moved away from The Beatles. As a wee boy in the Sixties everyone seemed to sound like the Fab Four to me. I think it would be fair to say that Oasis copied quite a bit from the Merseyside quartet when they kicked off in the Nineties. I've got a fair range of tastes concerning music and I have to admit I enjoy the soothing tones of Frank Sinatra and Dean Martin. I once went to see the Rat Pack tribute show when I was in Las Vegas one year and it was a smashing evening. Those guys were something else.

Actresses? I bet you can't see this one coming – I am a massive fan of Bette Davis. I realise she was no Angelina Jolie, but I thought she was magnificent in just about everything she did. They used to say you couldn't take your eyes off Humphrey Bogart when he was on screen. Away from the cameras, he was an unassuming bloke who wore a wig, wasn't not too tall and had buck teeth. However, he had a presence as soon as those cameras started to roll. It's the same for me with Bette Davis. As she got older, she played so many character roles and who could forget her starring alongside Joan Crawford in *Whatever Happened To Baby Jane*? Legend has it that the two of them hated the sight of each other in real life, so, possibly, they didn't have to put too much into their roles as feuding sisters for this particular movie! *All About Eve, Hush Hush Sweet Charlotte, The Anniversary, Jezebel* – all classics, as far as I am concerned. So good, in fact, even I could sit through them again. Believe me, that's saying something.

I enjoy a good laugh as much as the next guy and one television programme that always gets me grinning is *Blackadder*. Rubber-faced Rowan Atkinson is wonderful in the title role. It really is side-splittingly funny and the characters are marvellous. What about Baldrick and his cunning plans? I reckon he would have made a first-rate manager! I thought all four series were top-notch material and Ben Elton is obviously a very talented comedy writer. An American import I have a laugh at is *Everybody Loves Raymond*. It hasn't quite got the following of *Cheers*, *Two-And-A-Half Men*, *King of Queens* and so on, but it's good, witty stuff. Raymond is a sportswriter in the series and I've probably met a few like him on my travels through the game. I pop in some mornings and watch it at my Ma's. She enjoys it as much as I do. *Only Fools And Horses* does it for me, too. Mind you, I've got a few mates from Possil who could buy and sell Del Boy. You better believe it!

If you struggled to guess my favourite actress, you've got two chances – slim and none – of coming up with one of the guys I rate as a top comedian. I've seen this bloke live and I think he is hilarious, an absolute riot. Ladies and gentlemen – or, as he might say, ladels and jellyspoons – please put your hands together for . . . Ken Dodd! I went to see him in Glasgow not so long ago and he was exceptionally funny. He's value for money, too. He was on the stage for about four hours and really gave a first-rate performance. Just ask anyone who was there that night. He put so much drive and energy into his performance that I had to remind myself that I was watching a bloke who was over eighty-years-old. A quote that has often been attributed goes along these lines, 'If I get a hard audience they are not going to go away until they laugh. Those seven laughs a minute, I have got to have them.' Just the seven, Ken? Of course, The Big Yin himself, Billy Connolly, is right up there, too. What an impact he made on the comic circuit when he first came to prominence in the Seventies. He's got a unique brand of humour and I suppose he can take it as a compliment that so many other comedians try to mimic him. There's only one Billy Connolly, though.

I can't mention favourite things and leave out Partick Thistle, can I? You may have gathered by now that I am a bit of a Celtic fan. That's an understatement on the scale of saying it can be a tad damp in Scotland in the summer! However, the Jags have a special place in my heart, too. I've got so many fond memories of Firhill and, as I have said elsewhere, the supporters have always been superb to me. Trust me, it is always appreciated and their kindness will never be taken for granted. Nor will it ever be forgotten. Another club I've always had a soft spot for are Manchester United. There is a special aura about them and I suppose it is a bit ironic that I made my one and only appearance for Celtic against them at Old Trafford. It's known as the Theatre of Dreams and I can see why. I wouldn't have minded a few more games in that sort of company. If you're reading this, Sir Alex, I've still got my boots and I am a free agent!

AFTERWORD
BY JIM DUFFY

Chic Charnley was a firework. During his playing days he was ready to go off at any point. If provoked, he would react in an explosive manner. However, he could also put on a terrific, eye-catching, spectacular display. He could dazzle and he could take your breath away with his touches, his vision and his passing. Chic possessed a spontaneous spark and if you had taken that away from him you would have deprived him of his strongest weapons and greatest gifts. Sometimes it worked for him and sometimes it didn't. One thing that is not up for debate, though, is the fact he had the God-given ability to do the most outrageous things with that ball at his left peg.

Everyone still talks about the David Beckham goal against Wimbledon years ago when the former Manchester United man lobbed the ball from his own halfway line over the head of the unfortunate Neil Sullivan, the former Scotland international goalkeeper. The TV cameras were there to capture the impudent effort and we are still watching replays to this day. Listen, I've seen Chic go even better than that, but, unfortunately, there was no television coverage at that particular game, so you will just have to take my word for it. And that of the few thousand who were there to witness it, too.

Chic was playing for Hibs in a League Cup tie against Alloa and we were shooting up the infamous Easter Road slope. He received a pass in his own centre circle and, without hesitation or even looking up, he struck this sweet left foot shot goalwards. It was uncannily accurate as it sailed onwards and, suddenly, the poor opposition keeper realised he was in trouble. He scrambled back as fast as he could, but there was no stopping this effort from Chic. It zeroed in on its target and the ball was nestling in the back of the net before anyone realised what was going on. It was no fluke, either. I had seen Chic do the same thing on at least three or four occasions and you need to be a special talent to see these opportunities and sum them up so quickly. He had the awareness to know precisely the location of the goalkeeper and the guile to carry out the manoeuvre. Obviously, that was one of Chic's strengths. He was no ordinary player, that's for sure. A guy with decent talent might think about doing something so adventurous and off the cuff, but won't be able to execute the move because he lacks the craft of a Chic Charnley.

I played alongside Chic at Partick Thistle, of course, and I was his manager at Dundee and Hibs as well. No wonder my hair has fallen out! Chic never held any surprises for me; he was from Possil and I was brought up in Maryhill, just up the road. So, we both had the same working-class backgrounds and I think that was a good, solid foundation for us to forge our friendship. We each knew where the other was coming from and there were never any hidden agendas. We spoke the same language. Mind you, it could be laced with a certain F-word on the odd occasion, as I recall.

One time we came close to falling out was when I was gaffer at Dundee and had just signed Chic from Dumbarton. I was still a young manager and realised there was still a lot I had to learn at that level. One Monday morning Chic telephoned to tell me he would have to miss training. 'You better have a good excuse, Chic,' I said. He replied, 'I've been on the toilet all night. I've got the worst diarrhoea ever. I can shit through the eye of a needle.' What my good mate didn't realise was that I knew he had been out with his Possil pals all weekend. That would have been good for a hangover or two. 'Get your arse up here,' I said. 'Stack up with toilet paper in your car if you have to, but I want to see you at training today. If not, I'll fine you a week's wages and that will be the most expensive shite you've ever had!' Chic duly turned up and we had a wee chat afterwards. That was the end of the matter.

By the way, I've got to give Chic a compliment here when we talk about training. Simply put, he was one of the best I have ever seen. He really put his back into it. There are a few individuals who merely go through the routine, but hardly break sweat. They pay for that attitude on matchday when you see some of them ready to pass out during a particularly gruelling encounter. Not Chic. He was a fit specimen, believe me, and had a great engine. Remember he was thirty-four years old when he was with me at Hibs and I think the Easter Road fans were fortunate enough to see him at his best. As Chic is well aware, I had to persuade the board into letting me sign him. Once again, that bloody, contorted reputation went before him. However, once I had underlined that we would be getting an absolute bargain and there was no financial risk as far as wages were concerned, they gave me the go-ahead. I bet they're glad they did.

Chic was magnificent for Hibs. Just ask the fans. A lot of them had been brought up watching the likes of Peter Cormack, Colin Stein, Pat Stanton, Peter Marinello, Alex Cropley and the like. The older ones could go back to the Fifties and recall stories about the Famous Five, Gordon Smith, Eddie Turnbull, Lawrie Reilly, Willie Ormond and Bobby Johnstone. They even had a guy called George Best there at a late stage in his career. They were supporters who like to be entertained and that was music to the ears of Chic. He liked his football laced with class and style. He was a character and I think the game lacks that type of personality today. He was a born entertainer and he had the skills to back up his outlook. I've felt like giving him a round of applause when he has done something extraordinary with a ball – and that was just in training! Even without an audience, Chic would do something special. It was second nature to him.

I backed the campaign to get him into the Scotland international squad at the time we were both at Hibs. Why not? There were guys who seemed to be automatic choices back then who were not on a par with Chic. Around that time he was wearing his sensible head. He knew I wouldn't tolerate any mischief. Yes, there was a bit of leeway between the two of us, but I never played favourites. He could have the odd Monday morning off just so long as I was told in advance. He could be going to a wedding or function with his mates and I knew what to expect. Something that kicked off on Saturday evening could often come to a halt on Monday morning – without a pitstop. It only happened two or three times in the season and, to be fair to Chic, he didn't play on it. He was great in the dressing room, too. He has a natural sense of humour and, more often than not, he is a bubbly character. I realise his commitment to winning as well as entertaining could sometimes become a bit overheated. He was never slow to tell a colleague what he thought of his performance. That firework thing again.

Chic's speed of thought was quite awesome. He was always plotting a move or two ahead of other individuals. You could see that by the way he would take quick free kicks. There was never any hesitation. He would grab the ball and despatch it where he thought it would do most damage to the opposition. His deadball kicking must have terrified rivals teams. He had the craft to hit the spot with amazing regularity and outstanding accuracy. I don't think I've seen anyone better at hitting long, raking passes. He could hit them from any range and distance and, nine times out of ten, they were spot on. He could switch play entirely and thump them across the pitch, he could wallop seventy-yard diagonal passes and, basically, our opponents just didn't know what to expect next. He could assess situations in a split-second and that is a wonderful gift.

Chic didn't score a lot of goals – that wasn't his role in the team, anyway – but when he did notch one you could be sure it would be memorable. What is it they say these days? He wasn't a great goalscorer, but he was the scorer of great goals? That just about sums up Chic. I've got a sneaky feeling that he might have mentioned his fabulous effort against his beloved Celtic elsewhere in this book. I was the Hibs boss that day, of course, and I immediately realised I had just witnessed a strike that will live forever in my memory bank. Chic didn't break stride that day as he seized on a bad pass from Henrik Larsson, making his Celtic debut. You knew in an instant there was only one destination for that ball as Chic raced forward – the back of the net. He struck it as sweetly as any shot I have ever seen and their goalkeeper, Gordon Marshall, hadn't a chance of keeping it out of the net. The entire movement and the execution of the effort summed up Chic perfectly. Anticipation, power and precision. It was a deadly combination. Even better, it was the winning goal. Chic earned his bonus that day.

I was brought up watching Celtic, my first club, in the Eighties and, of course, I knew all about the exploits of the likes of Bobby Murdoch and Bertie Auld in midfield. Back then, Celtic mainly played 4-2-4 with four across the back, two in midfield and two wide players and two strikers. That put a helluva onus on Bobby and Bertie in the middle of the park. I've watched television footage and I never once witnessed or detected that responsibility threatening to overpower either of them. They revelled in the task. Pouring through footage of these guys helped me appreciate the role of players operating in the engine room of the team. Chic was built for such a position because of his lightning thought processes, his ability to size things up in an instant and be able to hit passes that defied belief.

How much would Chic be worth in today's transfer market? If you are talking purely in terms of Scotland, I would have to place him in the £2 million-plus category. That's where you would have to start the bidding, as far as I am concerned. Take Kevin Thomson, for example. He cost Rangers £2 million when he signed from Hibs and, absolutely no disrespect to the lad, he is a defensive midfielder. Chic, on the other hand, is a creator and I think you have to pay extra for those qualities. It's easier to break things down than set things up and that was Chic's forte. Players with that sort of ingenuity are few and far between, unfortunately.

Of course, our friendship has extended beyond the mere borders of football. We play tennis against each other quite a lot, but, rather alarmingly, I keep hearing reports that I am on the receiving end of a sound thrashing on most occasions. Not true! I may be no Andy Murray, but I can certainly take care of Chic. I might have to get video evidence to stifle his boasts once and for all!

If we were both playing football today, he would be one bloke I would be more than happy to sit beside in the dressing room and call a team-mate. Chic Charnley. My type of player. My type of guy.